BIOTECHNICAL AND SOIL BIOENGINEERING SLOPE STABILIZATION

BIOTECHNICAL AND SOIL BIOENGINEERING SLOPE STABILIZATION

A Practical Guide for Erosion Control

Donald H. Gray
Professor of Civil and Environmental Engineering, University of Michigan

Robbin B. Sotir
President, Robbin B. Sotir & Associates, Inc.

A Wiley-Interscience Publication

JOHN WILEY & SONS, INC.

New York • Chichester • Brisbane • Toronto • Singapore

This text is printed on acid-free paper.

Library of Congress Cataloging in Publication Data:

Gray, Donald H.
Biotechnical and soil bioengineering slope stabilization: a practical guide for erosion control/Donald H. Gray and Robbin B. Sotir.
p. cm.
Includes bibliographical references and index.
ISBN 0-471-04978-6 (cloth : alk. paper)
1. Slopes (Soil mechanics) 2. Soil stabilization. 3. Soil erosion. 4. Soil-binding plants. I. Sotir, Robbin B. II. Title.
TA710.G6286 1995
624.1′51363—dc20 96-10211
CIP

Printed in the United States of America

10 9 8 7 6 5 4 3 2 1

PREFACE

Biotechnical and soil bioengineering stabilization provide attractive, cost-effective, and environmentally compatible ways to protect slopes against surficial erosion and shallow mass movement. This guidebook discusses the general principles and attributes of biotechnical/soil bioengineering stabilization and describes specific soil bioengineering measures that can be employed on slopes, such as live staking, live fascines, brushlayering, branchpacking, live crib walls, and slope gratings. The conjunctive use of plants and earth-retaining structures or revetments is also described. This biotechnical approach includes plantings on slopes above low toe-walls, on benches of tiered retaining walls, and in the frontal interstices, or openings of porous retaining structures, such as crib walls, gabions, and rock breast walls. It also entails the use of vegetation in porous hard armor revetments, such as rock riprap, gabion mattresses, and articulated blocks. The book describes recent developments with biotechnical ground covers or "reinforced grass" systems, which include the use of nets, mats, and other types of structural/mechanical reinforcement to improve the establishment and performance of grass cover on steep slopes or temporary waterways.

Biotechnical and Soil Bioengineering Slope Stabilization distills more than a decade of experience in this subject on the part of both authors into a useful reference handbook. Numerous illustrations from actual field applications and stabilization projects supplement the text. In addition carefully selected and well-documented case studies have been included to show how various soil bioengineering methods have been chosen for particular site conditions. We also include helpful background information on the nature of soil erosion and mass movement, the role and function of slope vegetation in the stability of slopes, and techniques for the selection, establishment, and maintenanceof appropriate vegetation.

Biotechnical and Soil Bioengineering Slope Stabilization is intended primarily as a reference handbook for practicing professionals. Information in the book should prove of value to practitioners in such diverse fields as geotechnical engineering, geology, soil science, forestry, environmental horticulture, and landscape architecture. Although oriented toward professional practice, it is written in such a way that it can be understood by students, laypersons, and other interested parties as well. Analytical or somewhat technical material in some of the chapters can be skimmed over without loss of continuity or utility. Lastly, the book can be used as a reference text in college-level courses, extension courses, and workshops whose course content includes such topics as erosion control, slope stability, watershed rehabilitation, and land restoration.

We would like to acknowledge the assistance of several persons who helped in the preparation and review of the book. Jade Vogel, Karen Heiser, and Mark Huber, with the firm of Robbin B. Sotir & Associates prepared many of the drawings and assisted with compilation of the manuscript. Barbara Roberts helped to edit chapters 6, 7, and 8. Critical reviews of chapter 9 were provided by Deron Austin and Marc Theisen of Synthetic Industries, Inc. and by Tim Lancaster of North American Green, Inc.

DONALD H. GRAY
ROBBIN B. SOTIR

October 1995

CONTENTS

1 Introduction to Biotechnical Stabilization

1.0 THE BIOTECHNICAL/SOIL BIOENGINEERING APPROACH TO SLOPE PROTECTION AND EROSION CONTROL

Biotechnical stabilization and soil bioengineering stabilization both entail the use of live materials—specifically vegetation. While both approaches share this common attribute, they are characterized by subtle but important differences as well.

Biotechnical stabilization utilizes mechanical elements (or structures) in combination with biological elements (or plants) to arrest and prevent slope failures and erosion. Both biological and mechanical elements must function together in an integrated and complementary manner. Engineering principles of statics and mechanics are used to analyze and design most conventional slope protection systems, while principles of plant science and horticulture are invoked to select, propagate, and establish suitable plant materials for erosion control purposes. A biotechnical approach requires that both disciplines be integrated. Knowledge and awareness of important considerations from each of these major disciplines are necessary for successful implementation. Accordingly, this book will introduce the engineer to some aspects of horticulture, and the plant specialist to some concepts from mechanics.

Biotechnical stabilization can be characterized by the conjunctive use of live vegetation with retaining structures and revetments (see Figures 1-1 and 1-2). It is also characterized by the combined use of vegetation with structural-mechanical elements in ground cover systems (see Figure 1-3). Many new types of biotechnical ground covers have been developed in the past decade to improve the erosion resistance of grassed or vegetated surfaces. These ground covers are typically manufactured and delivered to a site as "rolled erosion control products"—a form that greatly facilitates placement. "Reinforced grass," a term coined by Hewlett et al. (1987), refers to a grass surface that has been artificially augmented with an open structural coverage (netting, mulch blanket, turf reinforcement mat, interlocking concrete blocks, etc.) to increase its resistance to erosion above that of grass alone and to improve the growth and establishment of the grass itself.

Soil bioengineering, on the other hand, can be regarded as a specialized area or subset of biotechnical stabilization. Soil bioengineering is somewhat unique in the sense that plant parts themselves, that is, roots and stems, serve as the main structural and mechanical elements in a slope protection system. Live

Figure 1-1. Vegetated, concrete crib wall retaining structure. Flowering plants have been introduced in the frontal openings or bays of the wall.

Figure 1-2. Vegetated, tiered, or stepped-back retaining wall. Flowering shrubs (*Forsythia spp.*) have been planted on the horizontal steps.

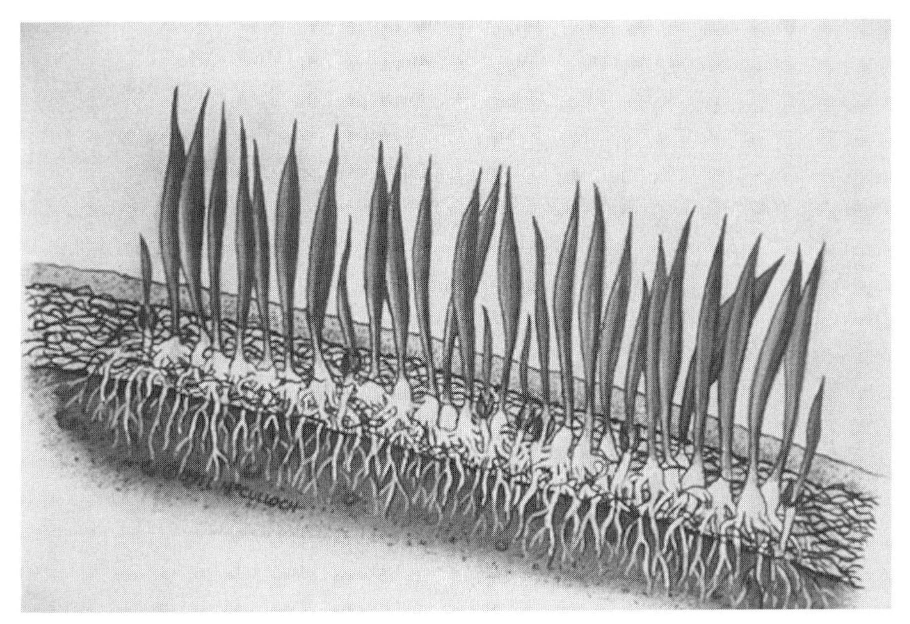

Figure 1-3. Vegetated, turf reinforcement mat. Topsoil and grass seed are introduced in an open, three-dimensional web or network of fibers. (Used with permission of Synthetic Industries, Inc.)

cuttings and rooted plants are imbedded in the ground in various arrangements and geometric arrays (see Figures 1-4 and 1-5) in such a way that they serve as soil reinforcements, hydraulic wicks (or drains), and barriers to earth movement. Soil bioengineering treatments provide sufficient stability so that native vegetation and surrounding plants can gain a foothold and eventually take over this role. Successful implementation of soil bioengineering stabilization likewise requires some knowledge of the factors governing the mass and surficial stability of slopes. If living plants and imbedded plant parts are to serve as soil reinforcements and drains, then some understanding is also required about the hydraulic and mechanical effects of slope vegetation.

1.1 HISTORICAL DEVELOPMENT

Live plants and other natural materials have been used for centuries to control erosion problems on slopes and along river banks in different parts of the world. These natural materials and methods of erosion control became less popular with the arrival of the Industrial Revolution. The age of machines and the development of concrete and steel technology encouraged the use of rigid, inert construction materials in engineered projects. These materials allowed for exact geometric measurements and time invariant designs suited to precise hydraulic

Figure 1-4. Landslide slip stabilized by live staking. Live willow stakes have been inserted into the ground on 2-foot centers in a triangular spacing pattern. Cuttings have rooted and leafed out.

and stress calculations. These inert, "hard" systems also promised initially to be more durable, cheaper, and safer.

In the United States the trend toward the abandonment of vegetation and natural structures for erosion control and slope protection was relatively rapid and pronounced. The low cost of energy, the relatively high cost of labor, and the wide distribution and abundance of raw materials needed in the fabrication of steel and concrete encouraged this trend. Europe experienced a similar trend; however, a few practitioners continued to use and improve both live and mixed construction methods. By the 1930s a number of professionals in various disciplines developed successful techniques employing the basic concepts of soil bioengineering. These techniques included the use of green willow as a live construction element, the planting of dry stone wall joints with woody cuttings, and crib wall construction with vegetative inclusions as an integral component of the wall. These techniques and methods have been described in detail by Hugo Schiechtl (1980), a foremost practitioner and proponent of soil bioengineering.

Soil bioengineering has been practiced widely with considerable success in Europe. A professional association for the promotion and advancement of soil bioengineering, Gesellschaft fur Ingenieurbiologie, has been active in Germany for several years. This association holds annual meetings to exchange information and to present technical papers on soil bioengineering. The only comparable forum to be held in the United States was a three-day workshop sponsored

(*a*)

(*b*)

Figure 1-5. Unstable slope stabilized by live fascines. Bundles of willow cuttings were placed in shallow trenches on contour, creating a series of benches or barriers across the slope. (*a*) After construction. (*b*) Three months later the imbedded stems have rooted and leafed out.

by the National Science Foundation in the summer of 1991. This workshop brought together leading soil bioengineering practitioners and other experts to assess the state-of-the-practice and to identify research needs. Position papers and a list of needed research tasks were presented in a Workshop Proceedings (National Science Foundation, 1991).

1.2 RATIONALE FOR BIOTECHNICAL APPROACH

Biotechnical and soil bioengineering stabilization offer a cost-effective and attractive approach for stabilizing slopes against erosion and shallow mass movement. These approaches capitalize on the advantages and benefits that vegetation offers for erosion control and slope protection. Other attributes and characteristics of biotechnical and soil bioengineering stabilization are discussed later in the book. The value of vegetation in civil engineering in general and the role of woody vegetation in particular for stabilizing slopes has gained considerable recognition in recent years (Greenway, 1987; Coppin and Richards, 1990; Gray, 1994). Woody vegetation affects the hydrology and mechanical stability of slopes in several ways. The stabilizing role of woody vegetation for protecting slopes against mass movement is examined in detail in Chapter 3. A few examples of this protective role, namely, buttressing action and root reinforcement, are depicted in Figures 1-6 and 1-7.

Figure 1-6. Slope buttressing by deep-rooted ponderosa pine tree. Unbuttressed part of slope on left has failed. Mendocino National Forest, California.

Figure 1-7. Sand dune permeated by roots of woody dune vegetation. Roots reinforce the soil and bind particles together, thereby increasing the coherence and stability of sand slopes.

Several biotechnical methods enhance landscaping opportunities and options on vertical and sloping surfaces. Suitable plants can be incorporated into openings and around structures in a variety of ways. Retaining structures and revetments can be designed and built with grace, elegance, and respect for visual sensitivity and natural settings. We are no longer restricted to using only stark, artificial, and inert materials. Instead, retaining walls and structures can be designed and constructed to combine both inert structural elements and living vegetation in a pleasing and functional manner.

The advantages and opportunities inherent to a soil bioengineering approach have been recognized belatedly in the United States. Kraebel (1936) was among the first to apply these live construction techniques to solve serious erosion problems on fill slopes on mountain roads in Southern California. The USDA Natural Resources Conservation Service (formerly known as the Soil Conservation Service) employed soil bioengineering techniques for lake bluff and streambank protection in the 1940s (USDA Soil Conservation Service, 1940; Edminster et al., 1949). More recently the California Department of Transportation (White and Franks, 1978) experimented with these methods on cut and fill slopes along highways in the Sierra Nevada. The U.S. National Park Service (Weaver and Sonnevil, 1981) has also employed soil bioengineering methods in their watershed rehabilitation program in Redwood National Park. Much of this earlier work is described and summarized elsewhere (Gray and Leiser, 1982).

In spite of its relatively embryonic state in North America, there is no question about the widespread interest in and increasing application of soil bioengineering technology in the United States. The USDA Natural Resources Conservation Service added a chapter to its Engineering Fieldbook on soil bioengineering for upland slope protection and erosion control (USDA Natural Resources Conservation Services, 1992). Short courses on soil bioengineering and biotechnical slope protection have been offered on a fairly regular basis in the United States during the past decade, most notably by the Universities of Michigan and Wisconsin. Soil bioengineering has been adopted as an attractive, cost-effective, and environmentally sound alternative for a variety of applications ranging from the stabilization of both cut and fill slopes along highways to streambank protection work. Recent examples of successful applications of soil bioengineering solutions to stabilization of highway cut and fill slopes are described by Gray and Sotir (1992). Several of these applications and case studies are documented and described in this book.

1.3 HEDGEROWS AS QUINTESSENTIAL BIOTECHNICAL STRUCTURES

Plants and structures can be used together to construct beautiful and functional living walls, medians, and barriers. Nowhere is this more evident than along the Cornish country lanes in southwest England, which are often lined with flowering hedgerows. Cornish stone hedges are splendid examples of living, man-made structures and communities–literally small ecosystems that host their own distinctive collection of flora and fauna (Bryson, 1993). Similar walls and hedgerows are also found in northern England, Scotland, and Normandy. The Normandy hedgerows, known locally as bocages, consist of low earthen dikes topped by a growth of trees and shrubbery. Some of these hedgerows date back to Roman times, when they were constructed by Gallic peasants to mark the boundaries of fields. The Normandy hedgerows bedeviled Allied tanks and armored units in World War II. They provided ideal defensive positions and concealment. The main problem facing armored units was to break through the hedges into the enclosed pastures so that they could support the infantry at close range.

The Cornish hedgerow is essentially an earth-filled stone wall. The wall is constructed with dry laid stones surrounding a central, earthern core, as shown schematically in Figure 1-8. It is this earthen core that gives the Cornish wall or hedge its unusual properties and attributes. During construction live cuttings of various plant materials are intentionally inserted through openings between stones into the earthen core. Eventually the cuttings root and leaf out. Ultimately, vegetation growing in the wall can become so well established that the wall takes on the appearance of a hedge or "living wall," as shown in Figure 1-9. Dowdeswell (1987) has published detailed descriptions and instructions for hedgerow construction, including information on the most suitable plant species for inclusion in the wall. Some hedgerows ultimately support not only plants

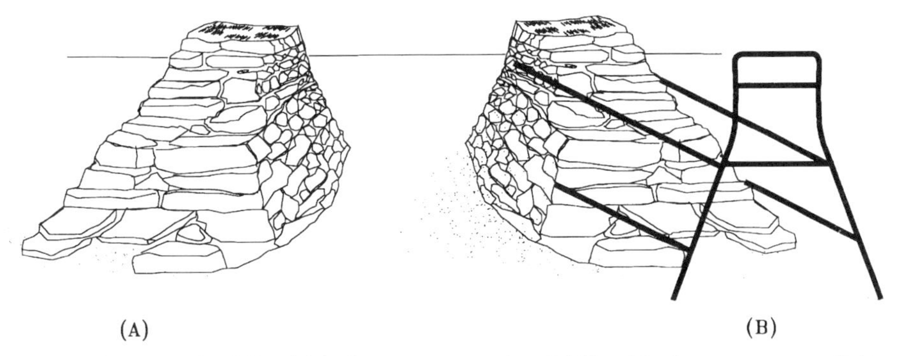

Figure 1-8. Example of Cornish hedgerow construction. (*a*) The side slopes are set at a slight batter to minimize the effects of erosion. (*b*) An iron frame may be used to improve uniformity in height and cross section. (From Dowdeswell, 1987.

and shrubs but also large mature trees (see Figure 1-10). Counterintuitively, the tree roots do not pry the wall apart; instead they literally flow and wiggle around the rocks, binding the entire rock wall together into a coherent, unitary mass. This attribute of plant roots, which has been termed "edaphoecotropism" (Vanicek, 1973), is discussed further in Chapter 4. Hedgerows are quintessential

Figure 1-9. View of fully vegetated Cornish hedgerows. Transplants and cuttings have rooted in the earthen core of dry-laid stone walls.

Figure 1-10. Cornish earth-filled stone wall or hedgerow supporting mature tree. Instead of prying the wall apart, the roots of the tree bind stones together into a coherent, unitary mass.

biotechnical structures that embody both the visually pleasing and functional characteristics of such structures.

1.4 EXTENT AND SEVERITY OF SOIL EROSION AND MASS MOVEMENT

The twin problems of erosion and mass movement are widespread and costly. They affect and disrupt agriculture, transportation routes, mining and forestry operations, water supply and storage, and urban construction sites. Damages from both these land degradation processes run into hundreds of millions of dollars annually in the United States. The conditions under which they occur, their location, and specific estimates of this damage are reviewed briefly next.

1.4.1 Soil Erosion

As a rule, regions with highly erodible soils, high relief, sparse vegetation, and strong winds, and with a dry climate but with occasionally intense storms exhibit the highest erosion losses. Human activities frequently intensify or accelerate rates of erosion, particularly if they entail stripping or removing vegetation; tilling, grading, or otherwise disturbing the soil; and increasing or concentrating runoff.

The total amount of soil eroded from the land and delivered to waterways is very large. Estimates of annual sediment yields in the United States range up to at least 2 billion tons/year (Brandt, 1972). Of the total amount eroded, about one-fourth to one-third is transported to the oceans, with the rest being deposited in flood plains, river channels, lakes, and reservoirs. The impact of this sediment on water quality is very significant; sediment is the number one pollutant in our nation's waterways in terms of magnitude and adverse impact. According to a recent National Water Quality inventory (U.S. Environmental Protection Agency, 1992), siltation and nutrients (nitrogen and phosphorus) from erosion impair more miles of rivers and streams than any other pollutant.

The principal sources of sediment include erosion from natural or undisturbed land, cultivated land, urban construction sites, highways and roadsides, streambanks, and mining or resource extraction operations. Typical erosion rates from those sources are summarized in Table 1-1. Erosion rates range from a

TABLE 1-1. Erosion Rates Associated with Different Land Uses

Sediment Source	Erosion Rate (tons/mile^2/year)	Geographic Location	Comment
Natural	32–192	Potomac River Basin	Native cover
	200	Pennsylvania and Virginia	Natural drainage basin
	320	Mississippi River Basin	Throughout geologic history
	13–83	Northern Mississippi	Forested watershed
	25–100	Northwest New Jersey	Forested and undeveloped land
Agricultural	200–500	Eastern U.S. piedmont	Farmland
	320–3,840		Established as tolerable erosion
	1,030	North Mississippi	Pasture land
	12,800	Missouri Valley	Loess region
	13,900	Northern Mississippi	Cultivated land
	10,000–70,000		Continuous row crop without conservation practices
Urban	140–2,300	Washington, D.C., area watersheds	Range in rates as urbanization increases
	1,000	Washington, D.C., area	750-mile^2 area average
	500	Philadelphia area	Area average
	50,000	Kensington Maryland	Land undergoing extensive grading
	1,000–100,000		Small urban construction sites
Highway right-of-way	36,000	Fairfax County, Virginia	Construction on 179 acres
	50,000–150,000	Georgia	Cut slopes

Source: U.S. Environmental Protection Agency (1973).

low of 15 tons/mile^2/year for natural or undisturbed areas to a high of 150,000 tons/mile^2/year for highway construction sites. These erosion rates underscore the fact that human activities (e.g., farming, mining, timber harvesting, and road building) tend to accelerate erosion rates greatly over background or predisturbance levels.

The economic costs of erosion and sedimentation are substantial. Direct costs include damage to property from either sediment accretion or loss of soil. Sediment removal costs alone may range from \$7 to \$68/yard^3 (1973) dollars), depending upon whether the sediment is removed from streets, sewers, or basements (U.S. Environmental Protection Agency, 1973). Indirect costs and damages of erosion and sedimentation probably exceed direct costs but are much harder to assess. These damages are summarized in Table 1-2. The U.S. Army Corps of Engineers, for example, annually dredges approximately 400 to 500 million yards^3 of sediment from the nation's rivers and harbors to keep them navigable. Conservative estimates place the loss of reservior capacity from siltation at 1 million acre-feet/year.

The U.S. Army Corps of Engineers (1978) has estimated that out of approximately 3.5 million miles of streams (7 million bank-miles), a total of approximately 8 percent, or about 575,000 bank-miles, are experiencing erosion to some degree. Of the total bank-miles experiencing erosion, some 142,000 bank-miles were reported to have serious erosion. While this degree of erosion occurs on only 2 percent of the 7 million bank-miles in the nation, it resulted in an estimated total damage of \$200 million (1978 dollars) annually.

1.4.2 Mass Movement

Slope failures or mass movement can occur anywhere steep slopes and slide-prone or weak materials are encountered. The presence of water in the slope also plays a critical role, both in increasing driving forces and in reducing resistance to sliding. Slope failures affect and disrupt transportation routes, urban development, mining, and timber harvesting operations. These activities are very often themselves a major cause of mass movement.

Individual slope failures are usually not as spectacular as certain other nat-

TABLE 1-2. Damages Caused by Erosion and Sedimentation

Damages on Land	Damages in Streams
Agricultural productivity loss	Flooding
Roadway, sewer, basement siltation	Eutrophication
Drainage disruption	Increased treatment costs
Undermining of pavements and foundations	Siltation of harbors and channels
Dam failure casued by piping or overtopping (earth dams)	Loss of wildlife habitat and disruption of stream ecology
Gullying of roads	Lost recreational value
Adverse aesthetic impact	Adverse aesthetic impact

ural hazards such as earthquakes, major floods, and tornadoes. On the other hand, slope failures are collectively more widespread, and total financial loss from slope movements exceeds that of most other geologic hazards (Alfors et al., 1973; Fleming and Taylor, 1980). The costs of slope failures include both direct and indirect losses. Direct costs are associated with actual damages to installations or property (Figure 1-11). Indirect costs may include: (1) loss of

Figure 1-11. Destruction of hillside homes by mudslide. Photo taken in 1976. (Photo courtesy of Los Angeles Department of Building and Safety.)

tax revenues and reduced real estate values, (2) loss of productivity of agricultural or forest lands, (3) impairment or degradation of water quality, and (4) loss of industrial productivity caused by disrupted transportation systems. A photograph of a large landslide that resulted in discharge of hundreds of tons of sediment into the Navarro River in Northern California is shown in Figure 1-12. This landslide was associated with timber harvesting operations in the steep slopes above the river, a connection that is explored further in Chapter 3. This landslide, which also resulted in temporary damming of the Navarro River, is an excellent example of indirect costs associated with large mass movements or slope failures. As with erosion costs, indirect costs of slope failures are harder to evaluate, but they may be larger and more significant in many cases than direct costs.

With regard to transportation routes, a survey conducted by the Federal Highway Administration indicated that approximately $50 million was spent annually to repair landslides on the federally financed portion of the national highway system (Chassie and Goughnor, 1976). Distribution of the direct costs of major landslides for 1973 by Federal Highway Administration region within the United States is shown in Figure 1-13. The cost for an individual region was based on both landslide risk and the amount of highway construction in the area. In addition the costs represent a single year; the average cost for a particular region could vary significantly from the given costs.

These costs may represent only the tip of the iceberg. Significantly, the

Figure 1-12. Large landslide, associated with timber harvesting operations, that dammed the Navarro River in Northern California. (Photo provided by Nicholas Wilson.)

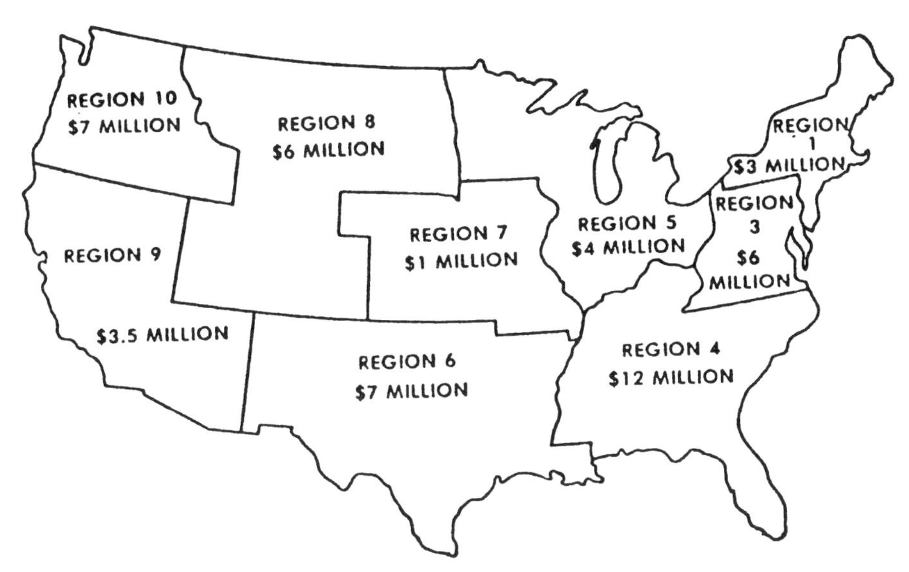

Figure 1-13. Cost of landslide repairs to federal-aid highways in the United States for 1973. (From Chassie and Goughnor, 1976.)

national highway system includes federal and state highways, but does not include most county and city roads, and streets or roads built by other government agencies. The U.S. Forest Service, for example, manages one of the largest transportation systems in the world. This agency authorizes the construction of some 7000 to 8000 miles of secondary or "low volume" forest roads each year and has maintenance responsibility for approximately a quarter million miles of existing roads (Howlett, 1975). Most of this system is located in steep, mountainous terrain comprised of frequently unstable, erodible, and fragile soils. As a result, erosion and mass movement pose constant problems. Costs of maintaining and repairing damage to this road system were estimated by the U.S. Forest Service at $82 million in 1975.

Total annual costs of landslides to highways and roads in the United States are difficult to determine precisely because of the aforementioned considerations and because of uncertainty in determining: (1) costs of smaller slides that are routinely handled by maintenance crews, (2) costs of slides on nonfederal-aid routes, and (3) indirect costs that are related to landslide damage. If these factors had been included, Chassie and Goughnor (1976) reckoned that $100 million would have been a conservative estimate of total annual cost of landslide damage to highways and roads in the United States.

Based on all the preceding estimates plus indirect costs and estimated damages to facilities not classed as roads and building, Schuster (1978) estimated direct and indirect costs of slope failures or mass movements in the United States to be in excess of $1 billion/year.

1.5 SCOPE AND ORGANIZATION OF BOOK

It should be noted at the outset that this book confines itself primarily to methods and techniques for protecting upland slopes against surficial erosion and mass movement. Upland slopes include natural slopes, ski slopes, embankment fills, highway and railroad cuts, landfill slopes, pipeline right-of-ways, mine spoil areas and tailings, gullies and ravines, and coastal dunes and bluffs. This represents a huge domain of problems and potential applications that are best treated in a single book. Streambank erosion problems are not specifically addressed. To be certain, many of the methods and techniques described herein are also applicable for streambank protection, but this topic lies largely outside the intended scope of this guidebook. Consideration of streambank erosion would require additional discussion and material on fluvial geomorphology. A reach of streambank cannot be treated in isolation without examining all the other factors controlling the dynamic stability of the river system as a whole. Accordingly, the reader is cautioned about applying any of the methods described herein to streambanks.

The book starts out in Chapter 2 by examining the two principal ways in which slopes are degraded, namley, by surficial erosion on the one hand and by mass movement on the other. Efforts to control or prevent this degradation are not likely to be effective unless the nature of the problem is well understood. It is essential, therefore, to review the mechanics of surficial erosion and mass movement, the factors that control them, and methods for predicting their occurrence.

Biotechnical and soil bioengineering rely heavily on the use of live plant material. Accordingly, Chapter 3 delves into the role and influence of slope vegetation on erosion and mass movement. Various mechanical and hydraulic influences of vegetation are examined in detail. Methods are presented there as well for maximizing the utility and value of vegetation while minimizing limitations and liabilities.

The principles of biotechnical stabilization and the main attributes and characteristics of soil bioengineering on the one hand and biotechnical stabilization on the other are discussed in Chapter 4. An attempt is made there to classify all the different approaches and methods to slope protection and erosion control in a systematic manner. This chapter also discusses the role and impact of slope grading and examines compatibility problems, if any, between engineering and biological elements of biotechnical structures.

The structural components or elements of biotechnical systems are discussed in Chapter 5, including basic requirements for internal and external stability of retaining walls, breast walls, and revetments.

The vegetative components are discussed in Chapter 6. Guidelines are presented therein for preliminary site analysis, species selection, site preparation, planting techniques, and maintenance. Emphasis is placed in this book on the collection, handling, and installation of live cuttings.

Soil bioengineering methods such as live staking, live fascines, brushlayers, branchpacking, and so on, are described in considerable detail in Chapter 7, with guidelines for their selection, construction, and aftercare. A number of specific applications or case studies are also cited in Chapter 7. These case studies include a highway cut stabilization, a major gully repair, and a high riverbank or bluff stabilization.

Biotechnical techniques enhance opportunities and options for landscaping on vertical and sloping surfaces and for softening the stark appearance of inert structural support systems. Guidelines for the integrated or combined use of vegetation with structural retaining walls, breast walls, and revetments are presented in Chapter 8.

Descriptions of biotechnical or composite ground covers and guidelines for their selection are presented in Chapter 9. These ground covers include mulch and erosion control blankets, turf reinforcement mats, erosion control revegetation mattresses, geocellular containment systems, and articulated block systems.

The book concludes with Chapter 10, which includes a discussin of new developments and likely future directions in the field of biotechnical stabilization. Anchored geonetting systems are a good example. Unlike most ground cover systems, an anchored geonet is tensioned and pulled down tightly on the ground surface by means of earth anchors, thereby imparting a normal or confining stress to the soil. The attributes of this system make it an attractive alternative to hard armor protection of sandy coastal landforms such as sand dunes, beaches, and bluffs. The "bio" in biotechnical stabilization is not restricted solely to flora or plant materials. Interesting developments with microscopic biological organisms or cultures as possible soil stabilizers are on the horizon. These and other new developments are described in Chapter 10 as well.

1.6 REFERENCES CITED

Alfors, J. T, et al. (1973). Urban geology: Master plan for California. *California Division of Mines and Geology Bulletin No. 198*, 112 pp.

Brandt, G. H. (1972). An economic analysis of erosion and sediment control methods for watersheds undergoing urbanization. Final Report to Department of Interior, Office of Water Resources Research, Contract No. 14-31-0001-3392.

Bryson, B. (1993). Britain's hedgerows. *The National Geographic Magazine* **185,** no. 10 (September) 78–100.

Chassie, R. G. and R. D. Goughnor (1976). National highway landslide experience. *Highway Focus* **8**(1): p. 1–9.

Coppin, N. J. and I. Richards (1990). Use of Vegetation in Civil Engineering. Sevenoaks, Kent (England): Butterworths.

Dowdeswell, W. H. (1987). *Hedgerows and Verges.* London: Allen & Unwin.

Edminster, F. C., W. S. Atkinson, and A. C. McIntryre (1949). Streambank erosion control on the Winooski River, Vermont. *USDA Circular No. 837*, 54 pp.

Fleming, R. W. and F. A. Taylor (1980). Estimating the costs of landslide damage in the United States. *U.S. Geological Circular No. 832*, 21 pp.

Gray, D. H. (1994). Influence of vegetation on the stability of slopes, *Proceedings*, International Conference on Vegetation and Slopes, Institution of Civil Engineers, University Museum, Oxford, England. 29–30 September, pp. 1–23.

Gray, D. H. and A. T. Leiser (1982). *Biotechnical Slope Protection and Erosion Control.* New York: Van Nostrand Reinhold.

Gray, D. H. and R. Sotir (1992). Biotechnical stabilization of cut and fill slopes. *Proceedings*, ASCE-GT Specialty Conference on Slopes and Embankments, Berkeley, CA, June, Vol. 2, pp. 1395–1410.

Greenway, D. R. (1987). Vegetation and slope stability, in: *Slope Stability*, edited by M. G. Anderson and K. S. Richards. New York: Wiley.

Hewlett, H. W. M., L. A. Boorman, and M. E. Bramley (1987). Design of reinforced grass waterways. CIRIA Report No. 116, Construction Industry Research and Information Association, London, 118 pp.

Howlett, M. R. (1975). Managing a 200,000-mile road system: Opportunity and challenge, in: "Low-Volume Roads," *Transportation Research Board Special Report No. 160*, NAS-NRC, Washington, DC, pp. 53–61.

Kraebel, C. J. (1936). Erosion control on mountain roads. *USDA Circular No. 380*, 43 pp.

National Science Foundation (1991). *Proceedings*, Workshop on Biotechnical Stabilization, edited by D. H. Gray, The University of Michigan, Ann Arbor, MI, August.

Schiechtl, H. M. (1980). *Bioengineering for Land Reclamation and Conservation.* Edmonton, Canada: University of Alberta Press, 404 pp.

Schuster, R. L. (1978). Introduction, in "Landslides: Analysis and Control," ed. by Schuster, R. L. and Krizek, R. J., Transportation Research Board Special Report 176, NAS-NRC, Washington, D.C., pp. 1–9.

USDA Soil Conservation Service (1940). Lake bluff erosion control. Open file Report prepared by USDA Soil Conservation Service, Michigan State Office, Lansing, MI, 81 pp.

USDA Natural Resources Conservation Service (1992). Chapter 18: Soil bioengineering for upland slope protection and erosion reduction. Part 650, 210-EFH, *Engineering Field Handbook*, 53 pp.

U.S. Army Corps of Engineers (1978). The Streambank Erosion Control Evaluation and Demonstration Act of 1974: Interim Report to Congress. Department of the Army, Washington, DC.

U.S. Environmental Protection Agency (1973). Comparative costs of erosion and sediment control construction activities, EPA-430/9-73-016, U.S. Government Printing Office, Washington, DC.

U.S. Environmental Protection Agency (1992). National Water Quality Inventory: 1992 Report to Congress. USEPA Office of Water, EPA841-R-001.

Vanicek, V. 1973). The soil protective role of specially shaped plant roots. *Biological Conservation* **5**(3): 175–180.

Weaver, W. and R. A. Sonnevil (1981). Relative cost effectiveness for erosion control in Redwood National Park. *Proceedings*, Symposium on Watershed Rehabilitation in Redwood National Park and Other Coastal Areas, U.S. National Park Service, pp. 341–360.

White, C. A. and A. L. Franks (1978). Demonstration of erosion and sediment control technology: Lake Tahoe region of California. California State Water Resources Control Board, Final Report, Sacramento, CA, 393 pp.

2 Surficial Erosion and Mass Movement

2.0 INTRODUCTION

Hills and uplands form as the result of tectonic forces that warp the earth's crust. Plutonic rock masses can also push up through the crust, forming mountain ranges such as the Sierra Nevada. Volcanic activity may bring up molten rock from the earth's interior and deposit it on the surface to create mountains and volcanoes as well. These mountainous areas and uplands in turn are subject to degradation and wear by the twin processes of surficial erosion and mass movement. Man-made slopes, for example, cuts and embankment slopes, are subject to the same degradation processes.

In order to control or prevent this wearing or wasting away of the earth's surface, it is first necessary to understand these two processes of degradation and the factors that control them. While the two processes share many similarities, they also differ in important respects. Surficial erosion entails the detachment and transport of individual particles whereas mass movement entails the movement of relatively large, initially intact masses of soil and/or rock along critical failure surfaces. Gravity is the main driving force behind mass wasting; wind and running water are the principal agents of erosion. The role and function of vegetation also differ substantially between these two processes.

2.1 DEFINITIONS

2.1.1 Surficial Erosion

Surficial erosion or soil erosion is the removal of surface layers of soil by the agencies of wind, water, and ice. Soil erosion involves a process of both particle *detachment* and *transport* by these agencies. Erosion is initiated by drag, impact, or tractive forces acting on individual particles of soil at the surface. Weathering processes such as frost action and wet-dry cycling set the stage for erosion by breaking up rock into smaller particles and weakening bonds between particles.

The two most common types of erosion are rainfall and wind erosion. Rainfall erosion starts with falling raindrops themselves. When these drops impact on bare or fallow ground, they can dislodge and move soil particles a surprising distance. At the onset of runoff, water collects into small rivulets, which may erode very small channels called rills. These rills may eventually coalesce into larger and deeper channels called gullies. Gullying is a complex and destructive process; once started, gullies are difficult to stop. Bare or unprotected earth surfaces are the most vulnerable to all forms of surficial erosion.

Erosion may also occur along streambanks where the velocity of the flowing water is high and the resistance of the bank material often low. Continuous or prolonged periods of inundation coupled with a variety of other degradation processes, also make streambank protection a formidable task. Piping or spring sapping is yet another type of erosion, in this case caused by seepage and emergence of water from the face of an unprotected slope.

2.1.2 Mass Movement

Mass movement is a descriptive name for the downward and outward movement of slope-forming materials—natural rock, soils, artificial fills , or combinations of these materials. The terms "mass erosion" or "mass wasting" are sometimes used. Mass movements are popularly known as landslides. Strictly speaking, however, landslides or slides refer to a particular type of mass movement.

Unlike soil erosion, mass movement involves the sliding, toppling, falling, or spreading of fairly large and sometimes relatively intact masses. Mass movements have been classified into categories based largely on the type of movement and material involved (Varnes, 1978). A slide is a relatively slow slope movement in which a shear failure occurs along a specific surface or combination of surfaces in the failure mass. Slides are amenable to quantitative stability analysis by techniques of limiting force equilibrium and limit analysis.

2.1.3 Salient Characteristics and Differences

Many of the same slope, soil, and hydrologic factors that control surficial erosion also control mass movement (e.g., steepness of slope and shear strength of soil). The two processes differ, however, in some important respects. The salient characteristics of surficial erosion on the one hand and mass movement on the other are contrasted in Figure 2-1. Precipitation, a key factor directly affecting rainfall erosion, only affects mass movement indirectly via its influence on the groundwater regime at a site. In contrast, geologic conditions such as orientation of joints and bedding planes in a slope, can have a profound influence on mass stability but not on surficial erosion. Vegetation has an important influence on both erosion and shallow mass movement.

Different predictive techniques or models have been developed to determine soil losses from surficial erosion or the likelihood of catastrophic slope failure in the case of mass movement. Rainfall and wind erosion, for example, are controlled by a number of soil, climatic, and topographic factors, including intensity and duration of precipitation, ground roughness, length and steepness of slope,

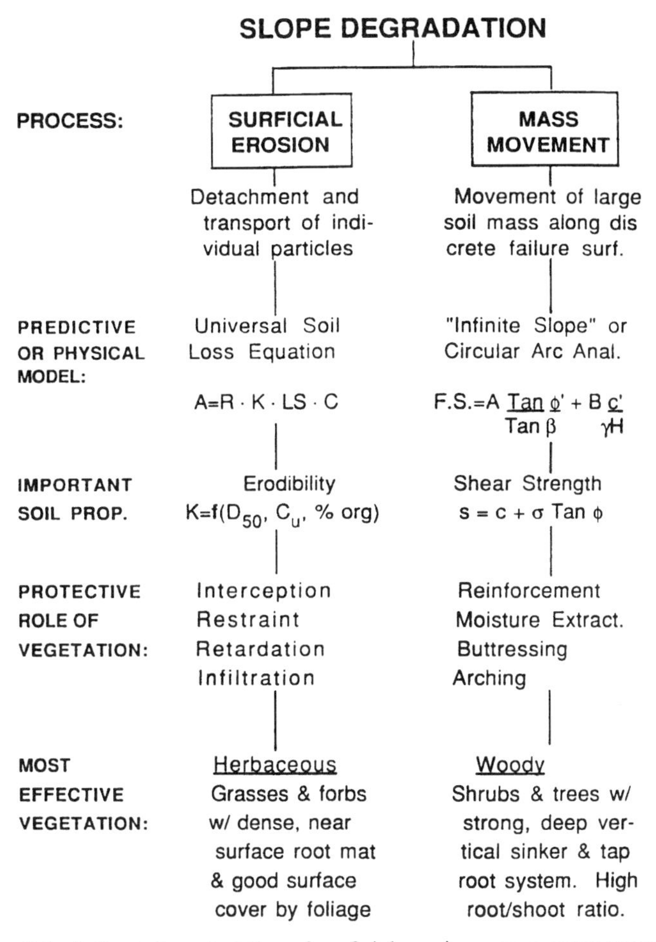

Figure 2-1. Salient characteristics of surficial erosion versus mass movement.

inherent soil erodibility, and type or extent of cover. All of these factors are taken into account explicitly in the universal soil loss equation (Wischmeier and Smith, 1978).

2.2 NATURE OF SURFICIAL EROSION

2.2.1 Agents and Types of Erosion

The primary agents of erosion include water, wind, and ice. These agents can scour and remove soil particles as a result of flowing past, impacting upon, or exiting from the surface of a soil. Each agent can also erode soil in quite different manners. Rainfall erosion, for example, can occur in the form of splash, sheet, or rill erosion. The various agents and types of erosion are summarized in Table 2-1.

Degradation processes often act jointly or in combination with one another. Glaciated landscapes, for example, may be the product of glacio-fluvial processes. Ground disturbed by creep or solifluction is more vulnerable to rilling and gullying. Groundwater piping likewise can trigger or promote gullying.

2.2.2 Mechanics of Erosion

Prevention and control of erosion depends on understanding the mechanics of the erosion process. Erosion is basically a twofold process that involves:

1. Particle detachment
2. Particle transportation

The forces acting on particles near a fluid-bed boundary or interface are shown schematically in Figure 2-2. Drag or tractive forces exerted by the flowing fluid are resisted by inertial or cohesive forces between particles.

Erosion is *caused* by: Drag or tractive forces (Water, wind, ice) = f {• Velocity • Discharge • Shape and roughness}

TABLE 2-1. Agents and Types of Erosion

Agent	Type of Erosion or Degradation Process	
Water	1. Raindrop splash	
	2. Sheet erosion	
	3. Rilling	
	4. Gullying	
	5. Stream channel erosion	
	6. Wave action	
	7. Piping and sapping	
Ice	1. Solifluction	
	2. Glacial scour	
	3. Ice plucking	
Wind	Wind erosion cannot be subclassified into "types"; instead it varies mainly by "degree."	
Gravity	1. Creep	These are usually classified under mass wast-ing, but they often act in conjunction with erosion
	2. Earth flow	
	3. Avalanche	
	4. Debris slide	

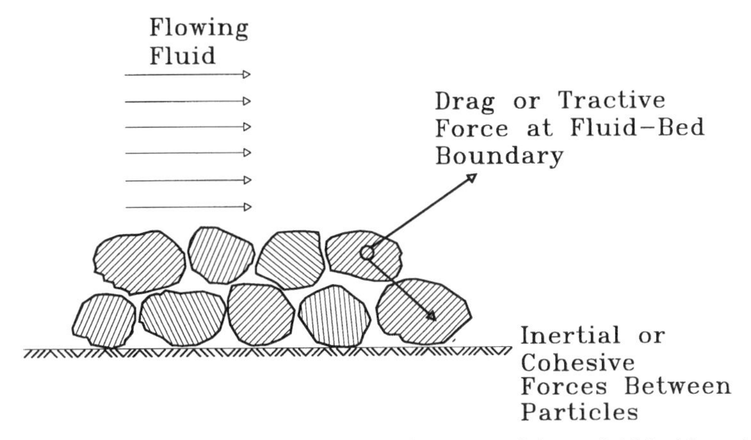

Figure 2-2. Schematic diagram of forces acting on particles at fluid-bed boundary.

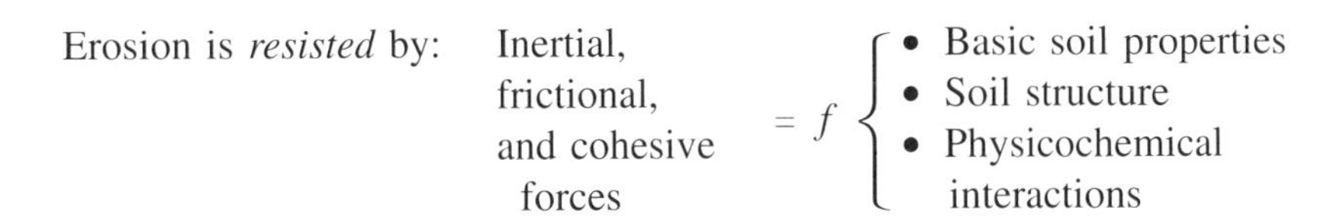

Erosion protection essentially consists of: (1) decreasing drag or tractive forces by decreasing the velocity of water flowing over the surface or by dissipating the energy of the water in a defended area, and (2) increasing resistance to erosion by protecting/reinforcing the surface with a suitable cover or by increasing interparticle bond strength.

2.3 PRINCIPAL DETERMINANTS OF EROSION

2.3.1 Rainfall Erosion

Rainfall erosion is controlled by four basic factors, namely, climate, soil type, topography, and vegetative cover. This relationship can be expressed schematically as follows:

Rainfall erosion = *f* {
- Climate—storm intensity and duration
- Soil—inherent erodibility
- Topography—length and steepness of slope
- Vegetation—type and extent of cover

The most important climatic parameters controlling rainfall erosion are intensity and duration of precipitation. Wischmeier and Smith (1958) have shown that the most important "single" measure of the erosion-producing power of a rainstorm is the product of rainfall energy times the maximum 30-minute rainfall intensity. Raindrops impacting on bare soil not only cause erosion; they also tend to compact the soil and decrease infiltration capacity.

The susceptibility of a soil to erosion is known as it "erodibility." Some soils (e.g., silts) are inherently more erodible than others (e.g., coarse, well-graded gravels). In general, increasing the organic content and clay size fraction of a soil decreases erodibility. Erodibility also depends upon such parameters as soil texture, antecedent moisture content, void ratio, exchange ions, pH, and composition or ionic strength of the eroding water. The dependence of soil erodibility on all these variables is summed up in Table 2-2.

There is no simple and universally accepted erodibility index for soils. Instead, various tests have been proposed for this purpose, including the SCS dispersion test (Volk, 1973), crumb test (Emerson, 1967), and pinhole test (Sherard et al., 1976). A suggested hierarchy of erodibility based on the Unified Soil classification system is:

Most Erodible ⟶ Least Erodible

ML > SM > SC > MH > OL ≫ CL > CH > GM > SW > GP > GW

where:

GW = well graded gravel
GP = poorly graded gravel
GM = silty gravel
SW = well graded sand
SM = silty sand
SC = clayey sand
ML = low plasticity silt
MH = high plasticity silt
CL = low plasticity clay
CH = high plasticity clay
OL = low plasticity organic soil

This erodibility hierarchy is simple, but based on gradation and plasticity indices of remolded or disturbed soils. Accordingly, it fails to take into account effects of soil structure, void ratio, and antecedent moisture content. Wischmeir et al. (1971) published an erodibility nomograph for use with the universal soil loss equation (see Section 2.5) that is based on easily measured soil properties.

TABLE 2-2. Soil Erodibility Trends

- Is low in well-graded, coarse gravels
- Is high in uniform silts and fine sands
- Decreases with increasing clay and organic content
- Decreases with low void ratios and high antecedent moisture content
- Increases with increasing sodium adsorption ratio and decreasing ionic strength of water

Topographic variables influencing rainfall erosion are: (1) slope angle, (2) slope length, and (3) size and shape of watershed. The influence or importance of length tends to increase as slopes become steeper. For instance, a doubling of slope length from 100 to 200 feet will only increase soil losses by 29 percent on a 6-percent slope, whereas the same doubling of slope length in a 20-percent slope will result in a 49 percent increase in soil loss. This is one of the reasons for benching or terracing and contour wattling long, steep slopes.

Vegetation plays an extremely important role in controlling rainfall erosion. Removal or stripping of vegetation by either human or natural agencies (e.g., wildfires) often results in accelerated erosion. Conversely, any measure that either preserves or enhances vegetation establishment will significantly retard soil loss and minimize erosion. The protective mechanisms of vegetation in preventing both surficial erosion and shallow mass movement are discussed at length in Chapter 3.

2.3.2 Wind Erosion

Wind erosion is controlled by the same basic factors that control rainfall erosion. The dependence of wind erosion on these factors can be expressed schematically in a similar fashion as follows:

Wind erosion = f {
- Climate—temperature, rainfall distribution wind velocity
- Soil—texture, particle size, moisture content, surface roughness
- Vegetation—type, height, and density of cover, seasonal distribution

These factors again can be expressed in terms of identifiable and measurable parameters as noted above. Unlike rainfall erosion, topographic parameters such as length and steepness of slope are relatively unimportant in the case of wind erosion. On the other hand, surface roughness and the presence of low barriers that act as wind breaks and sediment traps can be important. The climatic factors that most affect soil moisture are amount and distribution of rainfall, temperature, and humidity. Only relatively dry soils are susceptible to wind erosion. The most important characteristics of the wind are its velocity, duration, direction, and degree of turbulence. Wind can only pick up and carry in suspension dry soils with particle sizes primarily less than 0.1 mm, that is fine silt size material.

Wind erosion consists of three distinct phases: initiation of movement (detachment), transportation, and deposition. Soil movement by wind is initiated as a result of turbulence and velocity. The velocity required to start movement increases as the weight of particles increases. For many soils, this velocity is about 13 mph at a height of 1 foot above the ground. The velocity required

TABLE 2-3. Movement of Soil Particles by Wind

Mechanics of Movement	Size of Particles Moved (mm)	Percent of Soil Moved
Suspension	<0.1	3–38
Saltation (skipping and bouncing)	0.1–0.5	55–72
Surface creep (rolling and sliding)	0.5–1.0	7–25

Source: After Chepil (1945).

to sustain movement, once started, is less than that required to initiate movement. Similar relationships have been observed in the case of erosion by flowing water.

Laboratory studies by Chepil (1945) established that soil particles are transported by wind in the manner shown in Table 2-3. The major portion of soil particles transported by the wind occurs near the ground surface at heights under 3 feet (1 m). Approximately 62 to 97 percent of the total wind-eroded soil is transported in this zone near the surface, a fact that suggests the utility of installing relatively low barriers or windbreaks to filter and impede the movement of windborne soil. Vegetation readily serves this purpose in addition to its other control functions such as increasing surface roughness, slowing and deflecting the wind, and binding soil particles together. Because of these properties vegetation can be used very effectively in combination with fencing to trap drifting sand and build up dunes along beaches. Hardy, drought and salt tolerant plants, such as sea oats, beach grass, and bitter panicum, work well in this regard, as shown in Figure 2-3.

Figure 2-3. Dune vegetation helps to trap drifting sand and build foredunes shoreward of dry beach.

2.4 TYPES OF WATER EROSION

Water erosion manifests itself in many different ways. Rainfall erosion begins with raindrop splash and can progress ultimately to gullying and stream channel erosion. The salient characteristics of different types of water erosion are briefly reviewed next.

Raindrop Splash: Raindrop splash results from the impact of water drops falling directly on exposed soil particles or thin water surfaces covering the ground. Tremendous quantities of soil can be splashed into the air in this manner. On bare ground it has been estimated (Ellison, 1948) that as much as 100 tons/acre can be splashed into the air in a heavy storm. Splashed particles may move more than 2 feet vertically and 5 feet laterally on level ground. On steep slopes this splashing will cause a net downslope movement of soil.

Sheet Erosion: Sheet erosion is the removal of soil from sloping land in thin layers or sheets. From an energy standpoint raindrop erosion appears to be more important than sheet erosion because most raindrops have velocities of about 20 to 30 fps, whereas overland flow velocities are about 1 to 2 fps. Dry ravel and slope wash are forms of sheet erosion; the former occurs when surface layers of coarse-textured soil dry out and lose their apparent cohesion. The latter occurs when rainfall erodes without causing rilling or gullying. Sheetlike erosion is an important mechanism of slope retreat and source of sediment in cut slopes in granitic and andesite soils. Highway cuts in these soils often give the impression of being very stable (e.g., rills and gullies are absent) yet discharge tons of soil year after year into the roadside ditches (Howell et al., 1979).

Rilling: Rill erosion is the removal of soil by water from very small but well-defined, visible channels or streamlets where there is concentration of overland flow. An example of rill erosion at an urban construction site is shown in Figure 2-4. Rilling generally is more serious than sheet erosion because runoff velocities are higher in the rills or channels. Rilling is the form or erosion during which most rainfall erosion losses occur (Schwab et al., 1981). Rill erosion is most serious where intense storms occur in watersheds or sites with high runoff-producing characteristics and loose, shallow topsoil. Rills are sufficiently large and stable to be seen readily, but small enough to be removed easily by normal tillage and grading operations.

Gullying: Gullies are intermittent stream channels larger than rills. These channels carry water during and immediately after rains, and, unlike rills, gullies cannot be obliterated by normal tillage. Gullies tend to form where large volumes of runoff are concentrated and discharged onto steep slopes with erodible soils, e.g., undefended culvert outlets. Gullying is common in grasslands and is likely the principal form of erosion in steep, forested

Figure 2-4. Rill erosion at an urban construction site. Ann Arbor, Michigan.

watersheds. The dynamics of gully formation are complex and not completely understood. Various statistical models for predicting gully growth and development have been proposed (Beer and Johnson, 1963; Thompson, 1964). Gullies may not be as significant as rills in terms of total quantities of soil eroded, but they can be quite destructive in terms of damage to roadways, embankments, and watersheds as shown in Figure 2-5. They are also difficult to control and arrest. Effective gully control must stabilize both the channel bottom and headcuts. Continued downcutting of gully bottoms leads to deepening and widening, whereas headcutting extends the channel into ungullied headwater areas, and increases the stream net and its density by developing tributaries.

Stream Channel Erosion: Stream channel erosion consists of soil removal from stream banks and/or sediment scour along the channel bottom. Stream channel erosion should be considered separately from the rainfall-associated types of erosion discussed previously. A number of hydrologic/hydraulic and geomorphic variables govern the behavior of fluvial systems. These variables are in a dynamic equilibrium with one another. Investigators such as Leopold (1994), and Lane (1955) have determined several general relationships among these variables that are useful in analyzing stream activity. This balance or equilibrium will determine, for example, where material is likely to be eroded as opposed to deposited along a particular reach of stream.

Like gully erosion there are several processes acting along streams that

Figure 2-5. Hillside gully erosion caused by uncontrolled and concentrated discharge of runoff from parking lot.

are responsible for streambank and channel degradation (Keown et al., 1977). These include:

1. *Mass Wasting:* Slumping or sliding induced by undercutting, over-steepening, or impeded drainage during flood recession
2. *Flow Erosion:* Tractive stress imposed by flowing water
3. *Piping:* Seepage erosion as a result of bank drainage
4. *Wave Erosion:* Pumping action, pore pressure fluctuations, and wave run-up problems
5. *Freeze/Thaw Degradation:* Solifluction, volume instability, and impeded drainage

It is helpful to recognize and understand these processes when designing a prevention and control system. On the other hand, this classification of erosion processes is not particularly helpful in establishing the cause of stream channel and bank erosion. The cause must be linked to fluvial geomorphology and the play or relationship between hydraulic/hydrologic variables. From the stand-point of fluvial geomorphology, there appear to be three main actions or events (Keown et al., 1977; Leopold, 1994) that result in stream erosion:

1. *Widening:* Channel enlargement caused by increased streamflow and/or sediment discharges.

2. *Deepening:* Scouring of the channel bottom caused by increased flows and or changes in slope.
3. *Sinuosity Change:* Bank loss that occurs during and upon a change in planform or stream meander configuration. This bank loss is usually accompanied by accretion somewhere else along the affected reach.

Groundwater Erosion: Groundwater erosion is the removal of soil caused by groundwater seepage or movement toward a free face. Such erosion is commonly referred to as piping. The phenomenon is also known as spring sapping—literally the detachment and movement of soil particles at the point of emergence of a spring or seep in the ground. Piping occurs when seepage forces exceed intergranular stresses or forces of cohesion.

Pipes can form in the downstream side of earth dams (Sherard et al., 1972), gully heads, streambanks, and slopes where water exits from the ground. Once a pipe or cavity forms, it enlarges quickly because flow lines are attracted to areas of lower flow resistance, and this in turn results in further concentration of flow lines or flow net density in a positive feedback cycle.

2.5 SOIL LOSS PREDICTIONS

2.5.1 Historical Development

A semiempirical equation known as the universal soil loss equation (USLE) was developed by the USDA Agricultural Research Service in the early 1960s. A handbook that served as the main reference for this equation was first introduced in the mid 1960s (Wischmeier and Smith, 1965). A revised and definitive treatment of the USLE was subsequently published in the late 1970s as Agricultural Handbook 537 (Wischmeier and Smith, 1978). This relationship was developed originally for predicting erosion losses from cropland east of the Rocky Mountains. In subsequent years the USLE was modified and adapted to different regions of the United States (USDA Soil Conservation Service, 1972, 1977) and also for use at urban or highway construction sites (Israelson, 1980). These latter uses have stirred some controversy about the limits and applicability of the USLE and have prompted numerous efforts for further revision (Renard et al., 1991).

The USLE takes into account all the factors known to affect rainfall erosion, namely climate, soil, topography, and vegetative cover. It is based on a statistical analysis of erosion measured in the field on scores of test plots under natural and simulated rainfall. The annual soil loss from a site is predicted according to the following relationship:

$$A = R \cdot K \cdot LS \cdot C \cdot P \tag{2-1}$$

where: A = computed soil loss (e.g., tons) per acre for a given storm period or time interval;
R = rainfall factor;
K = soil erodibility value;
L = slope length factor;
S = steepness factor;
C = vegetation factor;
P = erosion control practice factor

2.5.2 Applications of USLE

Detailed procedures and charts for estimating soil losses and calculating values for the various parameters in the USLE are given in Gray and Leiser (1982) and Goldman et al. (1986). This information is not repeated herein; instead the objective will be to show which parameters have the greatest variability and effect on soil loss and which can be easily modified or changed.

The climatic factor (R) and the erodibility factor (K) only vary within one order of magnitude. These factors are essentially fixed for a given site and are not subject to change or modification. The cover factor (C) and the topographic factor (LS), on the other hand, can vary over several orders of magnitude, as shown in Tables 2-4 and 2-5 respectively. Moreover, unlike the other factors, the cover factor (C) and the topographic factor (LS) can be modified substantially to suit a particular erosion control objective. The topographic factor (LS) can be modified most easily by reducing slope length, that is by converting a long, steep slope into a series of short, steep slopes. This is commonly accomplished by benching or terracing. The use of live fascines or contour wattles (see Chapter 7) also achieves the same objective. The placement of fascines at

TABLE 2-4. Topographic Factors (*LS*) for Slopes

Slope	Slope	*LS* Values for Selected Slope Lengths *L* (feet(m))						
Ratio	Gradient	10	30	50	100	300	500	1000
(*H* : *V*)	*S* (%)	(3.0)	(9.1)	(15.2)	(30.5)	(91.0)	(152.0)	(305.0)
20 : 1	5	0.17	0.29	0.38	0.53	0.93	1.20	1.69
10 : 1	10	0.43	0.75	0.97	1.37	2.37	3.06	4.33
8 : 1	12.5	0.61	1.05	1.36	1.92	3.33	4.30	6.08
6 : 1	16.7	0.96	1.67	2.15	3.04	5.27	6.80	9.62
5 : 1	20	1.29	2.23	2.88	4.08	7.06	9.12	12.90
4 : 1	25	1.86	3.23	4.16	5.89	10.20	13.17	18.63
3 : 1	33.5	2.98	5.17	6.67	9.43	16.33	21.09	29.82
2.5 : 1	40	4.00	6.93	8.95	12.65	21.91	28.29	40.01
2 : 1	50	5.64	9.76	12.60	17.82	30.87	39.85	56.36
1.75 : 1	57	6.82	11.80	15.24	21.55	37.33	48.19	68.15
1.50 : 1	66.7	8.44	14.61	18.87	26.68	46.22	59.66	84.38
1.25 : 1	80	10.55	18.28	23.60	33.38	57.81	74.63	105.55
1 : 1	100	13.36	23.14	29.87	42.24	73.17	94.46	133.59

Source: Adapted from Israelson (1980).

TABLE 2-5. Vegetation Factor *C* Values for Pasture, Rangeland, and Idleland

Type and Height of Raised Canopy	Canopy Cover (%)	Canopy Type[a]	*C* Values for Selected Canopy and Ground Cover Conditions (Percent Ground Cover) 0	20	40	60	80	95–100
No appreciable canopy	——	G	.45	.20	.10	.042	.013	.003
		W	.45	.24	.15	.090	.043	.011
Canopy of tall weeds or short brush (0.5-m fall height)	25	G	.36	.17	.09	.038	.012	.003
		W	.36	.20	.13	.082	.041	.011
	50	G	.26	.13	.07	.035	.012	.003
		W	.26	.16	.11	.075	.039	.011
	75	G	.17	.10	.06	.031	.011	.003
		W	.17	.12	.09	.067	.038	.011
Appreciable brush or bushes (2-m fall height)	25	G	.40	.18	.09	.040	.013	.003
		W	.40	.22	.14	.085	.042	.011
	50	G	.34	.16	.09	.038	.012	.003
		W	.34	.19	.13	.081	.041	.011
	75	G	.28	.14	.08	.036	.012	.003
		W	.28	.17	.12	.077	.040	.011
Trees but no appreciable low brush low brush (4-m fall height)	25	G	.42	.19	.10	.041	.013	.003
		W	.42	.23	.14	.087	.042	.011
	50	G	.39	.18	.09	.040	.013	.003
		W	.39	.21	.14	.085	.042	.011
	75	G	.36	.17	.09	.039	.012	.003
		W	.36	.20	.13	.083	.014	.011

Source: USDA Soil Conservation Service (1978).

[a] G: Cover at surface is grass, grasslike plants, decaying compacted duff, or litter at least 2 inches deep.
W: Cover at surface is mostly broadleaf, herbaceous plants (as weeds) with little lateral-root network near the surface, and/or undecayed residue.

intervals of 10 feet, for example, in a 100-foot long slope, would reduce the soil loss to one-third of the original value. This reduction can be observed by examining the soil loss or *LS* factor values in Table 2.4 that correspond to a 100-foot long versus a 10-foot long slope for a given slope gradient.

The vegetative cover or *C* factor affects erosion via three separate and distinct, but interrelated, zones of influence, namely canopy cover, ground cover (vegetative cover in direct contact with the soil), and vegetative or crop residue at or beneath the surface. The effects of these three constituent influences can be observed in Table 2-5. For completely bare or fallow ground, the *C* factor is unity. The influence of ground cover is paramount compared to the type, extent, and condition of the canopy. This influence can readily be ascertained by comparing *C* factor values in the row versus column headings of Table 2-5. Even in the absence of appreciable canopy, the *C* factor value can drop to a value of only 0.003 when ground cover reaches 95 percent. This corresponds to almost a thousandfold reduction in erosion losses over the fallow or bare-ground case. None of the other variables or factors are amenable to management with such dramatic results as this one. Mulches and biotechnical ground covers (nets, erosion control blankets, turf reinforcement mats, etc.) provide initial erosion protection and enhance vegetative establishment and performance. They too can be assigned *C* factor values that are useful for evaluation and comparison, as shown in Table 2-6.

TABLE 2.6 Factors for Different Ground Covers

Type of Cover	*C* Factor	Soil Loss Reduction (%)
None	1.0	0
Native vegetation (dense, undisturbed)	0.01	99
Temporary seedings:		
90% cover, annual grasses, no mulch	0.1	90
Wood fiber mulch, 0.75 ton/acre, with seed	0.5	50
Excelsior mat, jute netting (slopes up to 2 : 1	0.3	70
Straw mulch		
1.5 tons/acre (3.4 t/ha), tacked down	0.2	80
4.0 tons/acre (9.0 t/ha), tacked down	0.05	95

Source: Goldman et al. (1986).

Mulches provide immediate protection on bare soils and may perform better than temporary seedings in certain instances. Straw is particularly effective as a mulch and ground cover (Kay, 1983). Straw that has been tacked down provides better protection (lower soil loss) than wood fiber mulch at equal application rates. Mulches and biotechnical ground covers are described and evaluated in greater detail in Chapter 9.

Factor *C* values tend to change with time following certain types of surface treatment such as mulching, seeding, and transplanting. For example, factor *C* values for grass may decrease from 1.0 (for fallow, bare ground) to about 0.01 between time of initial seeding and full establishment. In this case an average or weighted factor *C* value can be computed for a construction period using time as the weighting parameter and estimating *C* factor values for successive time intervals (Gray and Leiser, 1982).

The erosion control practice factor, or *P* factor, has been interpreted in several ways. These practices can be viewed as stuctural-mechanical, hydraulic, or grading practices that reduce the velocity of runoff and/or the tendency of water to flow directly downslope. Structural measures include grade stabilization structures (chutes, flumes, check dams, water ladders), level spreaders, diversions, and interceptor berms. Even sediment basins can be classified and evaluated as a form of control practice. Basins do not stop erosion; instead, they keep eroded material from leaving a site and causing off-side damage. Methods for determining and assigning factor P values for structural measures and basins are given elsewhere (U.S. Environmental Protection Agency, 1973).

In agricultural uses of the USLE, the *P* factor is often employed to describe different plowing and tillage practices. In construction site applications, on the other hand, *P* factor reflects the condition of the ground surface after grading operations. The ground surface condition or roughness can be affected by tractor treads and by raking, disking, or rolling, as shown in Table 2-7.

Some interesting trends can be observed. A slope left in a compacted and smooth condition is the most vulnerable to erosion; it is also the least amenable to vegetative establishment by seeding. Trackwalking up and down slope is

TABLE 2.7. Ground Surface Condition Factor *P* Values

Surface Condition	*P* Factor
Compacted and smooth	1.3
Trackwalked along contour	1.2
Trackwalked up and down slope	0.9
Punched straw	0.9
Rough, irregular cut	0.9
Loose to 12-inch (30-cm) depth	0.8

Source: From Goldman et al. (1986).

superior to trackwalking on contour because the orientation of the tractor cleats is parallel to the slope in the former. Cleat indentations oriented up and down slope can become the locus of rills, as shown in Figure 2-6. As noted in Table 2.7 changing the surface condition does not have a large effect on soil loss since all the factor *P* values are close to unity. On the other hand, the degree of surface roughness and orientation of indentations does have a significant influence on the lodgment of seed and establishment of vegetation on a slope, as shown in Figure 2-7.

Figure 2-6. Accelerated rill erosion as a result of trackwalking across the slope, which resulted in cleat indentations oriented up and down slope.

Figure 2-7. Lodgment of seed and establishment grass in cleat indentations oriented parallel to the slope.

2.5.3 Limitations of USLE

The USLE is an empirically based equation that predicts sheet and rill erosion from relatively small areas. Although it explicity takes into account all the factors known to affect rainfall erosion and is widely employed around the world, it has the following limitations:

- *It is Empirical.* Considerable judgment is required in assigning correct values to some of the factors in certain situations. This is particularly true in the case of the vegetation or cover factor.
- *It Predicts Average Annual Soil Loss.* The rainfall factor is based on the 2-year, 6-hour rainfall. Unusual storms or weather events during a particular year could produce more sediment than predicted.
- *It Does Not Predict Gully Erosion.* The USLE predicts soil loss from sheet and rill erosion, not erosion resulting from concentrated flow in large channels or gullies.
- *It Does Not Predict Sediment Delivery.* The USLE predicts soil loss, not sediment deposition. Soil lost from the upper portions of a slope or watershed will not automatically end up as sediment in a body of water. Instead it may collect on the lower portions of a slope or watershed.

In spite of its limitations, the USLE provides a simple, straightforward method of estimating soil losses, identifying critical areas, and evaluating the effectiveness of soil loss reduction measures. It provides a rational basis for sizing sediment basins and other sediment collection systems at construction sites. Significantly, the USLE provides an idea of the range of variability of each of the parameters, their relative importance in affecting erosion, and the extent to which each can be changed or managed to limit soil losses.

2.6 EROSION CONTROL PRINCIPLES

Surficial erosion can be controlled or prevented by observing some basic principles. These principles are universally applicable; they should be observed regardless of whether conventional or soil bioengineering treatments are contemplated. Erosion control principles are based on common sense, but are frequently violated in site development work. Many erosion control measures and products have been introduced over the years; they are more effective when applied in conjunction with the principles that are enumerated below:

1. Fit the development plan to the site. Avoid extensive grading and earthwork in erosion prone areas.
2. Install hydraulic conveyance facilities to handle increased runoff.
3. Keep runoff velocities low.
4. Divert runoff away from steep slopes and denuded areas by constructing interceptor drains and berms.
5. Save native site vegetation whenever possible.
6. If vegetation must be removed, clear the site in small workable increments. Limit the duration of exposure.
7. Protect cleared areas with mulches and temporary, fast growing herbaceous covers.
8. Construct sediment basins to prevent eroded soil or sediment from leaving the site.
9. Install erosion control measures as early as possible.
10. Inspect and maintain control measures.

Observance of these ten basic principles will greatly minimize erosion losses. Goldman et al. (1986) discuss and elaborate on their importance.

Some additional discussion is warranted here on the impact of grading practices on soil erosion. Both transportation corridors and residential developments in steep terrain require that some excavation and regrading be carried out to accommodate roadways or building sites. The manner in which this grading is planned and executed, and the nature of the resulting topography or landforms that are created affect not only the visual or aesthetic impact of the development but also the mass and surficial stability of the slopes and effectiveness of landscaping and revegetation efforts. A linear or planar slope will exhibit higher soil loss than a slope with a concave or decreasing gradient near the toe. Likewise, drainage channels brought down and across a slope in a curvilinear manner, which lengthens flow path and reduces gradient, are less susceptible to erosion than exposed channels brought directly down the face of a slope. Landform grading techniques (Schor, 1992, 1993) greatly mitigate erosion problems by observing these concepts and mimicking stable, natural slopes. Landform grading concepts are discussed further in Chapter 4.

2.7 NATURE OF MASS MOVEMENTS

2.7.1 Types of Slope Movements

Various schemes have been proposed over the years for classifying and describing slope movements (Sharpe, 1938; Varnes, 1958, 1978). Essential characteristics and attributes of mass soil movements are summarized in Table 2-8. The

TABLE 2-8. Basis for Classifying Mass Soil Movements

A. Materials

1. Ice
2. Rock (jointed, weathered, bedded, folded)
3. Soil (dry or saturated, sandy vs. clayey)

B. Velocity

1. Rapid (seconds → minutes): rockfalls, avalanches, earthflows
2. Intermediate (minutes → hours): debris slides, block glides, slumps
3. Slow (days → years): creep, solifluction, lateral spreading

C. Failure Mechanisms

1. Slides: Movement along well-defined sliding or shearing surface of largely intact block masses of earth and rock.
   a. *Planar:* Occurs in slopes where there is some geologic control, e.g., bedding planes, joints, colluvium mantle; also shallow slides (sloughing) in homogeneous sandy slopes.
   b. *Rotational:* Occurs in slopes composed of homogeneous, cohesive soils in which resistance to sliding is independent of depth; critical sliding surface tends to be an arc passing deeply under slope where shear resistance is lowest and shear stresses high.
2. Flows: Quasi viscous flow in which it is difficult to detect a distrinct sliding surface; motion dies out with depth; tends to occur in saturated soils (sands, silts, clays) with a high water content.
3. Falls: Falling mass of material loses coherent contact with stable, unmoving base; occurs in jointed brittle rock forming steep slopes.

classification system proposed by Varnes (1958, revised 1978) is perhaps the most useful and widely adopted. His system is based on two main variables:

1. Type of movement
2. Type of material

Types of movements or failure mechanisms are divided into five main groups: falls, topples, slides, spreads, and flows, as noted in Table 2-9. Materials are divided into two classes: rock and engineering soil. Soil is further divided into debris and earth. Debris is material that is transitional in gradation between rock and earth. An abbreviated version of Varnes (1958) classification system is presented in Table 2-9.

A comprehensive description of the various types of slope movements is outside the scope of this book. The interested reader should consult the original papers by Varnes (1958, 1978) for this purpose. On the other hand, it is important to understand and to be able to identify various types of slope movements, not only for purposes of avoiding unstable slopes but also for designing prevention and control systems. Methods useful for prevention and control of one type of slope failure (e.g., shallow, translational sliding) may be ineffectual against deep-seated rotational failures).

2.7.2 Causes of Slope Failure

The mass stability of slopes is governed by topographic, geologic, and climatic variables that control shear stress and shear resistance in a slope. Slopes fail when shear stress exceeds shear strength along a critical sliding surface. The factor of safety of a slope is defined as the ratio of shear strength to shear stress along a critical failure surface. The surface with the lowest ratio is the critical or failure surface. This surface also demarcates the boundary between stable and moving ground. The term slide implicitly specifies relative motion between the

TABLE 2-9. Abbreviated Classification of Slope Movements

Type of Movement		Type of Material		
			Engineering Soils	
		Bedrock	Predominantly Coarse	Predominantly Fine
Falls		Rock fall	Debris fall	Earth fall
Topples		Rock topple	Debris topple	Earth topple
Slides				
Rotational	Few units	Rock slump	Debris slump	Earth slump
Translational	Many units	Rock block slide	Debris block slide	Earth block slide
		Rock slide	Debris slide	Earth slide
Lateral spreads		Rock spread	Debris spread	Earth spread
Flows		Rock flow	Debris flow	Earth flow (soil creep)

Source: Adapted from Varnes (1978).

two. Methods of analyzing and computing the factor of safety of slopes against sliding are reviewed in Section 2.8.

Any variable or factor that increases shear stress or conversely that decreases shear strength will tend to cause slope movement. Causes of slope instability or failure have been grouped into these two categories by Varnes (1958), as illustrated in Table 2-10. Timely identification and recognition of these factors is the key to prevention and control of slope movements. Removal of lateral support by either natural or human agencies is probably the most common of all factors leading to instability. This observation suggests the use of buttress or retaining structures at the foot of slopes as a remedial measure. The addition of water to a slope may contribute simultaneously to an increase in stress and a decrease in strength. Water has been implicated as either the primary or a major controlling factor in 95 percent of all landslides (Chassie and Goughnor, 1976). Accordingly, drainage and diversion measures are without doubt among the most effective means of preventing and/or controlling slope failures. Landslide remedial measures and their effectiveness have been discussed by Zaruba and Mencl (1969), Hutchinson (1978), and Schuster (1992).

Vegetation can ameliorate many of the factors and conditions that cause instability. Woody vegetation growing on a slope can increase soil shear strength via root reinforcement, decrease soil moisture stress via evapotranspiration, and increase overall stability as a result of buttressing and soil arching action. The contribution and significance of slope vegetation to the mass stability of slopes are evaluated in detail in Chapter 3.

2.7.3 Indicators of Slope Instability

There are several visual indicators of hillside instability that are very useful for identifying areas of potential slope movement. Identification of unstable slopes

TABLE 2-10. Causes of Slope Failure

Increase in Shear Stress

1. Surcharging slope (structures and fills at top)
2. Removal of lateral support (cuts and excavations at toe)
3. Rapid changes in water level adjacent to slope ("sudden drawdown")
4. Increase in lateral stress (water filled cracks and fissures)
5. Earthquake loading (increases in horizontal or downslope driving forces)

Decrease in Shear Strength

1. Increased pore water pressure, which reduces effective stress (storm water infiltration into slope, uncontrolled discharge of water from drains, earthquake induced pure water pressure)
2. Presence of swelling clays (uptake of water with loss of intrinsic cohesion)
3. Weathering and physicochemical degradation (ion exchange, hydrolysis, solutioning, etc.)
4. Progressive failure by shear strain softening

or slopes with a high landslide potential can be made using certain topographic, vegetative, hydrologic, and geologic indicators, as summarized in Table 2-11. Most of these signs of hillside instability can be spotted by a visual reconnaissance in the field. They are signals for precautions or preventative measures that should be employed during and after slope disturbance in the area.

Peck (1967) stated a postulate that is useful for identifying potentially unstable slopes, namely if there is no evidence of old landslides in an area, then it is fairly unlikely that moderate construction activity will start a new one. On the other hand, if old landslides abound, then it is quite likely that even minor construction operations will activate an old slide or lead to a new one.

TABLE 2-11. Features Indicating Landslides or Areas with High Landslide Potential

Feature	Significance
1. Hummocky, dissected topography	Common feature in old and active progressive slides (slides with many individual components); slide mass is prone to gullying.
2. Abrupt change in slope	May indicate either an old landslide area or a change in the erosion characteristics of underlying material; portion with low slope angle is generally weaker and often has higher water content.
3. Scarps and cracks	Definite indication of an active or recently active landslide. Age of scarp can usually be estimated by the amount of vegetation established upon it; width of cracks may be monitored to estimate relative rates of movement.
4. Grabens or "stair step" topography	Indication of progressive failure; complex or nested series of rotational slides can also cause surface of slope to appear stepped or tiered.
5. Lobate slope forms	Indication of former earthflow or solifluction area.
6. Hillside ponds	Local catchments or depressions formed as result of feature 4 act as infiltration sources that can exacerbate or accelerate slope failure.
7. Hillside seeps	Common in landslide masses and areas with high landslide potential; can usually be identified by associated presence of phreatophyte vegetation, e.g., equisetum, cattails, alder, willows, in vicinity of seep.
8. Incongruent vegetation	Patches or areas of much younger or very different vegetation (e.g., alder thickets) may indicate recent landslides or unstable ground.
9. "Jackstrawed" trees	Leaning or canted trees on a slope are indicators of previous episodes of slope movement or soil creep.
10. Bedding planes or joints	Potential surface of sliding for translational slope movements.

2.8 SLOPE STABILITY PREDICTIONS

2.8.1 Approaches to Analysis

Numerous methods are available for predicting the stability of slopes and embankments. The two basic approaches are: (1) limit equilibrium analysis, and (2) deformation analysis. Most methods in use today fall into the former category. Limit equilibrium methods explicitly take into account the major factors that influence the shear stress and shear resistance of a slope (Table 2-10). In addition these methods are simpler to apply than deformation analysis. Limit equilibrium methods do not result in calculation of expected slope deformations; however, there are many instances in which precise deformations are of minor concern, provided the material essentially stays in place.

A detailed treatment of slope stability analysis lies outside the scope of this guidebook. The interested reader is referred to geotechnical texts on this subject (Lambe and Whitman, 1969; Morgenstern and Sangrey, 1978; Huang, 1983). The intent here is mainly to review basic concepts of stability analysis and to describe a few methods that are employed in practice. This review will serve: (1) to explain the combination of conditions that can result in slope failure; (2) to identify the relative importance to stability of different soil, slope, and hydrologic variables; and (3) to establish an analytical or quantitative basis for assessing the influence of vegetation on mass stability.

2.8.2 Limit Equilibrium Analysis

Limit equilibrium analysis is used to determine the factor of safety for a given slope; it can also be used to determine the effect on the stability of a slope of varying one or more parameters. A number of methods and procedures based on limit equilibrium principles have been developed for this purpose. Regardless of the specific procedure, the following principles (Morgenstern and Sangrey, 1978) are common to all methods of limit equilibrium analysis:

- *A Failure Surface or Mechanism is Postulated.* In the simplest case, idealized slopes are assumed to fail along planes or circular sliding surfaces, as shown schematically in Figure 2-8. More complex failure surfaces can also be proposed and analyzed when slope conditions are not uniform.
- *The Shearing Resistance Required to Equilibrate the Failure Mass is Calculated by Means of Statics.* The potential failure mass is assumed to be in a state of "limiting equilibrium," and the shear strength of the soil or rock in the failure mass is mobilized everywhere along the slip surface.
- *The Calculated Shearing Resistance Required for Equilibrium is Compared with the Available Shear Strength.* This comparison is made in terms of the factor of safety, which is generally defined as the factor by which the shear strength parameter must be reduced in order to bring the slope into a state of limiting equilibrium along a given slip surface.

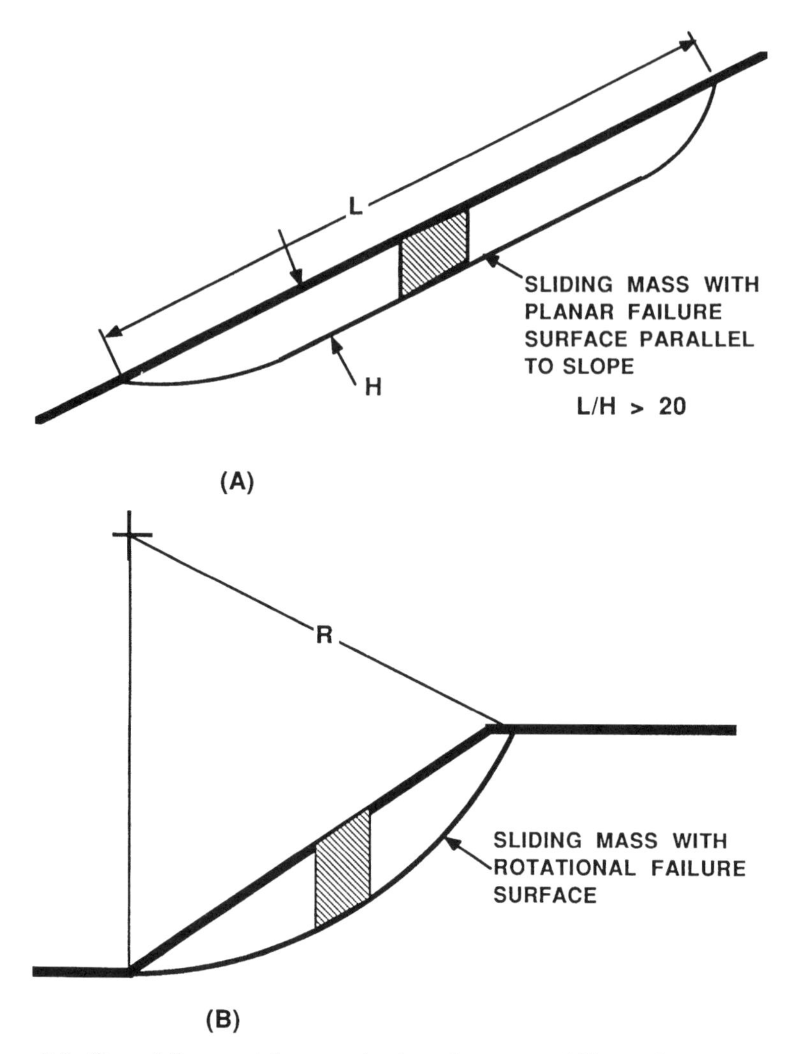

Figure 2-8. Slope failure models or mechanisms for mass stability analyses. (*a*) Translational failure, "infinite slope" model. (*b*) Rotational failure, circular arc model.

- *The Mechanism or Slip Surface with the Lowest Factor of Safety is Found by Iteration.* The surface with the lowest computed safety factor is the critical slip surface. If the location of the slip surface is predetermined or constrained by stratigraphic control (e.g., a weak clay seam, relict jointing surface, or contact between residual soil and underlying bedrock), other trials are usually unnecessary.

The essential requirements for conducting a slope stability analysis include the following:

- Accurate description of slope geometry
- Reliable soil properties (c, ϕ, γ)
- Correct definition of external loads, that is, surcharge, line loads, earthquake loads
- Correct description of slope hydrology, that is, phreatic surface (GWT) and seepage conditions
- Correct method of analysis

These constraints and requirements are sometimes difficult to observe in practice. Some common methods of analysis do not rigorously meet all requirements for static equilibrium. This is not as serious a limitation, however, as errors introduced by uncertainty in the proper choice and selection of input parameters, particularly values for shear strength of the soil (Singh, 1970). In spite of these drawbacks and problems, limit equilibrium analysis provides powerful insights into the factors and conditions governing the stability of slopes; it also provides a rational basis for assessing slope hazard and designing remedial measures.

2.8.3 Shear Strength Parameters

Determination of the factor of safety by limit equilibrium methods requires an estimate of the shear resistance that can be mobilized along the assumed failure surface. The shear strength of soil or unconsolidated rock is given by the Coulomb failure criterion:

$$s = c + \sigma \tan \phi \tag{2-2}$$

where:
s = shear strength of material;
σ = normal stress on the failure surface;
ϕ = angle of internal friction;
c = cohesion intercept

The angle of internal friction (ϕ) and the cohesion (c) are known as the shear strength parameters. They can be determined from various laboratory tests on representative samples of soil or, alternatively, back-calculated from analysis of a failed portion of a slope by assuming a factor of safety equal to unity.

An important consideration in slope stability analyses is whether to employ a total or an effective stress analysis. This decision determines what type of shear strength parameters must be used in the analysis. General rules and guidelines for selecting the appropriate type of parameters are discussed next and are also presented in Table 2-12 for clay slopes.

Total Stress Analysis: A total stress analysis using undrained shear strength parameters (c, ϕ) is limited to slopes where pore pressures are governed by

TABLE 2-12. Shear Strength Parameters for Clay Slope Stability Problems

	Cuttings		Natural Slopes	
Clay Type	Short Term	Long Term	Order of 100 Years	Order of 1000 Years
No preexisting slide				
Soft, normally consolidated, intact	$\chi \cdot c_u$		$c'\phi'$	
Lightly overconsolidated, intact		$c'\phi'$	$c'\phi'$	
Stiff, intact			$c'\phi'$	
Stiff, fissured	$f \cdot \chi \cdot c_u$	$r \cdot c', \phi'$	$c' = 0, \phi'$	$c' = 0, \phi' \rightarrow \phi_r'$
Jointed clays	$c' = 0, \phi'$			
Preexisting slide	$c'\phi_r'$	$c_r'\phi_r'$		$c_r'\phi_r'$

Source: From Skempton and Hutchinson (1969).

c_u = peak strength, undrained
$c'\phi'$ = peak strength parameters, drained
$c_r'\phi_r'$ = residual strength parameters ($c_r' = 0$)
χ = reduction factor for rate of testing, anisotropy, etc.
f = reduction factor fissures
r = time dependent reduction factor

total (external) stress changes and in which insufficient time has elapsed for significant dissipation of pore pressures. These conditions are characteristic of the so-called end-of-construction class of problems. A temporary cut made in a cohesive clay slope can be analyzed in this manner, provided it is not fissured and jointed (Table 2-12). A total stress analysis does not require a determination of pore pressure distribution in a slope, an important advantage of the analysis.

Effective Stress Analysis: When pore pressures are governed by steady state seepage conditions, or if long-term stability is a consideration, then stability analysis should be performed in terms of effective shear strength parameters. This is the usual condition for natural slopes in both soil and rock. All permanent cuts or fills also should be analyzed for long-term conditions to see whether these conditions control design (Table 2-12). Some jointed and fissured clays respond to drainage along dominant discontinuities so quickly that they should be analyzed in terms of effective stress regardless of the time of loading or unloading. Effective shear strength parameters (c', ϕ') may be obtained from results of either drained triaxial tests or undrained tests with pore pressure measurements (Bowles, 1970). An effective stress analysis requires that effective shear strength parameters be used and that the groundwater table position or pore pressure distribution in the slope be known from piezometric studies. If the pore pressure is unknown or cannot be determined, there is little point to an effective stress analysis; a total stress analysis should be employed instead.

2.8.4 Translational Slope Failures

A great number of slope failures can be characterized by shallow, translational sliding. These are also the types of slope failures most likely to be influenced by vegetation and by soil bioengineering treatment. Accordingly, we devote some attention to the analysis of translational failures. The reader is referred to other texts (Huang, 1983) for analysis of deeper seated, rotational failures. The stability of simple, natural slopes where all boundaries (ground surface, phreatic surface, and basal sliding surface) are approximately parallel can be modeled and analyzed by so-called "infinite slope" equations. In this analysis the slip surface is assumed to be a plane roughly parallel to the ground surface.This type of analysis is appropriate when sliding takes place such that the ratio of depth to length of the sliding mass is small, as noted in Figure 2-8. The following types of slopes or slope conditions meet the aforementioned criteria:

1. Loose products of weathering (residual soil) overlying an inclined bedrock contact
2. Inclined planes of stratification dipping downslope and underlain by stronger strata
3. Bedrock slopes mantled with glacial till or colluvium
4. Homogenous slopes of coarse-textured, cohesionless soil (sand dunes, sandy embankments, or fills)

In the first three slopes the surface of sliding is predetermined by stratigraphic control. In the fourth the slope failures are restricted to shallow, surface sloughing because shear strength increases steadily with depth.

Because of the geometry of an infinite slope, overall stability can be determined by analyzing the stability of a single, vertical element in the slope, as shown in Figure 2-9. End effects in the sliding mass can be neglected, and so too can lateral forces on either side of the vertical element, which are assumed to be opposite and equal. The factor of safety based on an infinite slope analysis for the conditions shown in Figure 2-9 is given by the following equation:

$$FS = \frac{[c'/\cos^2\beta\tan\phi' + (q_0 + \gamma H) + (\gamma_{\mathrm{BUOY}} - \gamma)H_w]\tan\phi'/\tan\beta}{[(q_0 + \gamma H) + (\gamma_{\mathrm{SATD}} - \gamma)H_w]} \tag{2-3}$$

where:

ϕ' = effective angle of friction;
c' = effective cohesion intercept;
β = slope angle of natural ground;
γ = moist density of soil;
γ_{BUOY} = buoyant density of soil;

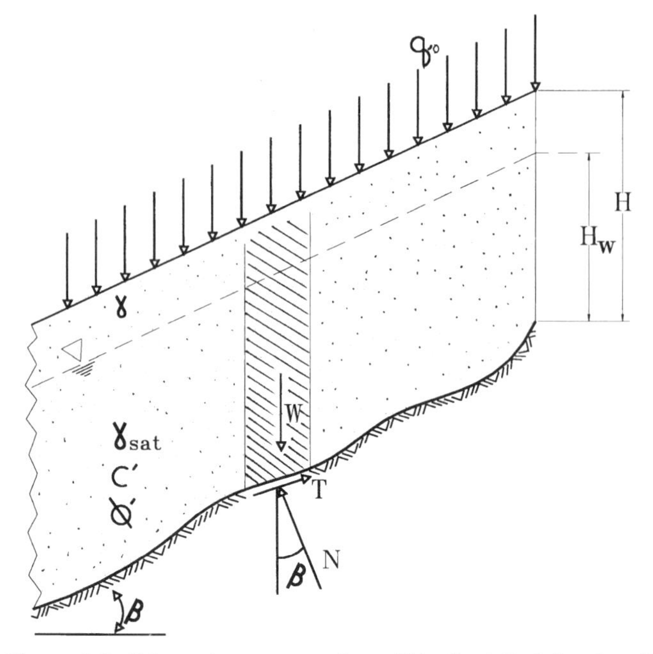

Figure 2-9. Schematic representation of idealized "infinite slope."

γ_{SATD} = saturated density of soil;
H = vertical thickness (or depth) of sliding surface;
H_w = piezometric height above sliding surface;
q_0 = uniform vertical surcharge stress on slope

This expression is quite general and takes into account the influence of surcharge (q_0), the presence of a phreatic surface or groundwater table in the slope (H_w), and the existence of cohesion (c'). The influence of root reinforcement in the soil, which affects only the cohesion and not the soil friction, can be accounted for by adding a "root cohesion" term (c_R) to the numerator of Equation 2-3, as explained in Chapter 3. When the groundwater table or phreatic surface is parallel to the slope, as shown schematically in Figure 2-9, seepage also occurs parallel to the surface. An alternative expression can be derived to account for groundwater movement or seepage in other directions, for example, vertically downward or emerging from the slope. These other directions may be important to consider as well, particularly in the case of slope vegetation and soil bioengineering systems that can modify seepage direction near the face of a slope. The influence of seepage direction on the factor of safety is evaluated in Chapter 3.

A number of particular or special cases of interest can be derived from the general expression (Equation 2-3) for an infinite slope. These special cases include the following:

Case i) Cohesionless Slope, No Surcharge ($c' = 0$; $q_0 = 0$)

$$FS = \frac{[\gamma(H - H_w) + \gamma_{\mathrm{BUOY}} H_w] \tan \phi' / \tan \beta}{[\gamma(H - H_w) + \gamma_{\mathrm{SATD}} H_w]} \tag{2-4}$$

This equation shows the influence of a rise in the phreatic surface in a shallow, cohesionless soil overlying an inclined bedrock contact. Note that the controlling factor is not the absolute rise but rather the ratio (H_w/H). In other words, the same rise in phreatic surface is more serious in a "thin" soil mantle than in a "thick" one.

Case ii) Saturated, Cohesionless Slope, No Surcharge ($c' = 0$; $q_0 = 0$; $H_w = H$)

$$FS = \{\gamma_{\mathrm{BUOY}} / \gamma_{\mathrm{SATD}}\} \frac{\tan \phi'}{\tan \beta} \tag{2-5}$$

But for most soils,

$$\gamma_{\mathrm{BUOY}} \simeq 0.5 \gamma_{\mathrm{SATD}}$$

Therefore,

$$FS = 0.5 \frac{\tan \phi'}{\tan \beta}$$

This equation yields the factor of safety for the worst case of complete saturation of a cohesionless infinite slope. The factor of safety is approximately one-half that of the dry case.

Case iii) Dry, Cohesionless Slope, No Surcharge ($c' = 0$; $q_0 = 0$; $H_w = 0$)

$$FS = \frac{\tan \phi'}{\tan \beta} \tag{2-6}$$

This equation shows that in a dry, cohesionless material the critical slope angle is equal to the angle of internal friction of the soil. If the material is end-dumped or side-cast in a loose condition, this angle will be equivalent to the angle of repose.

Case iv) Stable Slopes with Cohesion, No Surcharge ($q_0 = 0$; $F > 1$)

(a) *Dry Slope* ($H_w = 0$)

$$c_d / \gamma H = FS \cos^2 \beta \{\tan \beta - \tan \phi'\} \tag{2-7}$$

where: F = desired factor of safety;
c_d = required cohesion

(b) *Saturated Slope* ($H_w = H$)

$$c_d/\gamma H = FS \cos^2 \beta [\tan \beta - (\gamma_{BUOY}/\gamma_{SATD}) \tan \phi']\qquad (2\text{-}8)$$

These equations are useful for determining the amount of cohesion that must be present or developed, for example from root reinforcement, to achieve a desired factor of safety ($FS > 1$) for a given depth of sliding (H), slope angle (β), and friction angle (ϕ'). Contributions to cohesion from root reinforcement and methods for estimating root cohesion are described in Chapter 3. Note that the required cohesion is directly proportional to the thickness of the sliding mass.

Selection of Soil Parameters: Soil density should be determined in the field using standard procedures for field density testing (Bowles, 1970). Effective shear strength parameters c' and ϕ' should be used in an infinite slope analysis when groundwater and seepage conditions are taken explicitly into account. These shear strength parameters can be measured by reconstituting samples in the laboratory to their in-place density and performing a triaxial or direct shear test (Bowles, 1970). Shear strength parameters can also be obtained conveniently in the field by means of an in situ bore hole shear test (Wineland, 1975). An approximate estimate of friction angle can be obtained from gradation and density data using the nomograph shown in Figure 2-10. For many purposes this approximation is satisfactory in lieu of more expensive and time consuming lab strength tests.

Stability of Road Fills: Debris slides that originate from roadway fills are often characterized by movement along an approximately planar surface. One mode of failure is by shallow sloughing of the outside margins of a fill. This failure mode can be analyzed by the conventional infinite slope model, provided the thickness of the sliding mass is small relative to the slope length of the fill. Gonsior and Gardner (1971) suggest a minimum of 20 for the length/depth ratio.

The other mode of failure is sliding of the entire fill along the contact with the underlying natural ground. This mode is common in loose, side-cast fills on steep ground. The roadway in this case is supported part on cut and part on fill, as depicted schematically in Figure 2-11. Many low-volume roads are constructed in this manner. The assumptions of the infinite slope model are not strictly observed in this mode of failure. Instead a "sliding wedge" type of analysis seems more appropriate. Results of such an analysis show, however, that analogous equations describe the factor of safety.

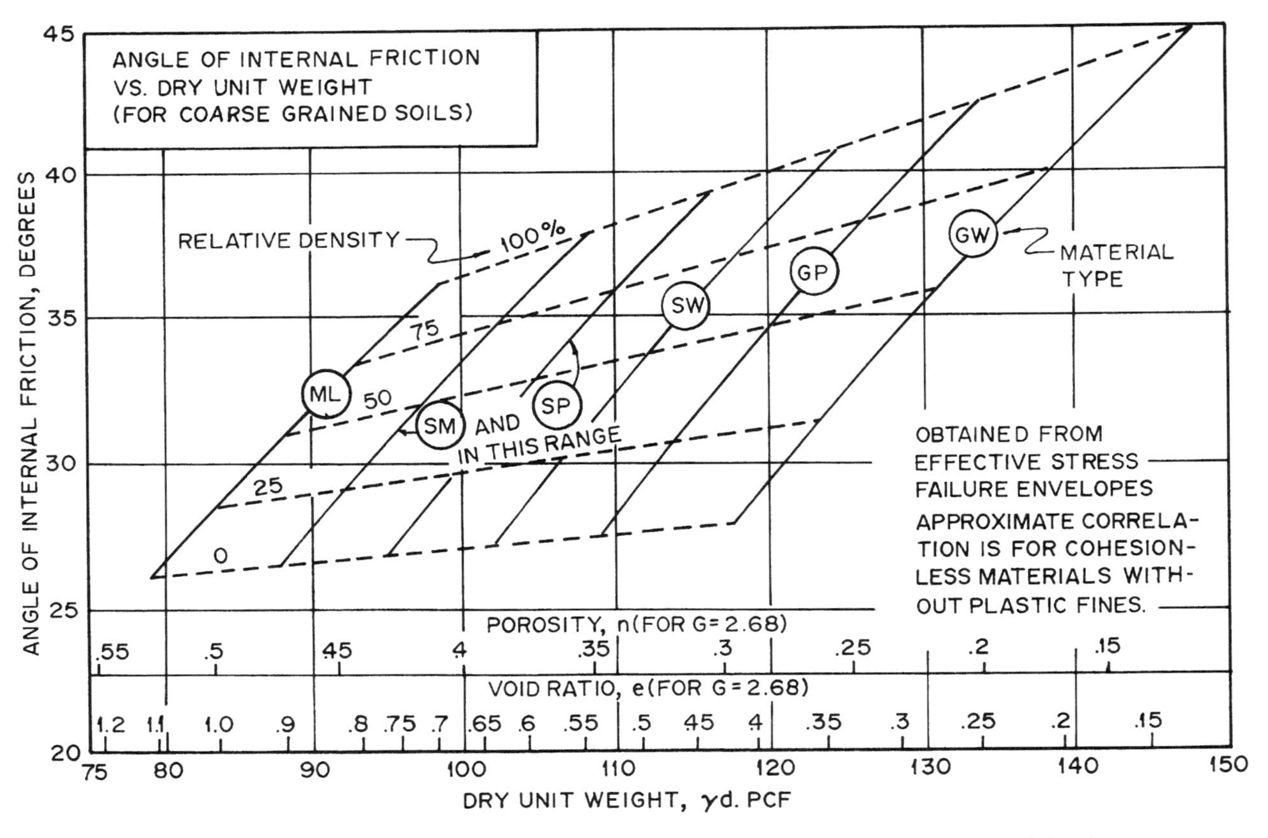

Figure 2-10 Correlation chart for estimating soil friction angles from graduation and density data. (From U.S. Department of Navy, 1971.)

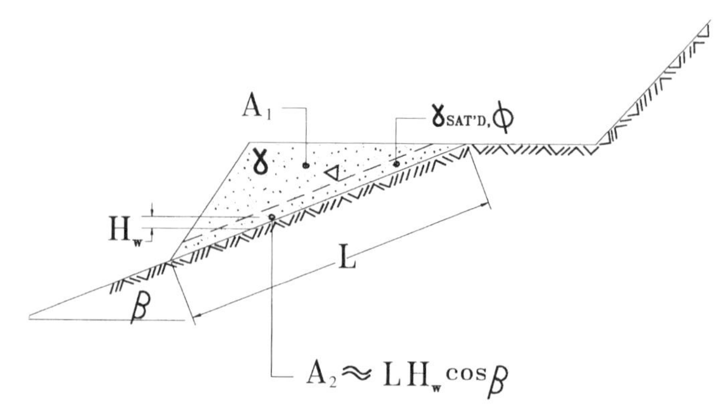

Figure 2-11. Schematic representation of roadway supported partly on a bench cut and partly on side-cast road fill. Saturated zone may develop at base of fill along contact with natural ground.

Case i) Dry Fill, No Cohesion ($c' = 0$; $H_w = 0$)

$$FS = \frac{\tan \phi'}{\tan \beta} \tag{2-9}$$

In this case both the "sliding wedge" and "infinite slope" model yield the same equation.

Case ii) Cohesionless Fill with Phreatic Surface near Contact In this case a thin, saturated zone develops at the base of the fill along the contact with the underlying natural ground, as shown schematically in Figure 2-11. In this case the factor of safety will be given by an equation analogous to the infinite slope equation (Equation 2-4), namely:

$$FS = \frac{[\gamma A_1 + \gamma_{BUOY} A_2] \tan \phi' / \tan \beta}{[\gamma A_1 + \gamma_{SATD} A_2]} \tag{2-10}$$

where: A_1 = cross-sectional area of fill above phreatic surface;
A_2 = saturated cross-sectional area of fill below phreatic surface

The areas A_1 and A_2 correspond to the heights $(H - H_w)$ and H_w in Equation 2-4. In some cases fill failures are more complicated and involve material beneath the fill that is activated by the additional fill weight. In those cases the failure surface can cut deeper than the contact with the ground, and the failure surface may be rotational.

2.9 CONTROL OF MASS WASTING

Measures to control or prevent mass wasting fall into two basic categories, namely: (1) measures that decrease shear forces, and (2) measures that increase shear resistance. These two approaches are antidotes to the factors that cause mass slope failures (see Table 2-10). Schuster (1992) has presented an excellent and comprehensive review of slope stabilization approaches and techniques.

Water is involved in the great majority of slope failures, because the presence of water both increases shear stresses and decreases shear strength. Accordingly, water drainage and diversion measures are extremely important in mass slope stabilization.

Soil bioengineering measures increase stability mainly by increasing shear resistance either from reinforcement of the soil mantle by roots and/or from interception of shallow slip surfaces by imbedded stems. The contribution to increased shear resistance by woody roots is discussed further in Chapter 3. Some soil bioengineering measures also improve stability by modifying the hydrologic regime in the soil, either by transpiring soil moisture or by acting as drains. Live fascines, elongated bundles of brush buried in shallow trenches, for example, can act as conduits to intercept and facilitate drainage. Live brushlayers inserted between lifts of soil near the face of a slope act as horizontal drains and improve stability by redirecting seepage flow direction in addition to mechanically reinforcing the soil.

2.10 REFERENCES CITED

Beer, C. E. and H. P. Johnson (1963). Factors in gully growth in deep loess area of Western Iowa. *Transactions ASAE* **6**(3): 237–240.

Bowles, J. E. (1970). *Engineering Properties of Soils and Their Measurement.* New York: McGraw-Hill.

R. G. Chassie and R. D. Goughnor (1976). National highway landslide experience. *Highway Focus* **8**(1): 1–9.

Chepil, W. S. (1945). Dynamics of wind erosion: I. Nature of movement of soil by wind. *Soil Science* **60:** 305–320.

Ellison, W. D. (1948). Erosion by raindrop. *Scientific American* (Aug.): 1–7.

Emerson, W. W. (1967). A classification of soil aggregates based on their coherence in water. *Australian Journal of Soil Research* **2:** 211–217.

Goldman, S. J., K. Jackson, and T. A. Bursztynsky, (1986). *Erosion and Sediment Control Handbook.* New York: McGraw-Hill.

Gonsior, M. J., and R. B. Gardner (1971). Investigation of slope failures in the Idaho Batholith. *USDA Forest Service Research Paper* INT-97, 34 pp.

Gray, D. H. and A. T. Leiser (1982). *Biotechnical Slope Protection and Erosion Control.* New York: Van Nostrand Reinhold.

Howell, R. B. et al. (1979). Analysis of short and long term effects on water quality for selected highway projects. Report No. FHWA/CA/TL-79/17, Final Report to California Department of Transportation, 245 pp.

Huang, R. (1983). *Stability of Earth Slopes*. New York: Van Nostrand Reinhold.

Hutchinson, J. N. (1978). Assessment of the effectiveness of corrective measures in relation to geologic conditions and types of slope movoement. *Bulletin at the International Association of Engineering Geologists* **16:** 131–155.

Israelson, E. (1980). Erosion control during highway construction—Manual on principles and practices. *Transportation Research Program Report No. 221*, NAS- NRC, Transportation Research Board, Washington, DC.

Kay, B. L. (1983). Straw as an erosion control mulch. *Agronomy Progress Report No. 140*, Univ. of California Davis Agricultural Experiment Station, Davis, CA.

Keown, M. P. et al. (1977). Literature survey and preliminary evaluation of streambank protection methods. *Technical Report* H-77-9, U.S. Army Waterways Experiment Station, Vicksburg, MS.

Lambe, T. W., and Whitman, R. V. (1969). *Soil Mechanics*. New York: Wiley.

Lane, E. W. (1955). The importance of fluvial morphology in hydraulic engineering. *Proceedings ASCE* **81**(745): 1–17.

Leopold, L. B. (1994). *A View of the River*. Cambridge, MA: Harvard University Press.

Morgenstern, N. R. and Sangrey, D. A. (1978). Methods of stability analysis, In: "Landslides: Analysis and Control," edited by R. L. Schuster and R. J. Krizek, *Transportation Research Board Special Report 176*, NAS-NRC, Washington, D.C., pp. 155–171.

Peck, R. (1967). Stability of natural slopes. *Journal Geotechnical Engineering* (ASCE) **93**(4): 437–451.

Renard, K. G., G. R. Foster, G. A. Weesies and J. P. Porter (1991). RUSLE: Revised universal soil loss equation. *Journal of Soil and Water Conservation*, **46:** 30–33.

Schor, H. (1992). Hills like nature makes them. *Urban Land* (Mar.): 40–43.

Schor, H. (1993). Landform grading: Comparative definitions of grading designs. *Landscape Architect and Specifier News* (Nov.): 22–25.

Schuster, R. L. (1992). Recent advances in slope stabilization. Keynote paper, Session G.3 (Stabilization & Remedial Works), *Proceedings*, 6th Intl. Conference on Landslides, Christchurch, New Zealand, Feb. 10–14.

Schwab, G. O., R. K. Frevert, T. W. Edminster, and K. K. Barnes (1981). *Soil and Water Conservation Engineering*, 3d ed. New York: Wiley.

Sherard, J. L., N. L. Ryker and R. S. Decker (1972). Piping in earth dams of dispersive clay. *Proceedings, Specialty Conference on Performance of Earth and Earth Supported Structures*, ASCE, Vol. 1, pp. 589–626.

Sherard, J. L. et al. (1978). Pinhole test for identifying dispersive soils. *Journal of Geotechnical Engineering* (ASCE) **102**(GT1): 69–85.

Singh, A. (1970). Shear strength and stability of man-made slopes. *J. of Soil Mechanics and Foundations Division* (ASCE) **96**(SM6): 1879–1890.

Skempton, A. W. and J. N. Hutchinson (1969). Stability of natural slopes and embankment foundations. *Proc., 7th International Conference on Soil Mechanics and Foundation Engineering*, Mexico City, Mexico, Vol. 1 pp. 291–340.

USDA Soil Conservation Service (1972). Procedures for computing sheet and rill erosion on project areas. Technical Release No. 51, Washington, DC., 14 pp.

USDA Soil Conservation Service (1977). Guides for erosion and sediment control in California. Miscellaneous Publication, Davis, CA, 32 pp.

USDA Soil Conservation Service (1978). Predicting rainfall erosion losses; a guide to conservation planning. USDA Handbook No. 537, Washington, DC.

U.S. Department of Navy (1971). *Design Manual DM-7 for Soil Mechanics, Foundations and Earth Structures*, Naval Facilities Engineering Command, Washington, DC.

U.S. Environmental Protection Agency (1973). Comparative costs of erosion and sediment control construction activities, EPA-430/9-73-016, U.S. Government Printing Office, Washington, DC.

Varnes, D. J. (1958). Landslide types and processes. In: "Landslides and Engineering Practice," edited by Eckel, *Highway Research Board Special Report No. 29*, NAS-NRC, Washington, D.., pp 20–47.

Varnes, D. J. (1978). Slope movements, types and processes. In: "Landslides, Analysis and Control," edited by R. L. Schuster and R. J. Krizek, *Transportation Research Board Special Report No. 176*, NAS-NRC, Washington, DC., pp. 11–33.

Volk, G. M. (1937). Method of determining the degree of dispersion of the clay fraction of soils. *Soil Science Society of America Proceedings* **2:** 432–445.

Wineland, J.D. (1975). Borehole shear device. *Proceedings, GT-ASCE Specialty Conference on In-Situ Measurement of Soil Properties*, Raleigh, North Carolina, Vol. 1, pp. 511–522.

Wischmeier, W. H., C. B. Johnson, and B. V. Cross (1971). A soil erodibility nomograph for farmland and construction sites. *Journal of Soil and Water Conservation* **26**(5): 189–193.

Wischmeier, W. H. and D. D. Smith (1978). Predicting rainfall erosion losses: a guide to conservation planning. *USDA Agricultural Handbook No. 537*, Washington, DC.

Zaruba, Q. and V. Mencl (1969). *Landslides and Their Control*, New York: Elsevier, and Prague: Academia.

3 Role of Vegetation in the Stability of Slopes

3.0 INTRODUCTION

Vegetation affects both the surficial and mass stability of slopes in significant and important ways. Various hydromechanical influences of vegetation, including methods for predicting and quantifying their magnitude and importance on stability, are described in this chapter. The stabilizing or protective benefits of vegetation depend both on the type of vegetation and type of slope degradation process. In the case of mass stability, the protective benefits of woody vegetation range from mechanical reinforcement and restraint by the roots and stems to modification of slope hydrology as a result of soil moisture extraction via evapotranspiration.

The loss or removal of slope vegetation can result in either increased rates of erosion or higher frequencies of slope failure. This cause-and-effect relationship can be demonstrated convincingly as a result of many field and laboratory studies reported in the technical literature. The findings of these studies are briefly reviewed in this chapter as well.

For the most part vegetation has a beneficial influence on the stability of slopes; however, it can occasionally affect stability adversely or have other undesirable impacts; for example, it can obstruct views, hinder slope inspection, or interfere with flood fighting operations on levees. A number of strategies and techniques are described at the end of this chapter to maximize benefits and minimize liabilities of plants. These include such procedures as the proper selection and placement of vegetation in addition to management techniques such as pruning and coppicing.

The right choice of plant materials is critical. A tight, dense cover of grass or herbaceous vegetation, for example, provides one of the best protections against surficial rainfall and wind erosion. Conversely, deep rooted, woody vegetation is more effective for mitigating or preventing shallow, mass stability failures. In a sense, soil bioengineering and biotechnical methods also can be viewed as strategies or procedures for minimizing the liabilities of vegetation while capitalizing on its benefits (see Chapters 7, 8, and 9).

3.1 INFLUENCE ON SURFICIAL EROSION

3.1.1 Stabilizing Functions

Vegetation plays an extremely important role in controlling rainfall erosion. Soil losses due to rainfall erosion can be decreased a hundredfold (USDA Soil Conservation Service, 1978) by maintaining a dense cover of sod, grasses, or herbaceous vegetation. The beneficial effects of herbaceous vegetation and grasses in preventing rainfall erosion are tabulated below:

- *Interception:* Foliage and plant residues absorb rainfall energy and prevent soil detachment by raindrop splash.
- *Restraint:* Root systems physically bind or restrain soil particles while aboveground portions filter sediment out of runoff.
- *Retardation:* Stems and foliage increase surface roughness and slow velocity of runoff.
- *Infiltration:* Plants and their residues help to maintain soil porosity and permeability, thereby delaying onset of runoff.

In the case of surficial erosion, herbaceous vegetation and grasses are more effective than woody vegetation because they provide a dense ground cover.

3.1.2 Vegetation Cover Factor

A good gauge of the influence of vegetation in preventing soil erosion can be obtained by examining the universal soil loss equation (USLE). The annual soil loss from a site is predicted according to the following relationship:

$$A = R \cdot K \cdot LS \cdot C \cdot P \tag{3-1}$$

where: A = computed soil loss (e.g., tons) per acre for a given storm period or time interval
R = rainfall factor
K = soil erodibility value
L = slope length factor
S = steepness factor
C = vegetation factor
P = erosion control practice factor

The USLE provides a simple, straightforward method of estimating soil losses and it provides an idea of the range of variability of each of the parameters, their relative importance in affecting erosion, and the extent to which each can be changed or managed to limit soil losses. The climatic (R), topographic

(*LS*), and erodibility (*K*) factors only vary within one order of magnitude. The vegetation or cover (*C*) factor, on the other hand, can vary over several orders of magnitude, as shown in Table 3-1. Moreover, unlike the other factors, the cover (*C*) factor can be readily decreased by the selection, method of installation, and maintenance of a particular cover system. Factor *C* values tend to change with time following certain types of surface treatment, such as mulching, seeding, and transplanting. For example, factor *C* values for grass may decrease from 1.0 (for fallow, bare ground) to about 0.001 between time of initial seeding and full establishment with a dense grass sod.

3.1.3 Recommended Vegetation

Under normal conditions, a dense cover of grass or herbaceous vegetation provides the best protection against surficial rainfall and wind erosion. A grass cover can be established by either seeding or sodding. Seed mixtures normally include grasses that germinate rapidly, such as rye or annual grass, to provide immediate short-term protection, and slower-growing perennial grasses that take more time to establish, but provide long-term protection. The opti-

TABLE 3-1. Cover Index Factor (*C*) for Different Ground Cover Conditions

Type of Cover	Factor *C*	Percent Effectiveness[a]
None (fallow ground)	1.0	0.0
Temporary seedings (90% stand):		
Ryegrass (perennial type)	0.05	95
Ryegrass (annuals)	0.1	90
Small grain	0.05	95
Millet or sudan grass	0.05	95
Field bromegrass	0.03	97
Permanent seedings (90% stand):	0.01	99
Sod (laid immediately)	0.01	99
Mulch:		
Hay, rate of application (tons/acre)		
0.5	0.25	75
1.0	0.13	87
1.5	0.17	93
2.0	0.02	98
Small grain straw, 2.0 tons/acre	0.02	98
Wood chips, 6.0 tons/acre	0.06	94
Wood cellulose, 1.5 tons/acre	0.10	90
Fiberglass, 1.5 tons/acre	0.05	95

Source: From USDA Soil Conservation Service (1978).

[a]Percent soil loss reduction as compared with fallow ground.

mum seed mix depends on soil, site, and climatic conditions. A horticulturist familiar with local conditions should be consulted for recommendations. Site preparation, mulching, and fertilization may also be required to insure germination and establishment (see Chapter 6 for guidelines).

3.2 INFLUENCE ON STREAMBANK EROSION

Streambanks and levees are subjected to erosion and scour by flowing water. The erosive power of flowing water increases with velocity. Slope vegetation can help to reduce this type of erosion in the following manner: above ground shoots bend over and cover the surface and/or reduce flow velocity adjacent to the soil/water interface, while below ground roots physically restrain or hold soil particles in place. The extent to which vegetation provides these benefits depends upon the surface area of vegetation presented to the flow and the flexibility of the stems. Dense grass swards and low shrubby species that extend numerous nonrigid branches and leaves into the flow (e.g., willows) are the most effective in this regard.

Some controversy exists about the wisdom of allowing woody vegetation to grow on levees, particularly on revetted sections. Objections that have been raised include loss of conveyance from increased roughness, difficulty of inspection, hindrance to flood-fighting operations, and alleged threats to structural integrity as a result of root penetration and subsequent piping. In response to these objections it should be noted that in large rivers, additional channel roughness will have a negligible effect on the stage of the design flood. The effects of vegetation on the structural integrity of sandy levees was investigated by Shields and Gray (1993). They conducted an extensive field study along a 6-mile reach of sandy channel levee adjacent to the Sacramento River near Elkhorn, California. Their study concluded that woody vegetation did not adversely affect the structural integrity of a levee. No open voids or conduits clearly attributable to plant roots were observed in the levee. On the contrary, the presence of plant roots reinforced the soil and increased the shear strength of the surface layers in a measurable manner.

In European practice, vegetation is often promoted as a means of stabilizing both streambanks and levee slopes. In Bavaria, West Germany, a common design practice is to construct widely spaced, vegetated levees. A mixture of plants, including reeds, grasses, and trees, is used with riprap and other standard engineering control measures to retard erosion (Keller and Brookes, 1984). Shields (1991) investigated the influence of woody vegetation growing in a structural, riprap revetment. His investigation showed that the frequency of revetment failure was actually lower in vegetated revetments as opposed to unvegetated sections. Vegetation helps to anchor the armor stones to the bank and increases their lift-off resistance. In addition, vegetated revetments provide riparian benefits and are less visually intrusive, as shown in Figure 3.1.

(*a*)

(*b*)

Figure 3-1. Contrasting views of streambank levees. (*a*) Rock riprap alone. (*b*) Vegetated riprap.

3.3 INFLUENCE ON MASS STABILITY

3.3.1 Hydromechanical Effects

Beneficial versus Adverse Effects: The protective role of vegetation on the stability of slopes has gained increasing recognition (Coppin and Richards, 1990). Greenway (1987) provides a good summary of the hydromechanical influences of vegetation as related to mass stability. These influences are depicted schematically in Figure 3-2 and tabulated according to whether they exert a beneficial or adverse effect.

Some mechanisms, for example, surcharge, may be either beneficial or adverse depending on soil and slope conditions, as discussed later in this chapter. The principal destabilizing mechanism is probably from windthrowing, which can cause local instability in a slope. Depletion of soil moisture by vegetation is a two-edged sword: on the one hand evapotranspiration leads to lower pore water pressures in a slope; on the other, soil moisture depletion can accentuate desiccation cracking, which in turn leads to higher infiltration capacity.

Beneficial Effects: The main beneficial effects of woody vegetation on the mass stability of slopes are listed below:

- *Root Reinforcement:* Roots mechanically reinforce a soil by transfer of shear stress in the soil to tensile resistance in the roots.
- *Soil Moisture Depletion:* Evapotranspiration and interception in the foliage can limit buildup of positive pore water pressure.
- *Buttressing and Arching:* Anchored and embedded stems can act as buttress piles or arch abutments to counteract downslope shear forces.
- *Surcharge:* Weight of vegetation can, in certain instances, increase stability via increased confining (normal) stress on the failure surface.

The most obvious way in which woody vegetation enhances mass stability is via root reinforcement. Extensive laboratory studies (Gray and Ohashi, 1983; Maher and Gray, 1990) on fiber-reinforced sands indicate that small amounts of fiber can provide substantial increases in shear strength. These findings have been corroborated by laboratory and field tests on root-permeated soils (Endo and Tsuruta, 1969; Waldron, 1977; Ziemer, 1981; Riestenberg and Sovonick-Dunford, 1983; Riestenberg, 1994; Nilaweera, 1994).

The soil buttressing and arching action associated with roots and the stems/trunks of woody vegetation are also important components of slope stabilization. In addition, evapotranspiration by vegetation can reduce pore water pressures within the soil mantle on natural slopes, promoting stability (Brenner, 1973). These beneficial effects have been analyzed in detail by Greenway

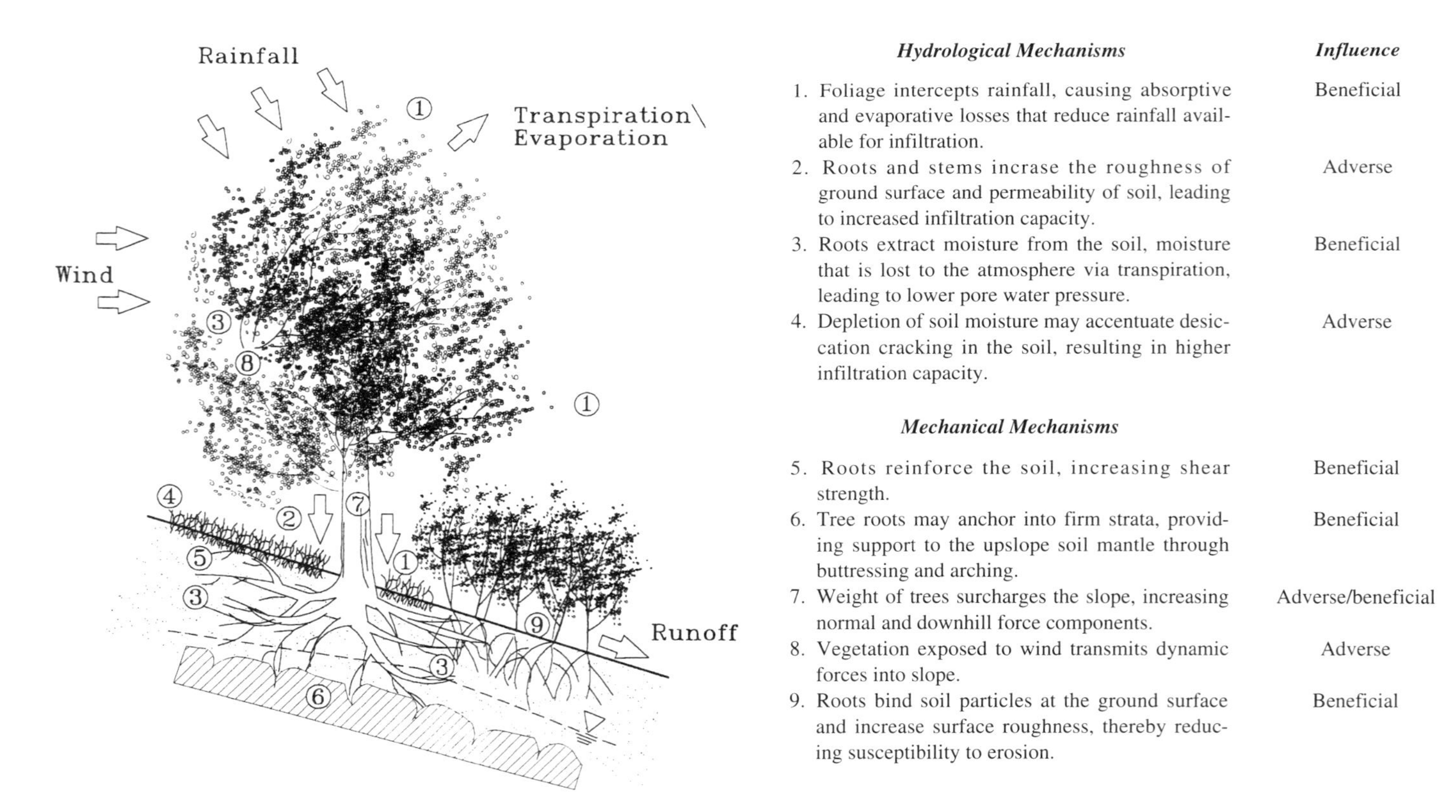

Hydrological Mechanisms	Influence
1. Foliage intercepts rainfall, causing absorptive and evaporative losses that reduce rainfall available for infiltration.	Beneficial
2. Roots and stems incrase the roughness of ground surface and permeability of soil, leading to increased infiltration capacity.	Adverse
3. Roots extract moisture from the soil, moisture that is lost to the atmosphere via transpiration, leading to lower pore water pressure.	Beneficial
4. Depletion of soil moisture may accentuate desiccation cracking in the soil, resulting in higher infiltration capacity.	Adverse
Mechanical Mechanisms	
5. Roots reinforce the soil, increasing shear strength.	Beneficial
6. Tree roots may anchor into firm strata, providing support to the upslope soil mantle through buttressing and arching.	Beneficial
7. Weight of trees surcharges the slope, increasing normal and downhill force components.	Adverse/beneficial
8. Vegetation exposed to wind transmits dynamic forces into slope.	Adverse
9. Roots bind soil particles at the ground surface and increase surface roughness, thereby reducing susceptibility to erosion.	Beneficial

Figure 3-2. Hydromechanical influences of vegetation on the mass stability of slopes. (From: Vegetation and slope stability by D. Greenway, in: *Slope Stability*, edited by M. G. Anderson and K. S. Richards. Copyright 1987. Reprinted by permission of John Wiley & Sons, Ltd.)

(1987) and Gray and Leiser (1982). Important findings are reviewed and summarized in succeeding sections of this chapter.

Detrimental Effects: The primary detrimental influence on mass stability associated with woody vegetation appears to be the concern about external loading and the danger of overturning or uprooting in high winds or currents (Nolan, 1984; Tschantz and Weaver, 1988). If a significantly sized root ball is unearthed during uprooting, it could reduce the stability of a cross section depending upon a tree's position on the slope. This problem is likely to be more critical for large trees growing on relatively small dams, levees, or streambanks. With regard to external loading, levee embankment slopes are generally shallow enough that the main component of the overburden weight may act perpendicular to, rather than parallel to, the failure surface, thereby increasing stability. However, the location of trees on the embankment must be considered in any slope stability analysis in order to ascertain the extent to which their weight might affect the balance of forces.

The problem of surcharge and windthrowing can be eliminated by proper plant selection or, alternatively, by the practice of coppicing. Planting shrubs or small trees, for example, with a large root-to-shoot ratio increases the belowground relative to the aboveground biomass, thereby minimizing problems associated with surcharge or windthrowing. Coppicing, a type of pruning procedure, also allows one to create an unobstructed view (a frequent reason for tree removal on slopes) by effectively reducing tree heights while retaining all the benefits provided by a tree's living root system. These mitigating practices are described in greater detail at the end of the chapter (see Section 3.7).

3.4 CONSEQUENCES OF VEGETATION REMOVAL

Given that woody vegetation growing on slopes reinforces soils and enhances stability, conversely, its removal should weaken soils and destabilize slopes. One of the earliest studies on this question was conducted by Bishop and Stevens (1964) in logged areas of southeast Alaska. Bishop and Stevens noted a significant increase in both frequency of slides and size of area affected by slides after clear-cut logging, as shown in Figure 3-3. They concluded that the destruction and gradual decay of interconnected root systems were the principal cause of increased sliding. Subsequent studies by other investigators (Wu, 1976; Megahan and Kidd, 1972; O'Loughlin, 1974; Swanston, 1974) have generally corroborated these findings.

Steep, metastable slopes underlain by weak rock and soils are particularly sensitive to disturbances created by road construction and timber harvesting. The practice of clear-cutting in particular warrants scrutiny in this regard. Clear-cutting is a silvicultural or tree-harvesting procedure in which all timber over a certain minimum diameter is felled and removed. This method is commonly

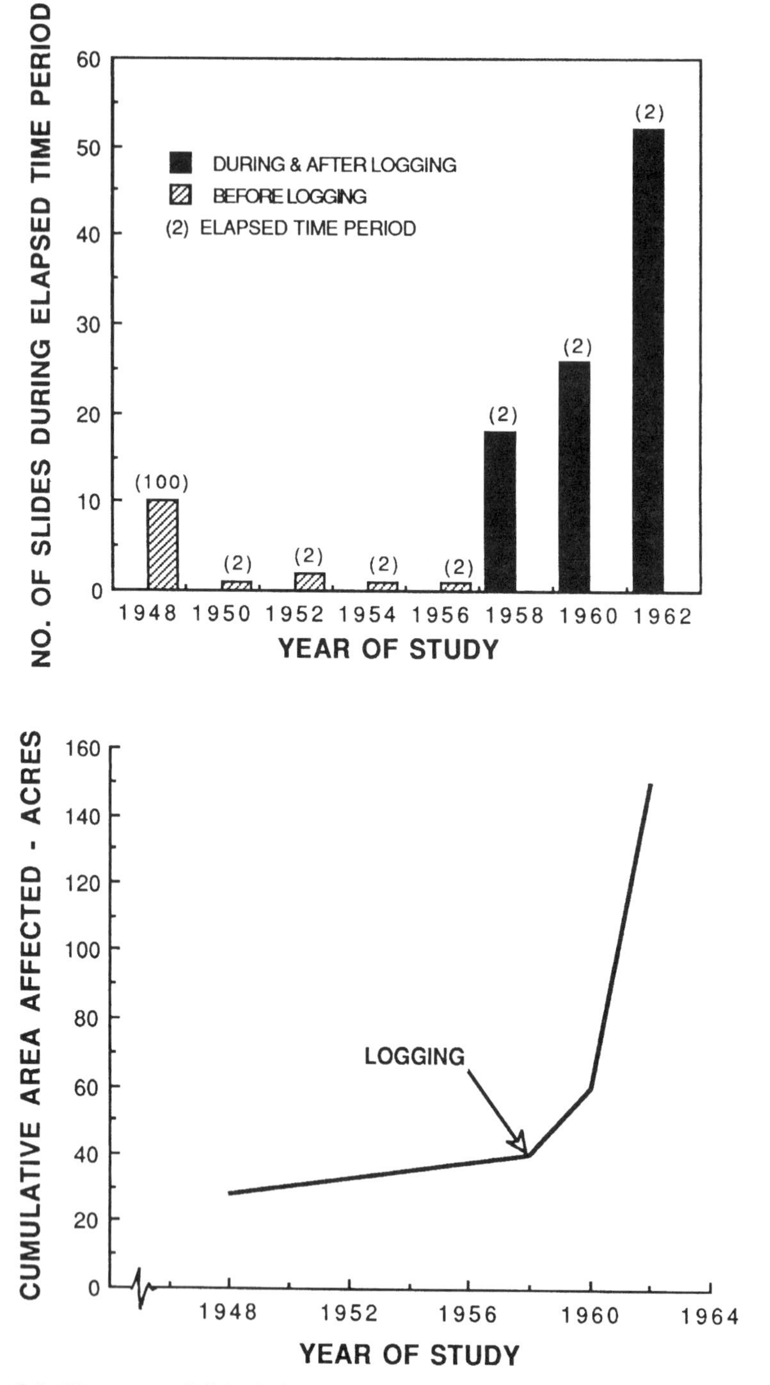

Figure 3-3. Frequency of slides before and after logging, Hollis, Alaska. (From Bishop and Stevens, 1964.)

Figure 3-4. View of clear-cut timber harvest site, Klamath National Forest, California.

employed to harvest timber in many parts of the world. Clear-cutting in mature, even aged stands of timber can result in denuded sites, prone to both surficial erosion and mass movement, as shown in Figure 3-4.

Amaranthus et al. (1985) investigated and compared the effects of timber harvesting on rates of mass erosion in seven different areas in the Pacific Northwest (U.S. mainland) with high rainfall rates. They estimated that the rates of mass erosion from debris slides in disturbed areas were on average some 100 times greater in road right-of-way areas than on undisturbed natural slopes (see Table 3-2). Their findings also showed that rates of erosion in harvested areas were on average five times greater than in undisturbed, vegetated sites.

TABLE 3-2. Mass Erosion Rates Estimated in Seven Pacific Coast Studies

Type of Terrain or Land Use	Average Annual Erosion Losses (yd^3/acre-yr)	Increase over Natural Background
Natural areas	0.17	——
Harvest areas	0.60	5×
Road rights-of-way	21.10	109×

Source: Adapted from Amaranthus et al. (1985).

3.5 ROOT MORPHOLOGY AND STRENGTH

3.5.1 Introduction

Woody vegetation affects shallow mass stability mainly by increasing the shear strength of soil via root reinforcement. The value of the root system in this regard will depend upon the strength and interface properties of the roots themselves and on the concentration, branching characteristics, and spatial distribution of roots in the ground. Root strength and architecture in turn are governed by the type of plant and by local soil/site conditions. Some way of assessing both these factors is required before the mechanical or reinforcing effect of woody plant roots on mass stability of slopes can be accounted for in a reliable and systematic manner.

3.5.2 Root Architecture

Structure Classification and Terminology: Specific terms have been adopted to describe the various parts of a tree root system, as noted in Figure 3-5. *Taproot* refers to the main vertical root directly below the bole of the tree, *sinker root* refers to vertical roots coming either from the bole or from laterals, and *lateral root* refers to roots growing from the central bole but in a horizontal orientation.

The overall shape or morphology of a tree root system can also be categorized. Three distinct forms have been recognized, namely, taproot, heartroot, and plateroot shapes, as shown schematically in Figure 3-6. Variants of these basic shapes may also occur. Morphology is controlled both genetically and by environmental conditions. The development of a particular root architecture in response to either of these factors dictates its contribution to slope stability. In general, root systems with strong, deeply penetrating vertical or sinker roots that penetrate potential shear surfaces are more likely to increase stability against shallow sliding. A high density or concentration of small-diameter fibrous roots is also more effective than a few large-diameter roots for increasing the shear strength of a root-permeated soil mass.

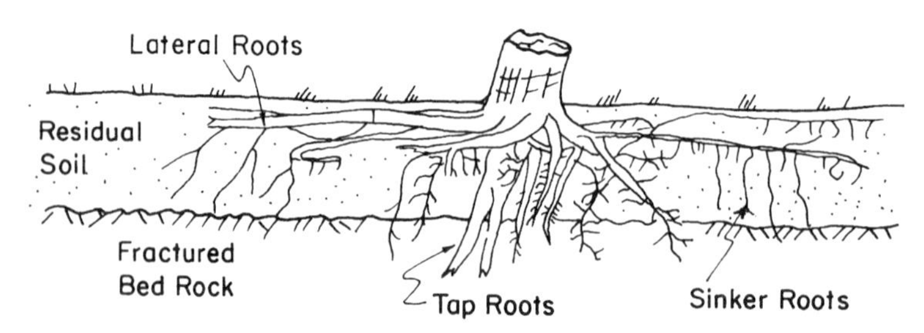

Figure 3-5. Main components of woody root system including lateral, tap, and sinker roots.

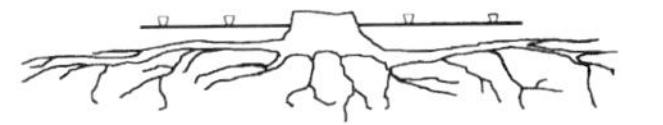

Plateroot

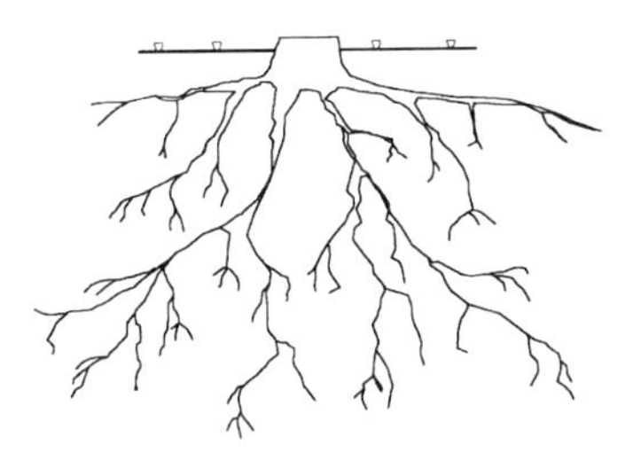

Heartroot

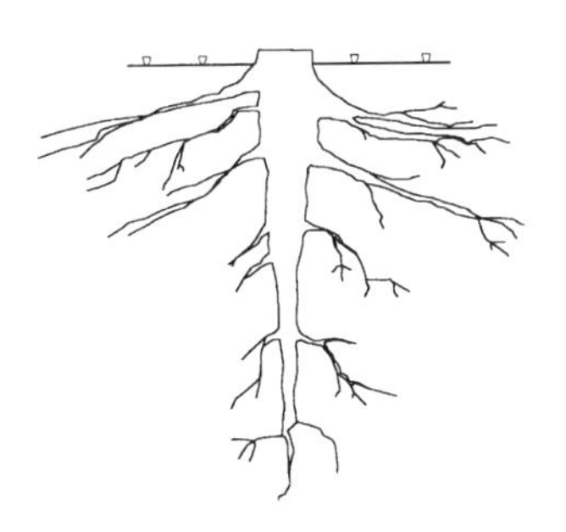

Taproot

Figure 3-6. Principal morphological shapes of woody root systems. (From Wilde, 1958.)

Depth and Distribution of Root Systems: Deeply penetrating vertical taproots and sinker roots provide the main contribution to the stability of slopes vis-à-vis resistance to shallow sliding. Mechanical restraint against sliding only extends as far as the depth of root penetration. In addition, the roots must penetrate across the failure surface to have a significant effect. The influence of root reinforcement and restraint for different slope stratigraphies and conditions is summarized in Figure 3-7. The most effective restraint is provided where roots penetrate across the soil mantle into fractures or fissures in the underlying bedrock (Type B) or where roots penetrate into a residual soil or transition zone whose density and shear strength increase with depth (Type C).

Because of oxygen requirements, the roots of most trees tend to be concentrated near the surface. As a rough rule of thumb the mechanical reinforcing or

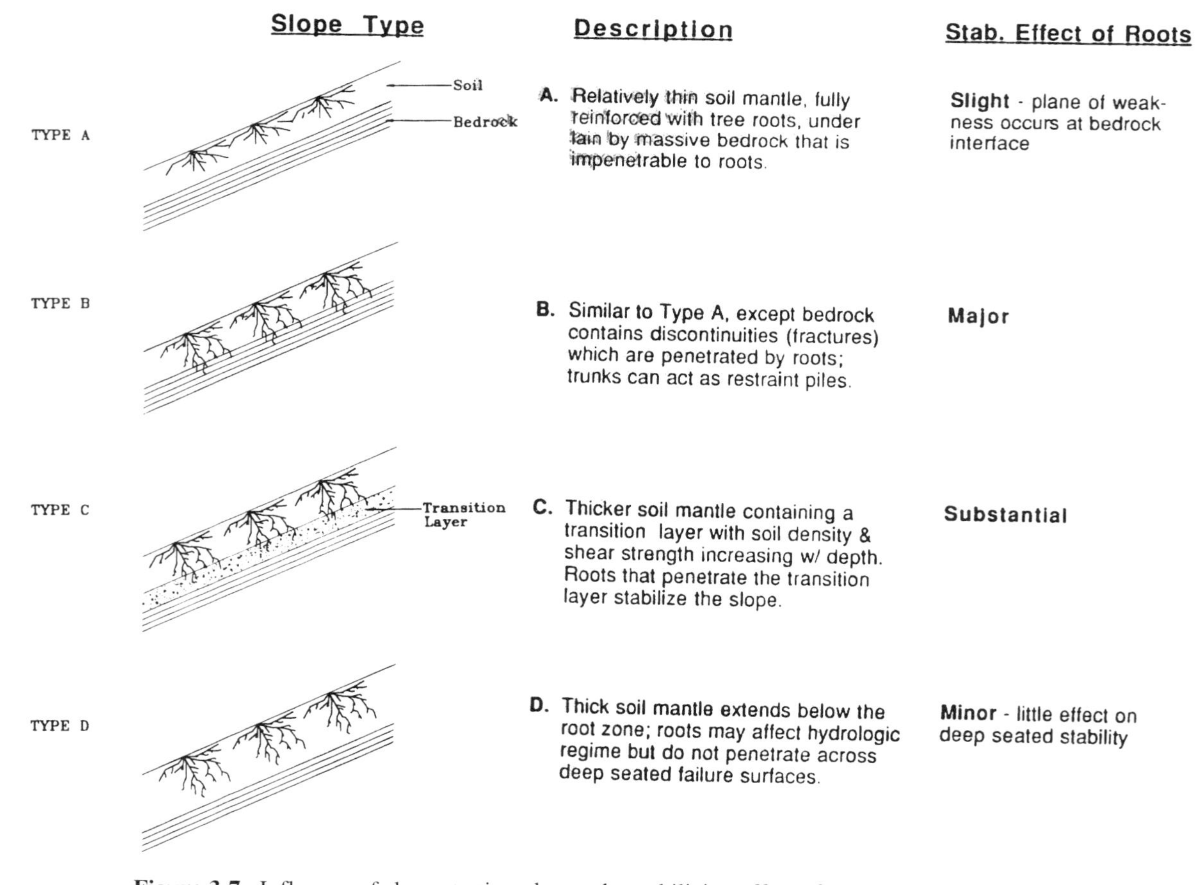

Figure 3-7. Influence of slope stratigraphy on the stabilizing effect of roots against slope failure. (Adapted from Tsukamoto and Kusuba, 1984.)

restraining influence of roots on a slope is probably limited to a zone about 5 feet (1.5 m) from the surface. Studies by Patric et al. (1965) in a loblolly pine plantation showed that 80 to 90 percent of the roots in their test plots were concentrated in the first 3 feet (0.9 m). The bulk of the near-surface roots were laterals; in contrast, roots below 3 feet were generally oriented vertically.

Root morphology studies require careful excavation and are difficult and expensive to undertake, particularly in the case of large mature trees. The root architecture and distribution of mature Monterey pine (*Pinus radiata*) and other tree species have been reported by Watson and O'Loughlin (1985, 1990). They presented diagrammatic plan and section views of the root system of a 25-year-old excavated Monterey pine tree. At age 25 years the main laterals extended up to 34 feet (10.4 m) from the bole. The vertical roots penetrated a maximum depth of 10 feet (3.10 m), but were about 7.8 feet (2.4 m) on average.

Root area ratios were measured as a function of depth in a sandy levee along the Sacramento River in California (Shields and Gray, 1993) for a variety of woody plant species. The term "root area ratio" refers to the fraction of the total cross-sectional area of a soil that is occupied by roots. Root area ratio versus depth curves from this study are shown in Figure 3-8 for two mutually perpendicular transects adjacent to a group of elderberry bushes. These distri-

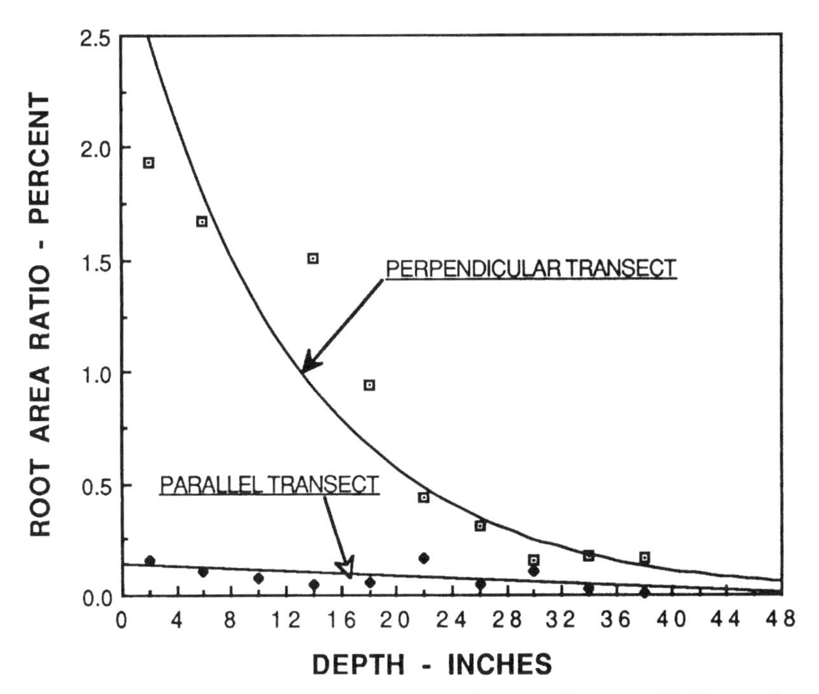

Figure 3-8. Root area ratio versus depth measured in mutually perpendicular trench excavations at elderberry site, Sacramento River channel levee.

bution curves are typical of root concentrations (or root biomass) as a function of depth. Few roots were encountered below a depth of 4 feet in either transect.

The stabilizing effect of roots is lowest when there is little or no penetration across the shear interface (Figure 3-7, Types A and D). However, even in these cases lateral roots can play an important role by maintaining the continuity of a root-permeated soil mantle on a slope. An interesting example of the value of lateral root reinforcement is shown in Figure 3-9. Clearing and grading along the top of a steep, knife-edge ridge for a pipeline right-of-way eliminated the continuity of the root-permeated/reinforced soil mantle across the ridge top. This in turn triggered a peculiar "zipper like" slope failure, in which a thin, forested soil mantle on either side of the ridge pulled apart and slid away when the lateral roots were cut.

Root Spread: Tree roots can spread out for considerable distances; in one reported instance (Kozlowski, 1971) roots of poplars growing in a sandy soil extended out 65 m. The extent of root spread is normally reported in relative multiples of the tree height or crown radius. Kozlowski (1971) cites 10-year-old pine trees growing on sandy soil with a root spread about 7 times the average height of the trees, probably an extreme case. More typical of root systems was the case of fruit trees growing on clay, which had roots extending 1.5 times the crown radius. Similar trees growing on loam extended 22 times, and those on

Figure 3-9. "Zipper type" slope failure along steep-sided knife ridge. Forested, root-permeated soil mantle pulled apart and slid away following grading and clearing of vegetation along ridge top for oil pipeline.

sand up to 3 times the crown radius. A useful rule of thumb is that a root system will spread out a distance at least equal to the 1.5 times the radius of the crown. The hydraulic influence of a tree, that is, significant soil moisture reductions caused by evapotranspiration, can be felt to a distance of at least 1 times the tree height, as shown in Figure 3-10. These findings have implications with regard to both slope stability and safe placement of structures adjacent to trees growing on compressible soils.

Factors Affecting Root Development: Root development and structure are affected initially by genetic disposition but ultimately are governed more by environmental and edaphic conditions (see Sutton, 1969). Henderson et al. (1983) have noted that root systems tend to grow wide and deep in well drained

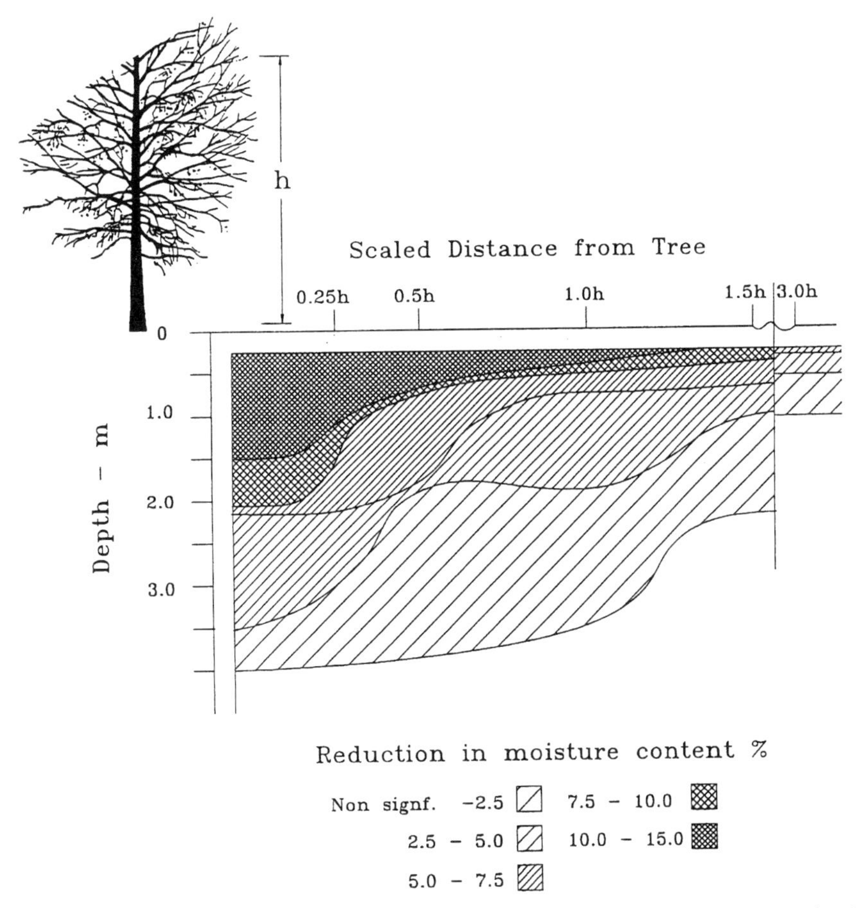

Figure 3-10. Reductions in soil moisture near a poplar tree (*Populus deltoides*) growing in Boulder clay. (From Biddle, 1983. Used with permission of the Institution of Civil Engineers.)

soils as opposed to developing a flat, platelike structure in a surface soil underlain by a more dense or rocky substratum.

The degree to which roots are able to penetrate underlying bedrock depends to a large extent on the nature and extent of discontinuities (i.e., joints and fractures) in the bedrock. Trees growing in shallow, coarse-textured soils developed on granitic bedrock, for example, can develop sinker and taproots that penetrate into fissures and fractures in the underlying bedrock, as shown in Figure 3-11. The overlying soil developed on granitic bedrock is often coarse and incapable of holding much moisture; consequently, roots seek out water in the fractures and fissures in the underlying bedrock. This adaptation in turn insures that the trees will be well anchored to the slope and that they will help to restrain movement of the soil mantle by a combination of buttressing and arching action (Gray and Leiser, 1982).

Root Structure/Distribution: Experimental Methods: Various methods for determining root structure and distribution are described in detail by Bohm (1979). Root area ratios or root biomass concentrations as a function of depth are required in order to estimate rooting contributions to soil shear strength. This ratio varies spatially in three dimensions, as shown schematically in Figure 3-12. The area ratio of greatest interest coincides with the critical sliding surface. Normally this surface is oriented parallel to the slope.

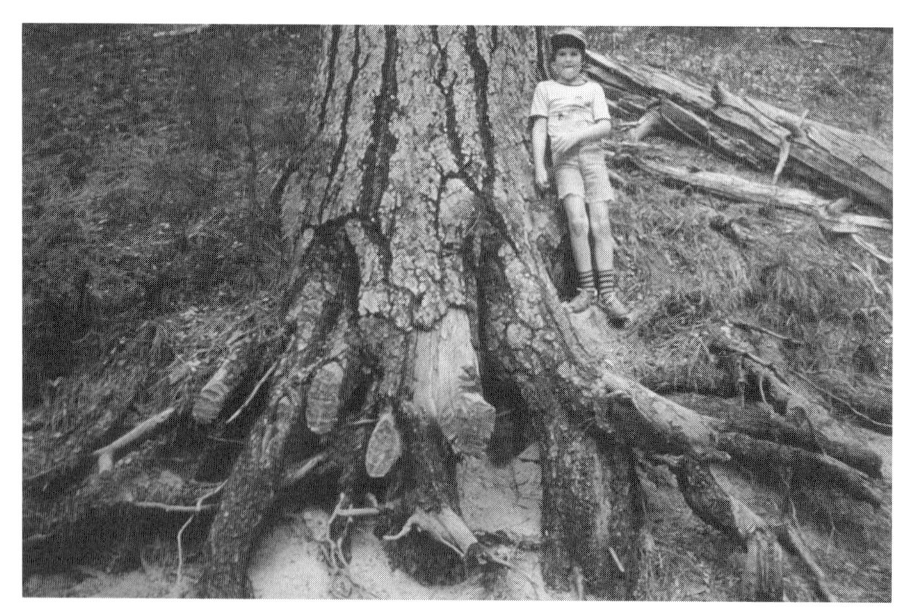

Figure 3-11. Root system of a western yellow pine (*Pinus ponderosa*) exposed in a road cut in granitic terrain showing development of extensive vertical root system that penetrates into fractures in the bedrock.

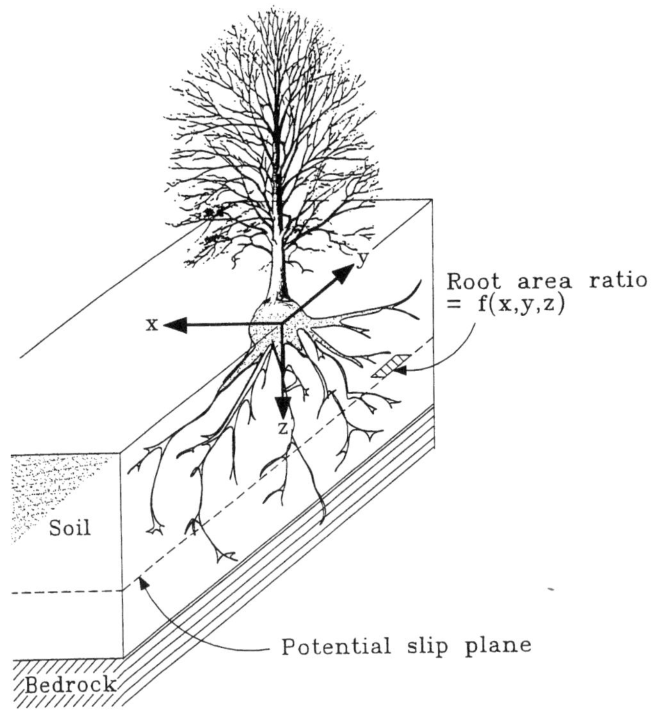

Figure 3-12. Schematic diagram showing variation in root area ratio within a slope. (From: Vegetation and slope stability by D. Greenway, in: *Slope Stability*, edited by M. F. Anderson and K. S. Richards. Copyright 1978. Reprinted by permission of John Wiley & Sons, Ltd.)

One approach is simply to recover large samples of root-permeated soil from various depths and measure the root biomass per unit volume at each depth by sieving the soil, recovering, and weighing the roots. Root biomass per unit volume can be converted to an equivalent root area ratio if the unit weight or density of the roots is known. Alternatively root area ratios can be measured directly in an excavated trench using the "profile wall" method (Bohm, 1979). In this method the roots exposed in the vertical side of the trench are carefully mapped by means of a gridded, acetate overlay, as shown in Figure 3-13. The root area ratios in this case lie in a vertical plane, and it is necessary to assume that area ratios are the same in all directions if the information is to be used for shear strength predictions in an "infinite slope" stability analysis (see Section 3.6.3).

The range in root area ratios measured using the profile wall method along two transects oriented perpendicular and parallel to the crest of a sandy levee along the Sacramento River in Northern California is listed in Table 3-3. The transects or trenches passed adjacent to groves of different tree species. Note that the maximum root area ratio was about 2 percent. This probably represents

Figure 3-13. Profile wall method used to determine root distribution and area ratio as function of depth. Exposed roots are mapped on gridded acetate overlay.

the upper limit for area ratios of root-permeated soils in field and laboratory tests. These root area ratios and their variation with depth can be used to obtain an estimate of the rooting contribution to shear strength with depth as explained in Section 3.6.3.

3.5.3 Root Strength

Factors Affecting Strength: Wide variations in tensile strength of roots have been reported in the technical literature, variations depending on species and on such site factors as growing environment, season, root diameter, and orientation. Greenway (1987) has compiled an excellent review of root strength and factors

TABLE 3-3. Maximum and Minimum Root Area Ratios Measured in Trench Excavations in a Sandy Channel Levee

Site Number	Site Description	Perpendicular Trench	Parallel Trench
2	Control	0.010–0.58	0.010–2.02
3	Dead oak	0.001–0.24	0.001–0.40
4	Live valley oak	0.060–0.83	0.008–0.13
5	Willow	0.004–0.34	0.001–0.32
7	Elderberry	0.070–1.10	0.006–0.16
8	Black locus	0.001–0.85	0.001–0.12

affecting it. With regard to the influence of seasonal effects, Hathaway and Penny (1975) reported that variations in specific gravity and lignin/cellulose ratio within poplar and willow roots produced seasonal fluctuations in tensile strength. Schiechtl (1980) observed that roots growing in the uphill direction were stronger than those extending downhill, as shown in Table 3-4 for several tree species.

Ranges in Root Tensile Strength and Modulus: Root tensile strengths have been measured by a number of different investigators, notwithstanding difficulties in conducting such tests. Nominal tensile strengths reported in the technical literature are summarized in Table 3-5 for selected shrub and tree species. Tensile strengths vary significantly with diameter and method of testing (e.g., in a moist or air dry state). Accordingly, the values listed in Table 3-5 should be considered only as rough or approximate averages. Nevertheless, some interesting trends can be observed in the tabulated strength values. Tensile strengths can approach 70 MPa but appear to lie in the range of 10 to 40 MPa for most species. The conifers as a group tend to have lower root strengths than deciduous trees. Shrubs appear to have root tensile strengths at least comparable to that of trees. This is an important finding because equivalent reinforcement can be supplied by shrubs at shallow depths without the concomitant liabilities of trees resulting from their greater weight, rigidity, and tendency for windthrowing. This could be an important consideration, for example, in streambank or levee slope stabilization. Willow species, which are frequently used in soil bioengineering stabilization work, have root tensile strengths ranging from approximately 14 to 35 MPa (2000 to 5000 psi).

It is important to recognize that root tensile strength is affected as much by differences in size (diameter) as by species. Several investigators (Turnanina, 1965; Wu, 1976; Burroughs and Thomas, 1977; Nilaweera, 1994) have reported a decrease in root tensile strength with increasing size (diameter). Roots are no

TABLE 3-4. Root Tensile Strength of Different Tree Species Showing the Influence of Orientation

	Tensile Strength (MPa)			Number of
Plant Species	Maximum	Minimum	Mean	Samples
Alnus incana				
Uphill	10.6	55.5	32.8	28
Downhill	6.9	56.2	28.3	10
Alnus japonica				
Uphill	12.5	90.5	42.0	24
Downhill	17.2	73.8	40.1	25
Pinus densiflora				
Uphill	30.9	71.2	47.6	6
Downhill	12.7	33.8	24.8	9
Horizontal	8.9	41.6	28.4	5

Source: From Schiechtl (1980). Used with permission of author.

TABLE 3-5. Nominal Tensile Strength of Selected Tree and Shrub Species

Species	Common Name	Mean Tensile Strength (MPa)
	Tree Species	
Abies concolor	Colorado white fir	11
Acacia confusa	Acacia	11
Alnus firma var. *multinervis*	Alder	52
Alnus incana	Alder	32
Alnus japonica	Japanese alder	42
Betula pendula	European white birch	38
Ficus benjamina	Banyan	13
Hevea braziliensis	Rubber tree	11
Nothofagus fusca	Red beech	32
Picea sitchensis	Sitka spruce	16
Picea abies	European spruce	28
Pinus densiflora	Japanese red pine	33
Pinus lambertiana	Sugar pine	10
Pinus ponderosa	Ponderosa (western yellow) pine	10
Pinus radiata	Monterey pine	18
Populus deltoides	Poplar	37
Populus euramericana 1488	American poplar	33
Pseudotsuga mensieii	Douglas fir (Pacific coast)	55
Pseudotsuga mensieii	Douglas fir (Rocky mountains)	19
Quercus robur	Oak	20
Sambucus callicarpa	Pacific red elder	19
Salix fragilis	Crack willow	18
Salix helvetica	Willow	14
Salix matsudana	Willow	36
Salix purpurea (Booth)	Purple willow	37
Tilia cordata	Linden	26
Tsuga heterophylla	Western hemlock	20
	Shrub Species	
Castanopsis chrysophylla	Golden chinkapin	18
Ceanothus velutinus	Ceanothus	21
Cytisus scoparius	Scotch broom	33
Lespedeza bicolor	Scrub lespedeza	71
Vaccinium spp.	Huckleberry	16

Source: Adapted from Schiechtl (1980). Used with permission of author.

different in this regard than fibers of other materials, which exhibit a similar trend. The variation in root tensile strength with root diameter for several tropical hardwood species is shown in Figure 3-14. Root tensile strengths vary from approximately 8 to 80 MPa for root diameters ranging from 2 to 15 mm. The tensile strengths of Rocky Mountain Douglas fir reported by Buroughs and

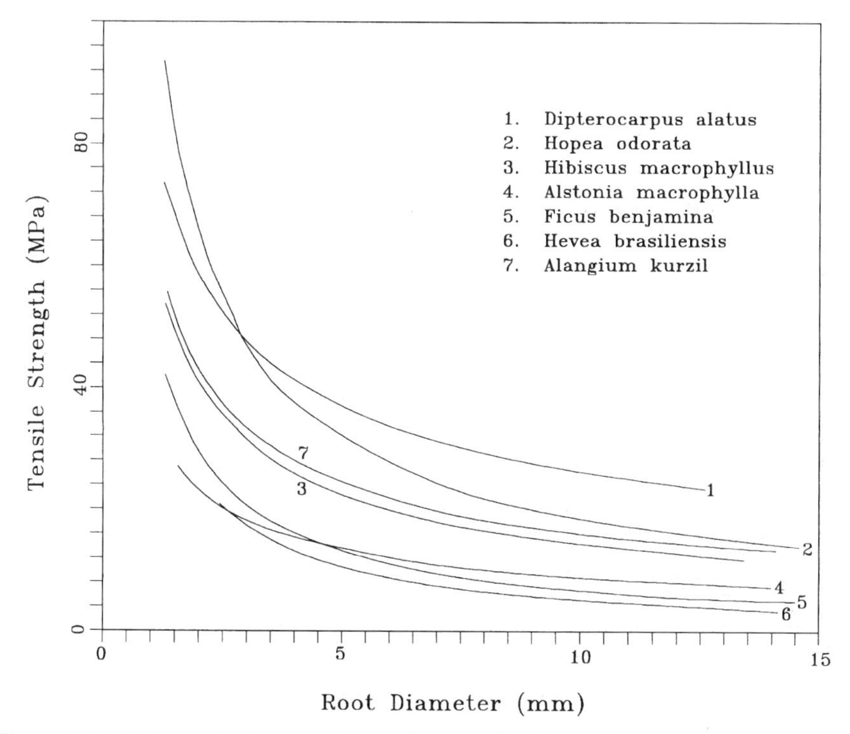

Figure 3-14. Relationship betveen rolt tensile strength and root diameter for several tropical hardwood species. Adapted from Nilaweera, 1994.

Thomas (1977) fall approximately in the midrange position of these plots. A decrease in root diameter from 5 to 2 mm can result in a doubling or even tripling of tensile strength.

Clearly, finer roots can contribute significantly to soil reinforcement and shear strength increase. Finer roots have the advantage of not only higher tensile strengths but also superior pullout resistance because they have higher specific surface areas than larger roots at equivalent area ratios. The relationship between root tensile strength and diameter can be expressed in the form of a simple logarithmic equation as follows:

$$T_r = nD^m \tag{3-2}$$

where: T_r = root tensile strength
D = root diameter
n and m = empirical constants for a given tree species

The values of n and m for the hardwood roots plotted in Figure 3-13 varied from 29.1 to 87.0 and from −0.76 to −0.45, respectively, for root diameters measured in millimeters and tensile strengths in MPa.

The tensile modulus of roots is also of some interest because in many cases the full tensile strength of the roots is not mobilized. Instead, the amount of mobilized tensile resistance will be a function of the modulus and amount of tensile strain or elongation in the root. Only limited data on tensile modulus of roots are available. Hathaway and Penny (1975) presented typical stress-strain curves for several species of poplar and willow. They tested root specimens, without bark, that had been air dried and then rewetted by soaking prior to testing. The ultimate breaking strains, Young's moduli, and tensile strengths measured in these tests are presented in Table 3-6.

Root Decay and Strength Loss: Woody vegetation improves the strength and stability of soil on steep slopes; conversely, its removal by felling or wildfire tends to decrease stability (see Section 3.4). The main reason for the loss of stability and increase in frequency of slope failures following felling is root decay and loss of strength. The smallest roots, which, as noted earlier, have the highest tensile strengths and best interfacial friction or pullout resistance, are the first to disappear after cutting or felling. There will be a period of time between cutting and regeneration of new growth when stability will gradually decrease, reach a minimum, and then increase again as new roots are established in the soil. The time to reach this minimum depends on the tree species, site conditions, and timing of reforestation efforts (Gray and Megahan, 1980).

Strength loss with time following cutting has been reported by a number of investigators. The decline of tensile root strength can be approximated by a negative exponential relationship (Ziemer and Swanston, 1977; O'Loughlin and Watson, 1979). The form of this relationship can be expressed as follows:

$$T_{rt} = T_{r0}e^{-bt} \tag{3-3}$$

where: T_{r0} = tensile strength of root wood sampled from live trees
T_{rt} = tensile strength of roots sampled from stumps cut t months before sampling

TABLE 3-6. Tensile Strength and Stress-Strain Behavior of Some Poplar and Willow Roots

Species	Clone	Tensile Strength (MPa)	Young's Modulus (MPa)	Ultimate Strain (%)
Poplar	*Populus* "I-78"	45.6	16.4	17.1
	Populus "I-488"	32.3	8.4	16.8
	Populus yunnamensis	38.4	12.1	18.7
	Populus deltoides	36.3	9.0	12.4
Willow	*Salix matsundana*	36.4	10.8	16.9
	Salix "Booth"	35.9	15.8	17.3

Source: From Hathaway and Penny (1975).

b = probability of decay

t = age of stump (time between felling and sampling)

The term e^{-b} is an expression of the strength decay rate; accordingly, the time for root strength to decline to half the living root strength is:

$$t_{0.5} = \log 0.5 / \log e^{-b} \quad (3\text{-}4)$$

where: $t_{0.5}$ = the root strength "half life" after felling

O'Loughlin and Watson (1979) measured the tensile strengths of *Pinus radiata* roots at different times after felling and for living trees; their results are listed in Table 3-7.

The mean tensile strength of *Pinus radiata* roots in this study decreased from about 18 to 3 MPa 29 months after cutting. The mean root diameter increased from 5.3 to 8.3 mm, reflecting the faster decay rate and disappearance of smaller roots. For *Pinus radiata* the strength decline curve from these data is:

$$T_{rt} = 19.0\, e^{-0.056t} \qquad (r2 = 0.95)$$

and the strength "half life" is:

$$t_{0.5} = 14.8 \text{ months}$$

Similar relationships have been reported for other tree species. Burroughs and Thomas (1977) determined the tensile strength of roots of Rocky Mountain and Pacific Coast species of Douglas fir as a function of both age after cutting and root diameter. A pronounced decrease in root tensile resistance with time was observed, with the decline being more rapid in the Pacific Coast species, as shown in Figure 3-15.

TABLE 3-7. Tensile Strengths of Radiata Pine Roots at Different Elapsed Times after Felling

Root Class	Mean Tensile Strength (MPa)	Mean Root Diameter (mm)	Number Tested
Living trees	17.6	5.3	188
Cut 3 months	14.4	5.6	105
Cut 9 months	12.3	6.2	134
Cut 14 months	11.0	6.8	140
Cut 29 months	3.3	8.3	59

Source: From O'Loughlin and Watson (1979).

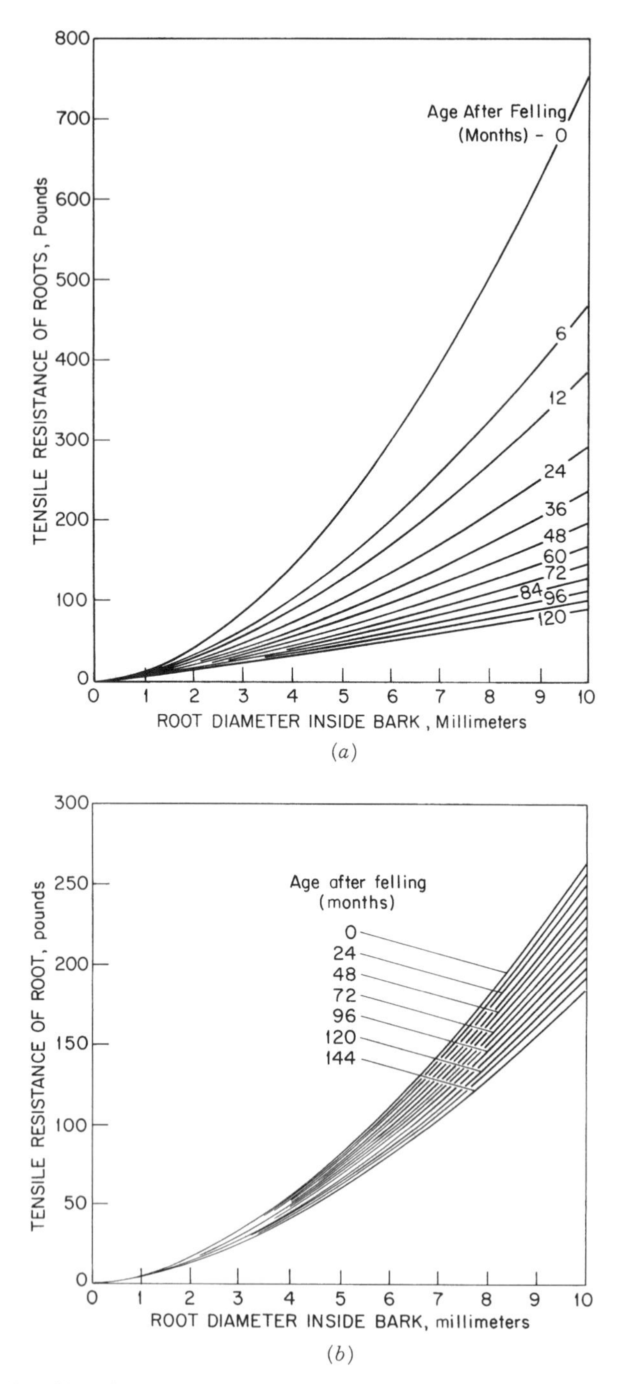

Figure 3-15. Tensile resistance of roots as a function of root diameter and time after felling. (*a*) Coastal Douglas fir. (*b*) Rocky Mountain Douglas fir. (From Burroughs and Thomas, 1977.)

3.6 ROOT/FIBER SOIL REINFORCEMENT

3.6.1 Force-Equilibrium Models

Important investigations have been carried out on a number of fronts during the past two decades, investigations that have greatly improved our understanding of root reinforcement of soils and the contribution of roots to slope stability. These studies include modeling of root-fiber soil interactions, laboratory testing of fiber/soil composites, and in situ shear tests of root-permeated soils.

Relatively simple and straightforward force equilibrium models (Waldron, 1977; Wu et al., 1979; Waldron and Dakessian, 1981) provide useful insights into the nature of root-fiber soil interactions and the contribution of root fibers to soil shear strength. More sophisticated models based on the deformational characteristics of fiber-reinforced composites (Shewbridge and Sitar, 1989, 1990) and statistical models that take into account the random distribution and branching characteristics of root systems have also been developed (Wu et al., 1988a, 1988b).

Root fibers increase the shear strength of soil primarily by transferring shear stresses that develop in the soil matrix into tensile resistance in the fiber inclusions via interface friction along the length of imbedded fibers. This process is shown schematically in Figure 3-16 for an imbedded fiber oriented perpendicularly to the shear surface. When shear occurs the fiber is defomed as shown. This deformation causes the fiber to elongate, provided there is sufficient interface friction and confining stress to lock the fiber in place and prevent

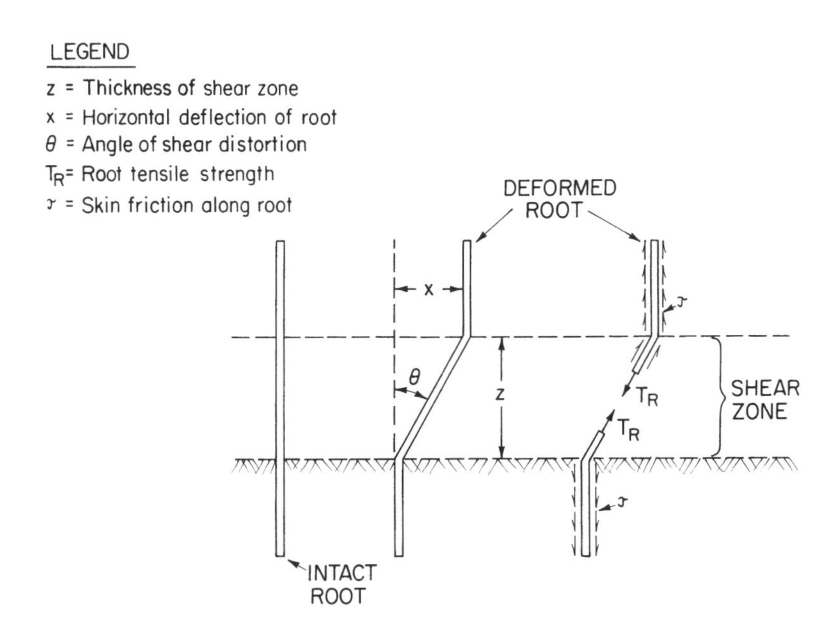

Figure 3-16. Schematic diagram of perpendicular root fiber reinforcement model.

slip or pullout. This fiber elongation mobilizes tensile resistance in the fiber. The component of this tension tangential to the shear zone directly resists shear, while the normal component increases the confining stress on the shear plane.

The assumption of initial fiber orientation perpendicular to the shear surface requires further discussion. Root fibers have many orientations and are unlikely to be oriented perpendicular to the shear failure surface. Furthermore, both theoretical analyses and laboratory studies (Gray and Ohashi, 1983) have shown that a perpendicular orientation is not the optimal orientation. Fibers oriented initially at an acute angle (<90 degrees) in the direction of maximum principal tensile strain result in the highest increase in shear strength. This orientation corresponds to the angle of obliquity $(45 + \phi/2)$, or approximately 60 degrees in most sands. Conversely, an oblique orientation with the shear surface (>90 degrees) can actually result in a shear strength decrease because the fibers initially go into compression rather than tension. The simple, perpendicular model is actually a very useful simulation because it yields an average estimate of all possible orientations. This finding is supported by both laboratory studies on sand/fiber mixtures (Gray and Ohashi, 1983) and by statistical studies of sands with randomly distributed fibers (Maher and Gray, 1990).

Based on this perpendicular model the increase in shear strength of the fiber/soil composite will be given by the following expression:

$$\Delta s = t_R[\sin\theta + \cos\theta \tan\phi] \qquad (3\text{-}5)$$

where: Δs = the shear strength increase
ϕ = the angle of internal friction of the soil
θ = the angle of shear distortion in the shear zone
t_R = the mobilized tensile stress of root fibers per unit area of soil

The mobilized tensile stress of root fibers (t_R) will depend upon the amount of fiber elongation and fixity of the fibers in the soil matrix. Full mobilization can occur only if the fibers elongate sufficiently and if imbedded root fibers are prevented from slipping or pulling out. The latter requires that the fibers be sufficiently long and frictional, constrained at their ends, and/or subjected to high enough confining stresses to increase interface friction. Accordingly, three different response scenarios are possible during shearing of a fiber-reinforced soil composite, namely, fibers break, stretch, or slip.

Fiber Break Mode: Shear strength increase from full mobilization of root fiber tensile strength requires calculation of the average tensile strength of the root fibers, T_R, and the fraction of soil cross section occupied by roots or (A_R/A). The mobilized tensile stress per unit area of soil (t_R) in this case is given by:

$$t_R = T_R(A_R/A) \tag{3-6}$$

The angle of shear distortion or angle (Figure 3-16) is given by:

$$\theta = \tan^{-1}(x/z) \tag{3-7}$$

where: x = the shear displacement
z = the shear zone thickness

The fraction of soil cross section occupied by roots, also termed the root area ratio, can be determined by counting roots by size class within a given soil as:

$$\frac{A_R}{A} = \frac{\sum n_i a_i}{A} \tag{3-8}$$

where: n_i = the number of roots in size class i
a_i = the mean cross-sectional area of roots in size class i

Accounting for the variation in root fiber tensile strength with root diameter mean tensile strength of roots (T_R) can be determined by:

$$T_R = \frac{\sum T_i n_i a_i}{\sum n_i a_i} \tag{3-9}$$

where: T_i = the strength of roots in size class i

By substituting Equation 3-6 into Equation 3-5, the predicted shear strength increase from full mobilization of root tensile strength will be given by:

$$\Delta S = T_R(A_R/A)[\sin\theta + \cos\theta\ \tan\phi] \tag{3-10}$$

The value of the bracketed term $[\sin\theta + \cos\theta \tan\phi]$ in Equation 3-10 is relatively insensitive to normal variations in θ and ϕ, so Wu et al. (1979) proposed an average value of 1.2 for this term. Equation 3-10 can then be simplified to:

$$\Delta S = 1.2T_R(A_R/A) \tag{3-11}$$

Thus the predicted shear strength increase depends entirely on the mean tensile strength of the roots and the root area ratio. This model assumes that the roots are well anchored and do not pull out under tension. The root fibers must be long enough and/or subjected to sufficient interface friction for this assumption to be satisfied. If a simple uniform distribution of bond or interface friction stress between soil and root is assumed, the minimum root length, L_{min}, required to prevent slippage and pullout is given by:

$$L_{min} = \frac{T_R D}{4\tau_b} \tag{3-12}$$

where: T_R = the root tensile strength
D = the root diameter
τ_b = the limiting bond or interface friction stress between root and soil

The bond stress between root fibers and soil can be estimated from the confining stress acting on the fibers and the coefficient of friction. For vertical fibers this bond stress varies with depth and can be calculated by:

$$\tau_b = z\gamma(1 - \sin\phi) f \tan\phi \tag{3-13}$$

where: z = the depth below the ground surface
γ = the soil density
ϕ = the angle of internal friction of the soil
f = the coefficient of friction between the root fiber and soil

The coefficient of friction between soil and wood ranges from 0.7 to 0.9. The rough texture and kinky shape of roots mean that their friction coefficients will likely lie closer to the high end.

Roots will generally exceed the length criteria given in Equation 3-12 except close to the ground surface where the confining stress and hence the bond stresses will be low. This claim is supported by field observations where a preponderance of broken roots, compared to roots that have pulled out, can be seen in landslide scars or failure surfaces.

Fiber Stretch Mode: Lack of sufficient fiber elongation coupled with strain compatibility requirements may prevent mobilization of root fiber tensile or breaking strength. In this case the calculation of the mobilized tensile strength (t_R) will be governed by the amount of fiber elongation and the fiber tensile modulus E_R. A force-equilibrium analysis yields the following expression for the mobilized tensile stress per unit area of soil:

$$t_R = k\alpha(A_R/A) \tag{3-14}$$

where:

$$k = (4z\tau_b E_R/D)^{1/2}; \qquad \alpha = (\sec\theta - 1)^{1/2} \tag{3-15}$$

where: E_R = the tensile modulus of the root fiber
z = the thickness of the shear zone
D = the fiber diameter
τ_b = the root/soil bond stress
θ = the angle of shear distortion

Equation 3-14 assumes a linear tensile stress distribution in the fiber, zero at the ends to a maximum value at the shear plane. A parabolic stress distribution would yield a slightly higher value (Waldron, 1977). By substituting Equation 3-15 into Equation 3-5, the predicted shear strength increase from mobilization of root tensile resistance from stretching will be given by:

$$\Delta S = k\alpha(A_R/A)[\sin\theta + \cos\theta\tan\phi] \qquad (3\text{-}16)$$

This expression reveals that shear strength increases vary inversely with the square root of the fiber diameter. Accordingly, at equal root area ratios, small diameter fibers will be more effective than large fibers.

Fiber Slip Mode: If the fibers are very short, unconstrained, and subject to low confining stresses, they will tend to slip or pull out when the soil/fiber composite is sheared. They will nevertheless continue to contribute a reinforcing increment. At incipient slippage, the maximum tension in a root fiber, T_N, is given by:

$$T_N = 2\tau_b L/D \qquad (3\text{-}17)$$

where: D = the root diameter
L = the root length in which the maximum stress occurs at the center

The shear strength increase or reinforcement from n slipping roots of one size class is given by:

$$\Delta S = \{\pi\tau_b nLD/2A\}[\sin\theta + \cos\theta\tan\phi] \qquad (3\text{-}18)$$

If there are j slipping root size classes with n_i roots in each size class, then the shear strength increase is given by:

$$\Delta S = \{\pi\tau_b/2A\}[\sin\theta + \cos\theta\tan\phi]\Sigma n_i L_i D_i \qquad (3\text{-}19)$$

Under field conditions roots occur in different sizes and lengths, and can have different tensile strengths and degrees of fixity. Accordingly, all three mechanisms may occur simultaneously. Waldron and Dakessian (1981) present procedures for systematically accounting for each. These models are idealizations of actual conditions, but they show what parameters are important and how they affect shear strength. Furthermore, the trends and relationships predicted by these simple force-equilibrium models have been validated by laboratory studies on fiber reinforced soils.

3.6.2 Laboratory and In Situ Tests

The presence of fibers (roots) in a soil increases the shear strength of the soil in ways predicted by the force-equilibrium models described in Section 3.6.1. Fiber reinforcement in a sandy, cohesionless soil is manifested in both the stress-strain behavior of the soil/fiber composite and in the failure envelopes, as shown in Figures 3-17 and 3-18, respectively.

Fiber reinforcement tends to increase the peak stress and reduce the amount of post-peak stress loss in dense soils. The failure envelopes in fiber reinforced sands tend to be bilinear. The initial part of the envelope is steep and then bends over and becomes parallel to the unreinforced envelope. The break point coincides with the critical confining stress. Below this stress the fibers tend to slip, while above this stress the fibers lock in and stretch. The breakpoint in the

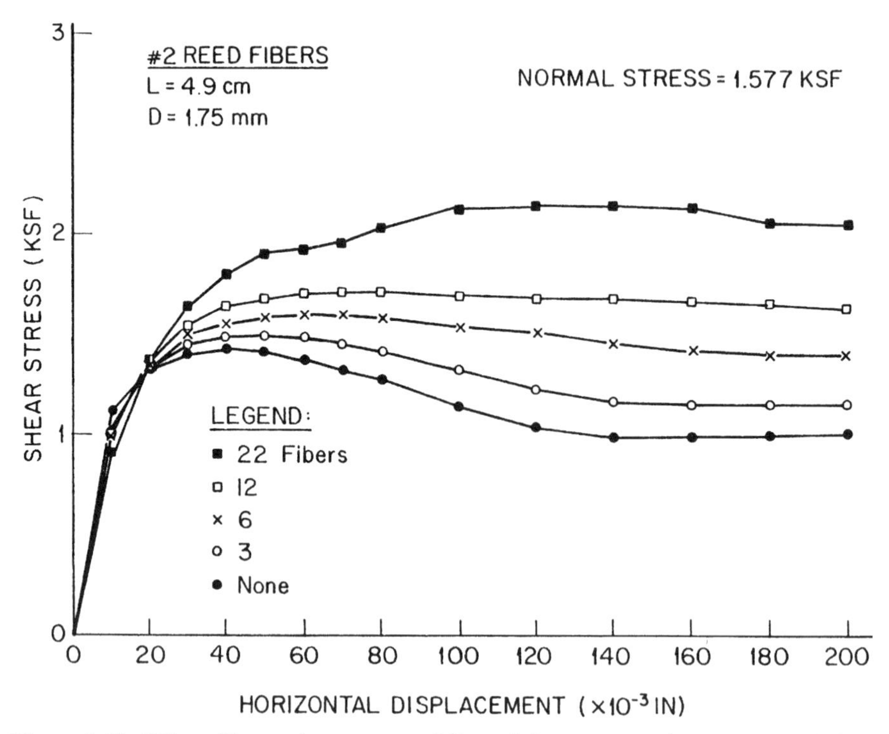

Figure 3-17. Effect of increasing amounts of fiber reinforcement on the stress-strain behavior of a dry sand in a direct shear test. Fibers oriented perpendicularly to the failure plane.

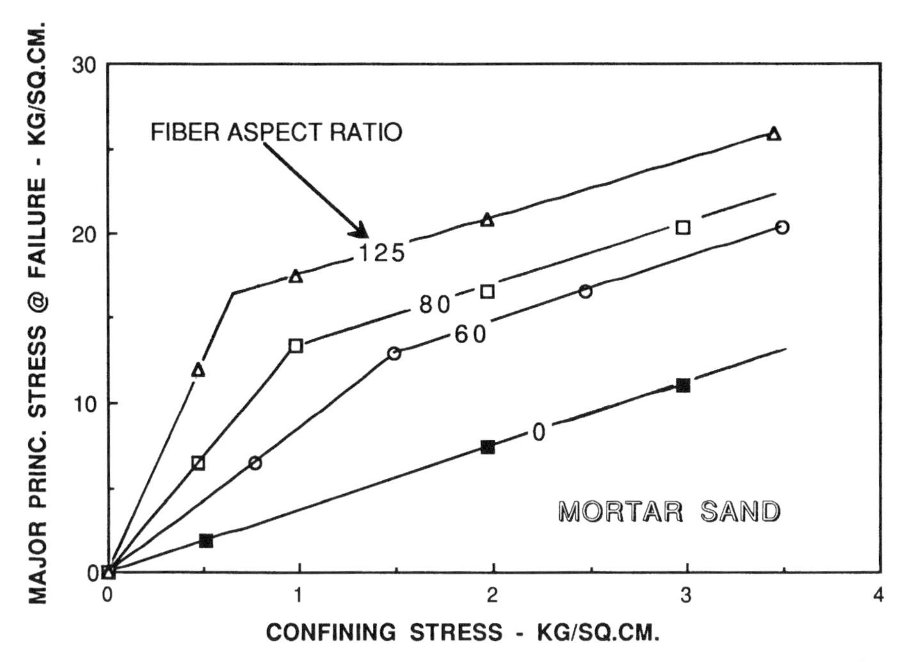

Figure 3-18. Effect of fiber additions on the failure envelopes of a well-graded, angular, dry sand reinforced with randomly distributed fibers with different length/diameter (aspect) ratios. Fiber/soil composites tested in triaxial compression.

envelope shifts to the left, that is, the critical confining stress is reduced, as the fiber length increases and the fibers lock in place more easily.

A bilinear or curvilinear failure envelope is a trademark of all fiber reinforced soils regardless of the type of test or reinforcement. Bilinear failure envelopes with a sharp, well-defined break in the envelope are particularly well manifested in angular, well-graded sands, as shown in Figure 3-18. Extrapolation of the second part of the envelope to the ordinate results in a cohesion intercept. In the case of dry, cohesionless sands reinforced with root fibers, this intercept defines a shear strength increase—sometimes referred to as the "root cohesion."

Laboratory tests show that the shear strength increase or root cohesion is proportional to the amount of fiber or root area ratio, as shown both in Figure 3-19 for a dry dune sand with oriented fibers tested in direct shear and in Figure 3-20 for the same sand with randomly oriented fibers tested in triaxial compression. This observation is consistent with predictions from the force equilibrium models in Section 3.6.1 and appears to hold at root area ratios up to 2 percent or weight concentrations up to 5 percent—the range of practical interest for most root-permeated soils.

Similar relationships have been noted in field or in situ tests on root- permeated soils. Endo and Tsuruta (1969) determined the reinforcing effect of tree roots on soil shear strength by running large scale, in situ direct shear tests on

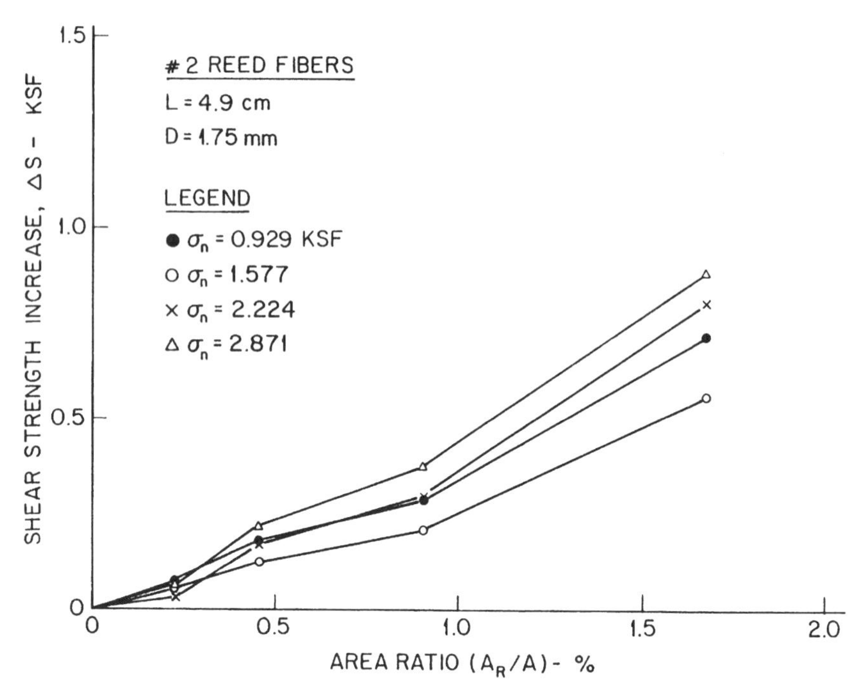

Figure 3-19. Shear strength increases versus fiber content for a dune sand with oriented fibers tested in direct shear at different confining stresses.

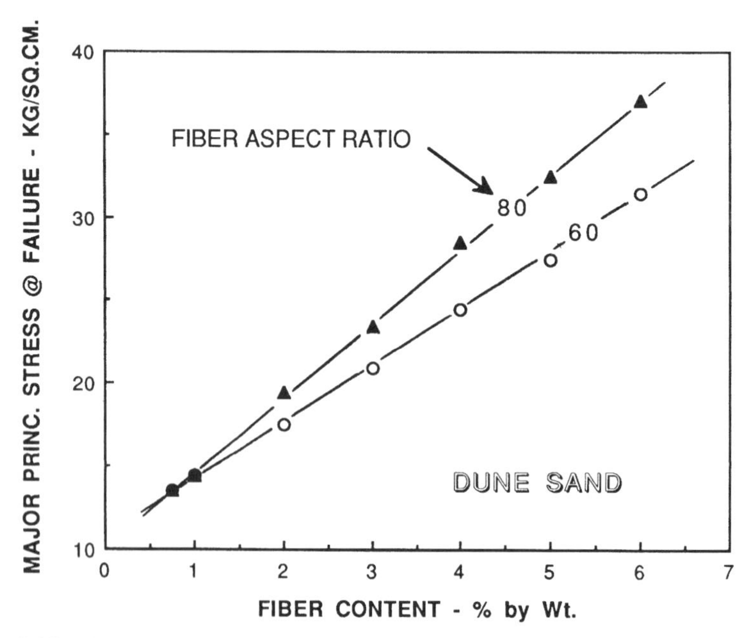

Figure 3-20. Shear strength increases versus fiber content for a dune sand tested in triaxial compression with randomly distributed fibers with different length/diameter (aspect) ratios.

soil pedestals of a clay loam containing live roots of young European alder trees (*Angus glutinosa*). A schematic diagram of their field test apparatus and procedure is shown in Figure 3-21. The shear strength of the soil they tested increased in direct proportion with the bulk weight of roots (or root biomass) per unit volume of soil. An empirical relation of the following form was obtained:

$$\Delta S = a(\rho_R + b) \tag{3-20}$$

where: ΔS = increase in shear resistance, kg/cm^2
ρ_R = biomass of roots per unit volume of soil, g/m^3
a, b = empirical constants ($a = 0.93 \times 10^{-5}$, $b = 53\ \text{g/m}^3$)

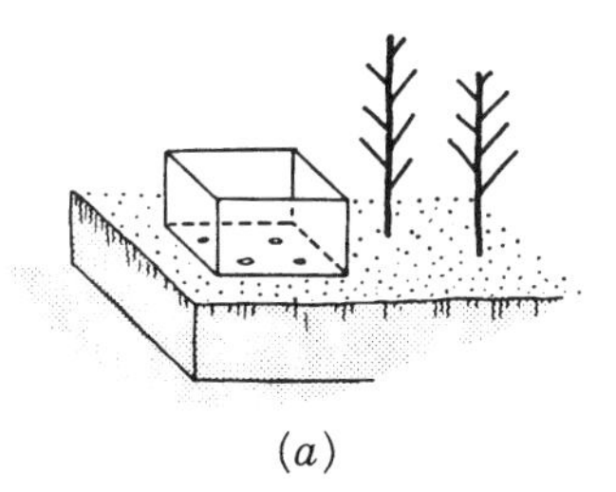

(*a*)

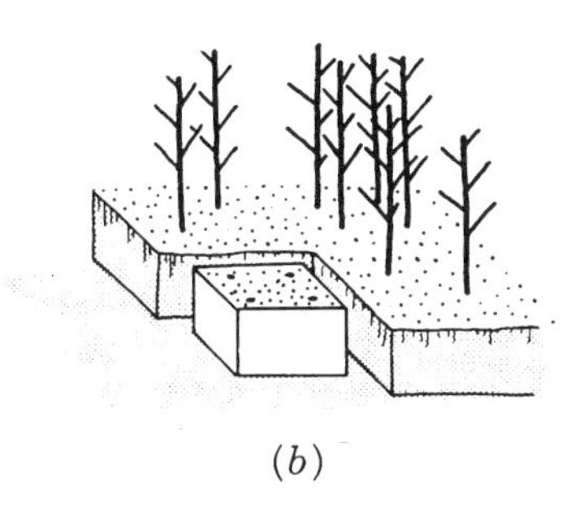

(*b*)

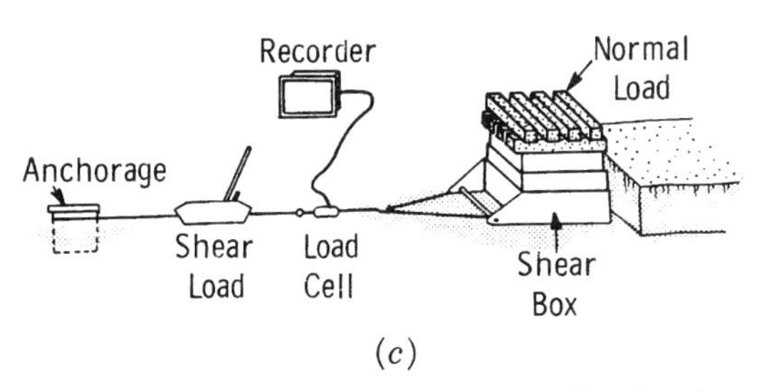

(*c*)

Figure 3-21. Schematic diagram of in situ shear tests on soil pedestals containing live roots of young nursery trees. (*a*) Soil pedestal guide box in place. (*b*) Soil pedestal excavated and exposed. (*c*) Shear box emplaced over pedestal. (From Endo and Tsuruta, 1969.)

This relationship is similar to the strength increase predicted by the theoretical, force-equilibrium model Equation 3-11. The root area ratio (A_R/A) can be substituted for the root biomass concentration (ρ_R), provided the former is multiplied by the unit weight of the root fibers (γ_F). Equation 3-20 can be written in the following equivalent form:

$$\Delta S = a\gamma_F(A_R/A) + ab \tag{3-21}$$

or

$$\Delta S = k_1(A_R/A) + k_2 \tag{3-22}$$

According to data from the study by Endo and Tsuruta (1969), the second term in Equation 3-22 can be neglected with little error for all root area ratios exceeding 0.1 percent. Equation 3-22 developed empirically from field test data is thus virtually identical in form to Equation 3-11, which was developed from a theoretical force-equilibrium model.

Ziemer (1981) conducted in situ direct shear tests on sands permeated with live roots of pine trees (*Pinus contorta*). The largest roots exposed in this shear cross section were under 17 mm. He also observed an approximately linear increase in shear strength with increasing root biomass, as shown in Figure 3-22.

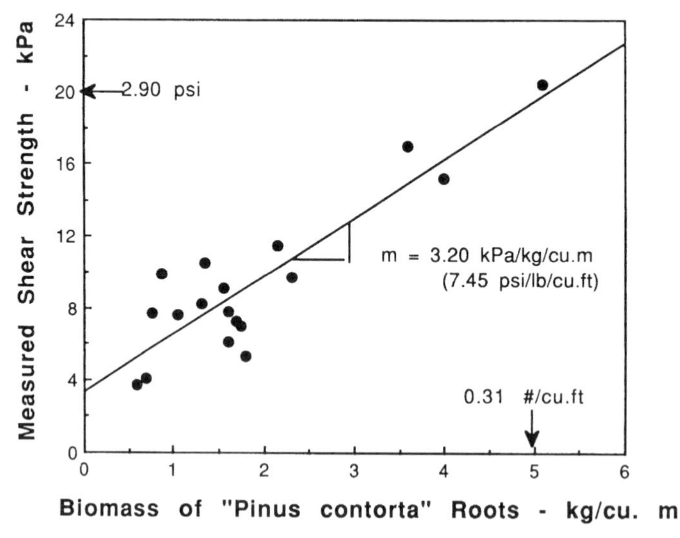

Figure 3-22. Results for in situ direct shear tests on sand permeated with pine roots. (From Ziemer, 1981. Roots and shallow stability of forested slopes. *International Association of Hydrological Sciences Publication No. 132*, pp. 343–361.)

Results of both Ziemer's in situ tests in sand using pine roots and Gray and Ohashi's laboratory tests in a sand using reed fibers are compared in Table 3-8. The shear strength increase response was very similar in both cases; the shear strength increase per unit fiber concentration ranged from 3.2 to 3.7 kPa/kg of root/m^3 soil (7.4 to 8.7 psi/lb of root/foot^3 of soil). This translates into an average value of 3.5 psi/percent root area ratio (assuming a root unit weight of 40 pcf).

These unit shear strength increases can be used to obtain order-of-magnitude estimates for rooting contribution to shear strength in stability analyses (see Section 3.6.3). All that is required is a measure of the root biomass at the depth of interest using the techniques described in Section 3.5.2. Nilaweera (1994) conducted extensive in situ direct shear tests on soil pedestals containing the roots of six different tropical hardwood trees. In addition, he measured the root tensile strength, root biomass as a function of depth, as well as the vertical and lateral uprooting resistance of each tree species. Results of his testing program are summarized in Table 3-9.

The test results for the hardwood trees are listed in order of decreasing shear strength increase per unit root biomass concentration in the soil, which ranged from 4 to 15 kPa/kg/m^3. These values are higher in general than those measured by Gray and Ohashi (1983) and Ziemer (1981) in their tests on fiber-reinforced sands (see Table 3-8). Only rubber tree (*Hevea braziliensis*) roots exhibited comparable values. The tensile strengths for the different roots shown in the first column are for 5-mm diameter roots, which represent a rough average for the different root size classes (see Figure 3-14). *Ficus benjamina* exhibited

TABLE 3-8. Summary of Root Fiber Contributions to Soil Shear Strength of Sands Based on Lab and *in situ* Tests

	Maximum Fiber or Root Biomass Concentration		Shear Stress Increase per Unit Fiber Concentration	
Fiber or Root System	Area Ratio (%)	Weight Concentration (lb root/ft^3 soil)	psi lb/ft^3	kPa kg/m^3
In situ tests on root-permeated sand:				
Tree roots (*pinus contorta*)	0.78	0.31	7.4	3.2
Vertical shear surface, coastal sand				
Live roots <17 mm diameter				
In situ direct shear test				
Lab tests on fiber-permeated sand:				
Reed fibers (*phragmites communis*)	1.70	0.68	8.7	3.7
Natural fibers: diameter = 2.0 mm				
Uniform sand				
Direct shear test				
Averages			8.1	3.5

TABLE 3-9. Tensile Strength and Influence of Hardwood Roots on Soil Shear Strength and Uprooting Resistance[a]

Species Name	Tensile Stress at 5-mm Diameter (MPa)	Maximum Root Volume (cm^2)	Uprooting Lateral (kN)	Uprooting Vertical (kN)	Shear Stress Increase (kPA)	Shear Stress Increase/Unit Biomass Concentration ($kPa/kg/m^3$)
Dipterocarpus alatus	33.6	95	61.6	1.81	1.64	15.5
Hopea oderata	28.5	90	43.9	1.94	1.29	12.2
Astonia macrophylla	13.5	200	40.8	2.46	1.51	9.5
Hibiscus macrophyllus	20.9	420	35.6	4.00	1.89	8.0
Ficus benjamina	12.7	500	88.0	4.67	2.10	6.2
Hevea braziliensis	10.7	120	25.8	1.36	0.83	4.3

Source: Adapted from Nilaweera (1994).

[a] Highest value (boxed); Lowest value (shaded).

the largest vertical and lateral uprooting resistance, in spite of its relatively low root tensile strength, simply because of its huge root volume or biomass. Lateral uprooting resistance was reasonably well correlated with shear strength increase from the presence of roots, and vertical uprooting resistance with root tensile strength, normalized by multiplying by the maximum root volume or biomass (see Figure 3-23).

Shear strength increase per unit root biomass concentration is plotted versus root tensile strength for the hardwood trees in Figure 3-24. A good linear correlation can be observed, which again supports the validity of the force-equilibrium model Equation 3-16, which postulates such a relationship.

3.6.3 Stability Analyses

The so-called "infinite slope" model is appropriate for analyzing shallow slides in which the failure surface is planar and parallel to the slope over most of its length. The infinite slope model is generally applicable for sandy slopes that are subject to shallow sloughing. Cohesionless soils are weakest near the surface where confining stress, and hence shear resistance, is lowest. The infinite slope model also is valid for analyzing slopes with a residual soil mantle underlain by a shallow bedrock contact—the most likely sliding surface.

The influence of groundwater must also be considered in stability analyses. Water plays an important role in both shallow mass wasting and surficial erosion processes, either by subsurface seepage or by overland flow. The presence of groundwater in a slope causes pore water pressures to develop, which can increase shear stress and decrease shear resistance. In addition, groundwater flow or seepage in a slope can exert destabilizing stresses depending on the

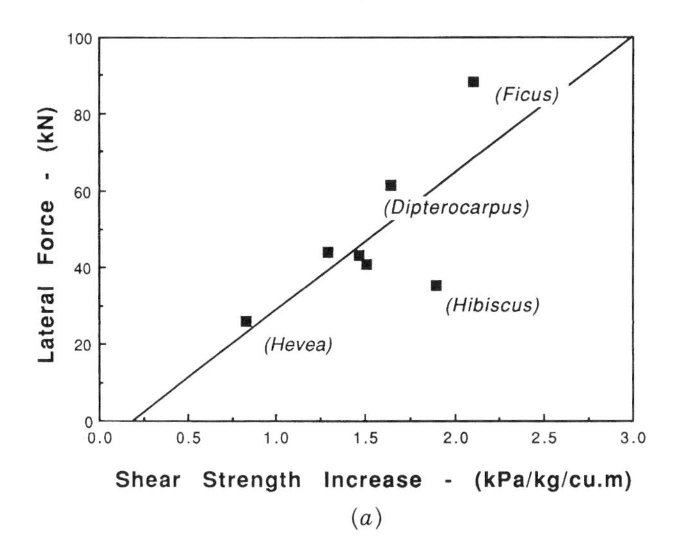

(*a*)

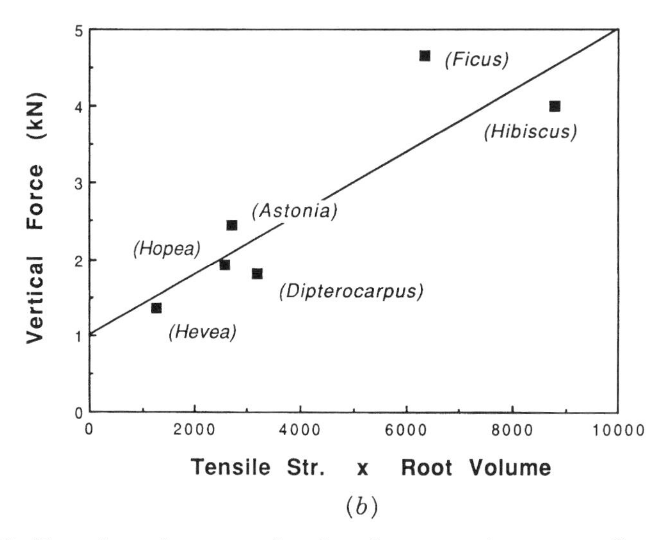

(*b*)

Figure 3-23. Uprooting resistance as a function of root strength parameters for various hardwood trees. (*a*) Lateral resistance versus shear strength increase. (*b*) Vertical resistance versus root tensile strength times root volume. (Adapted from Nilaweera, 1994.)

direction of flow. Groundwater flow or seepage that exits at or flows toward the free face of a slope presents a particular threat.

The factor of safety of an infinite slope that is subject to seepage can be computed as a function of vertical depth (H) and seepage direction (θ) with respect to a horizontal plane (see Figure 3-25). The factor of safety for this case is given by the following relationship:

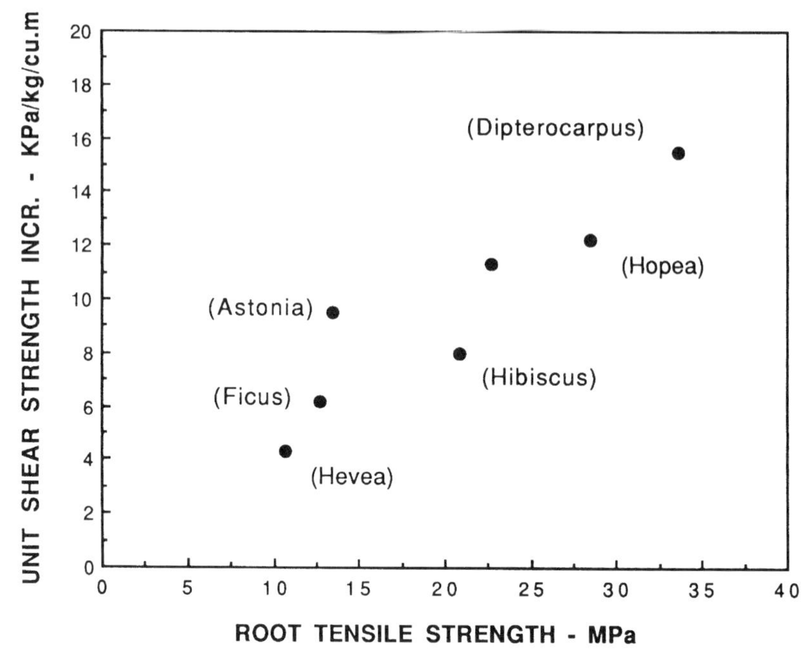

Figure 3-24. Shear strength increase per unit biomass concentration versus root tensile strength for different hardwood trees. (Adapted from Nilaweera, 1994.)

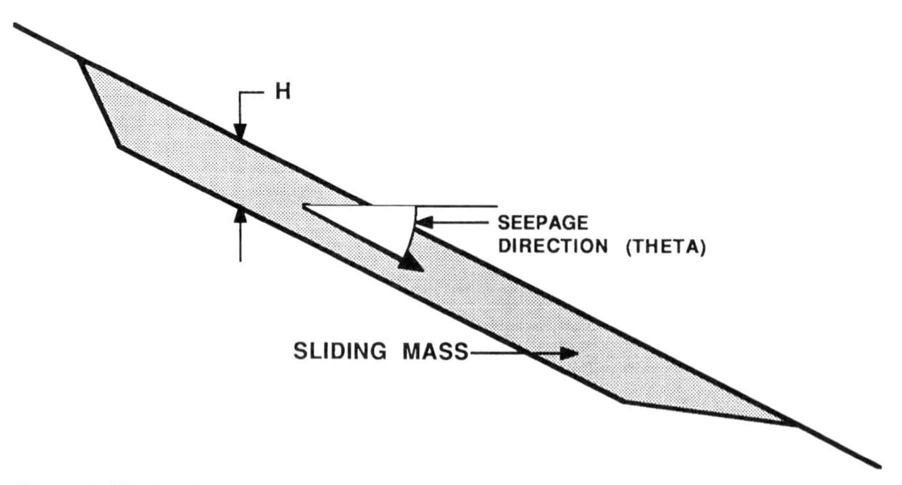

Figure 3-25. Infinite slope model for a cohesionless soil with seepage parallel to slope surface ($\theta = \beta$).

$$F = A[\tan(\phi)/\tan(\beta)] + B[(c + c_R)/\gamma H] \tag{3-23}$$

$$A = [1 - r_u/\cos^2(\beta)] \tag{3-24}$$

$$B = [1/\cos^2(\beta)\tan(\beta)] \tag{3-25}$$

$$r_u = [(\gamma_w/\gamma)\{1/(1 + \tan(\beta)\tan(\theta))\}] \tag{3-26}$$

where:
β = slope angle
ϕ = angle of internal friction
θ = seepage angle (with respect to horizontal)
H = vertical depth below surface
c = soil cohesion
c_R = root cohesion
γ = soil density
γ_w = density of water

Stability Without Roots: The infinite slope equation can be programmed into a spreadsheet and the factor of safety computed for different assumed values of soil friction (ϕ), soil cohesion (c), root cohesion (c_R), seepage direction (θ), depth (H), and slope angle (β). Results of an infinite slope analysis for a 2 : 1 slope with a friction angle of 34 degrees, no surcharge ($q_0 = 0$), and negligible cohesion ($c = 0$) are plotted in Figure 3-26. $\theta = 26.6$ degrees corresponds to the case of seepage parallel to the slope, whereas $\theta = 0$ degrees corresponds to

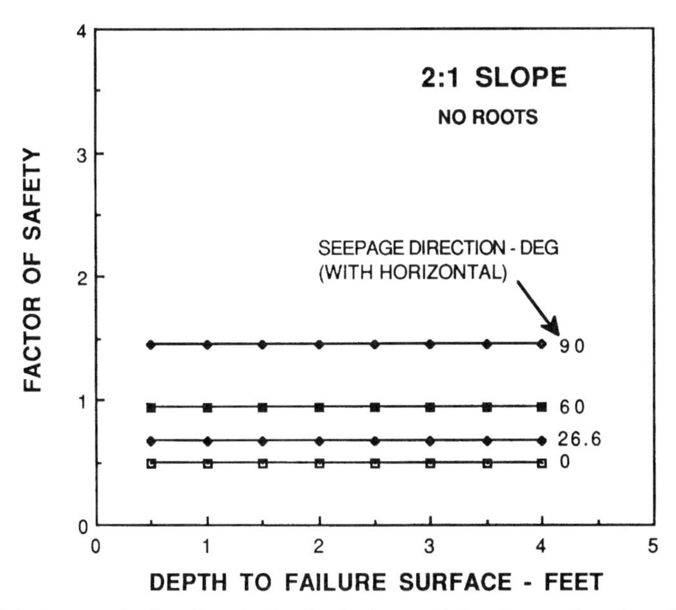

Figure 3-26. Factor of safety for a 2 : 1 cohesionless, infinite slope as function of depth and seepage direction. Soil friction angle = 34 degrees.

horizontal seepage toward the face of the slope. $\theta = 90$ degrees corresponds to vertical (downward) flow; *this condition yields a factor of safety identical to a dry slope.* The factor of safety is independent of depth when $c = 0$ and $q_0 = 0$. The factor of safety drops below one (failure) when the seepage parallels the slope ($\theta = 26.6$ degrees). Horizontal seepage ($\theta = 0$ degrees) results in the lowest safety factor, namely $F = 0.5$. Thus any seepage parallel to the slope, or worse yet, emerging from the slope, will result in sloughing and instability at the face.

Stability With Roots: The factor of safety calculations for the same slope shown in Figure 3-26 can be repeated taking into account the presence of roots. A root reinforcement factor determined from either laboratory or in situ shear strength tests (see Tables 3-8 and 3-9) can be used, along with an estimate of the root biomass concentration (or root area ratio), to compute a shear strength increase (or root cohesion) versus depth relationship. Methods of determining the root biomass or area ratio (RAR) as a function of depth were described previously (see Section 3.5.2). A root reinforcement factor of 3.2 psi/% RAR based on results of shear strength tests on root-permeated sands reported in Table 3-8 can be used to illustrate the calculation procedure. For illustrative purposes, the root area ratio versus depth relation shown in Figure 3.8, measured adjacent to elderberry shrubs growing in a sandy levee, was selected as well. The product of these two parameters yields the shear strength increase from the presence of roots (or root cohesion) at each depth. This root cohesion (c_R) was calculated and is shown tabulated in Table 3-10 for the same slope conditions shown in Figure 3-26.

The tabular results accounting for the influence of root cohesion on stability are plotted in Figure 3-27. Root reinforcement greatly enhances stability at shallow depths (>3 feet), even for the case of emergent or lateral seepage ($\theta < 26.6$

TABLE 3-10. Stability of Sandy Slope Against Shallow Sloughing or Face Sliding, Showing Influence of Variable Root Cohesion with Depth, 2 : 1 Slope

Depth to Failure Surface[a]	Saturated Soil Density	Soil Cohesion	Soil Friction Angle	Root Cohesion[b]	Safety Factor for Various Seepage Directions θ			
H	γ	C	ϕ	c_R	$\theta = 0$	$\theta = 26.6$	$\theta = 60$	$\theta = 90$
(feet)	(pcf)	(psi)	(degrees)	(psi)	(degrees)	(degrees)	(degrees)	(degrees)
0.5	118	0	34	0.38	2.76	2.91	3.18	3.66
1.0	118	0	34	0.33	1.46	1.61	1.88	2.36
1.5	118	0	34	0.28	1.03	1.18	1.45	1.93
2.0	118	0	34	0.23	0.81	0.96	1.23	1.17
2.5	118	0	34	0.19	0.68	0.83	1.10	1.58
3.0	118	0	34	0.14	0.60	0.75	1.01	1.50
3.5	118	0	34	0.09	0.54	0.68	0.95	1.44
4.0	118	0	34	0.04	0.49	0.64	0.91	1.39

[a] $A_r/A = 0.1333 - 0.013H$; based on results of root distribution with depth measured in a sandy levee (from Shields and Gray, 1993).

[b] c_R(psi) = 3.2 (A_r/A) (%); correlation based on direct shear tests on root/fiber/sand composites (Gray and Ohashi, 1983; Ziemer, 1981).

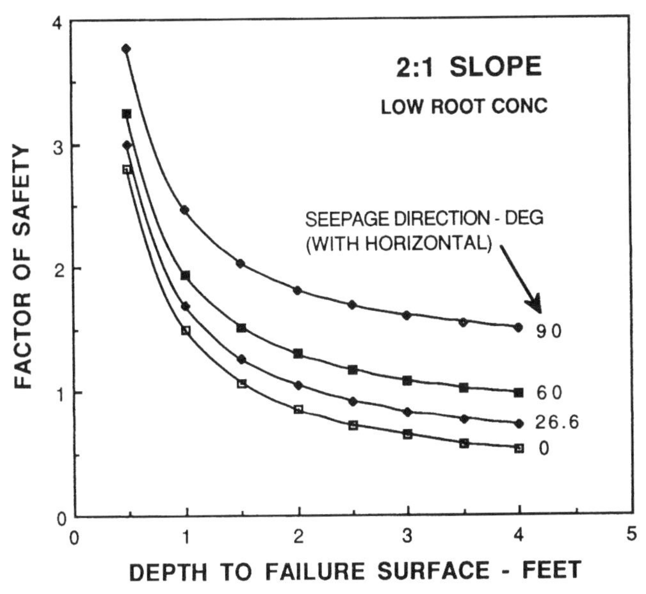

Figure 3-27. Influence of root reinforcement on the factor of safety of a 2:1 cohesionless, infinite slope as a function of depth and seepage direction. Soil friction angle = 34 degrees.

degrees). At greater depths (> 3 feet) the effect of root cohesion diminishes and the curves approach the same values for the case of an unreinforced, cohesionless slope (see Fig. 3-26). This follows both because there are fewer roots at greater depths and because the contribution of cohesion to the total shear resistance decreases with increasing depth.

Also note again the influence of seepage direction on stability with the factor of safety almost doubling at each depth as the seepage direction changes from a horizontal (emergent) to vertical (downward) direction. This finding is particularly significant in the case of brushlayer stabilization of slopes (see Section 7.5). Live brushlayers in a slope act as a series of horizontal drains that intercept seepage, either emerging from the slope or flowing parallel to the slope, and direct it downward toward the brushlayer drainage layers. This mechanism alone can increase stability significantly, in addition to reinforcement from adventitious rooting along the length of imbedded stems.

3.7 GUIDELINES FOR MAXIMIZING BENEFITS OF VEGETATION

3.7.1 General Observations

As noted earlier in this chapter, vegetation mostly has a beneficial influence on stability of slopes, but it can have detrimental or adverse effects as well (see Figure 3-2). Fortunately, a number of strategies and procedures can be adopted to maximize the benefits of vegetation while minimizing its liabilities. These strategies include selection of the appropriate species for particular site conditions and stabilization objectives, placement or location of vegetation in the right places, and management of the vegetation to mitigate any undesirable characteristics. The latter includes such procedures coppicing (essentially a pruning technique), thinning, burning, weeding, and fertilization.

3.7.2 Species Selection

Vegetation should be selected for desired stabilization objectives and be compatible with soil and site conditions. The latter includes consideration of soil type, water availability, nutrient status, soil pH, climate, possible browsing pressure, regulations governing the use of exotic or nonnative species, and so on. These factors are discussed at greater length in Chapter 6.

Certain types of plants are intrinsically better suited than others for specific stabilization objectives. Woody vegetation is stronger and deeper rooted than herbaceous plants and grasses and provides greater mechanical reinforcement and buttressing action at depth. Accordingly, woody plants are superior for mass stability. Grasses and herbaceous vegetation, on the other hand, grow close to the surface and provide a tight, dense ground cover. They tend to be superior, therefore, in intercepting rainfall and preventing surficial erosion. Shrubs are not as deep rooted as trees nor can they be expected to provide as much buttressing restraint. On the other hand, shrubs are more flexible, have less aboveground biomass, and exert less surcharge on a slope. They may be preferable, accordingly, in riverbank and levee stabilization, where these attributes would be advantageous. Table 3-11 lists the relative advantages and disadvantages of different plant types for various engineering functions and applications.

3.7.3 Placement Strategies

Several different placement or locational strategies can be invoked to maximize the utility of slope plantings and minimize possible problems. One of the main objectives raised to vegetation on slopes is that it obstructs views and hinders access. These objections have been raised both by home owners living on hillsides and by inspectors examining river levees. These problems can be addressed by pruning and coppicing techniques, which are described in the next section. They can also be addressed by placement of vegetation on a slope according to its height and shape or density of the crown foliage. Smaller shrubs should be grown near the top of the slope and larger trees placed near the bottom, as shown schematically in Figure 3-28. This simple procedure will improve views from the top, eliminate weight from the top of the slope, and put maximum buttressing restraint and reinforcement near the base, where it is most needed. In the case of river levees plants can be located in such a way to meet both stabilization objectives and to create relatively clear fields of view

TABLE 3-11. Suitability of Plant Types for Different Engineering Functions and Applications

Type	Advantages	Disadvantages
1. Grasses	Versatile and cheap; wide range of tolerances; quick to establish; good dense surface cover	Shallow rooting; regular maintenance required
Reeds and sedges	Establish well on riverbanks, etc.; quick growing	Hand planting expensive; difficult to obtain
2. Herbs	Deeper rooting; attractive in grass sward	Seed expensive; sometimes difficult to establish; many species die back in winter
Legumes	Cheap to establish; fix nitrogen; mix well with grass	Not tolerant of difficult sites
3. Shrubs	Robust and fairly cheap; many species can be seeded; substantial ground cover; deeper rooting; low maintenance; many evergreen species	More expensive to plant; sometimes difficult to establish
4. Trees	Substantial rooting; some can be seeded; no maintenance once established	Long time to establish; slow growing; expensive
Willows and poplars	Root easily from cuttings; versatile, many planting techniques; quick to establish	Care required in selecting corrective type; cannot be grown from seed

Source: From Coppin and Richards (1990). *Use of Vegetation in Civil Engineering*, CIRIA Book 10. Used with permission of Construction Industry Research and Information Association.

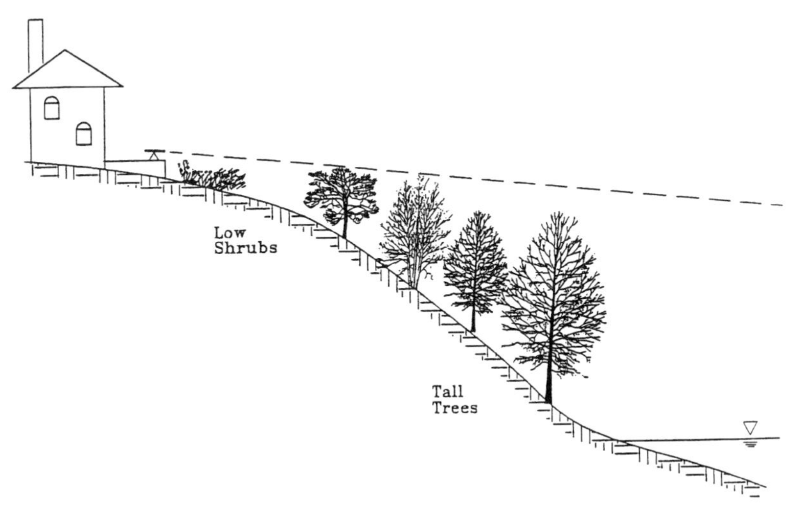

Figure 3-28. Schematic illustration of planting arrangement on a slope to retain views and to maximize slope protection.

for inspection and access purposes. A possible planting scheme to achieve these dual goals is shown in Figure 3-29.

Another approach is to locate vegetation in conformance with "landform grading" practices (Schor, 1980, 1992). Landform grading replicates the irregular shapes of natural, stable slopes. Landform graded slopes are characterized by

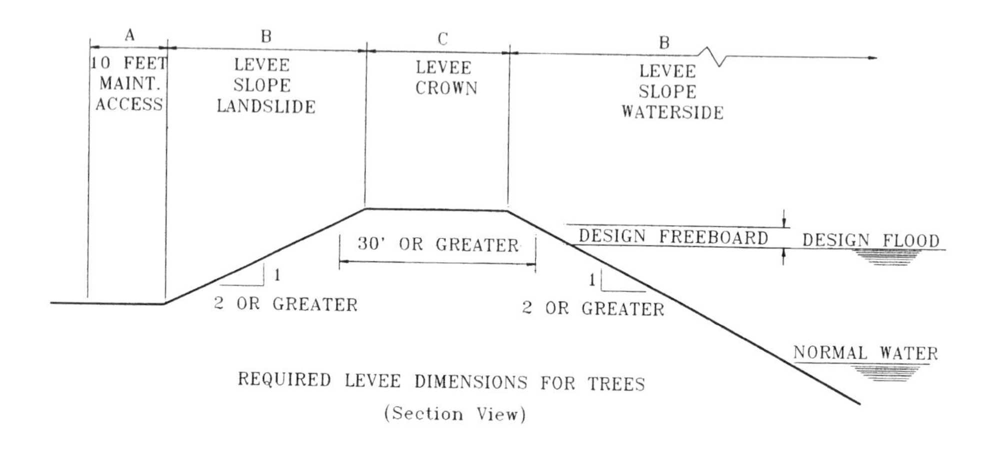

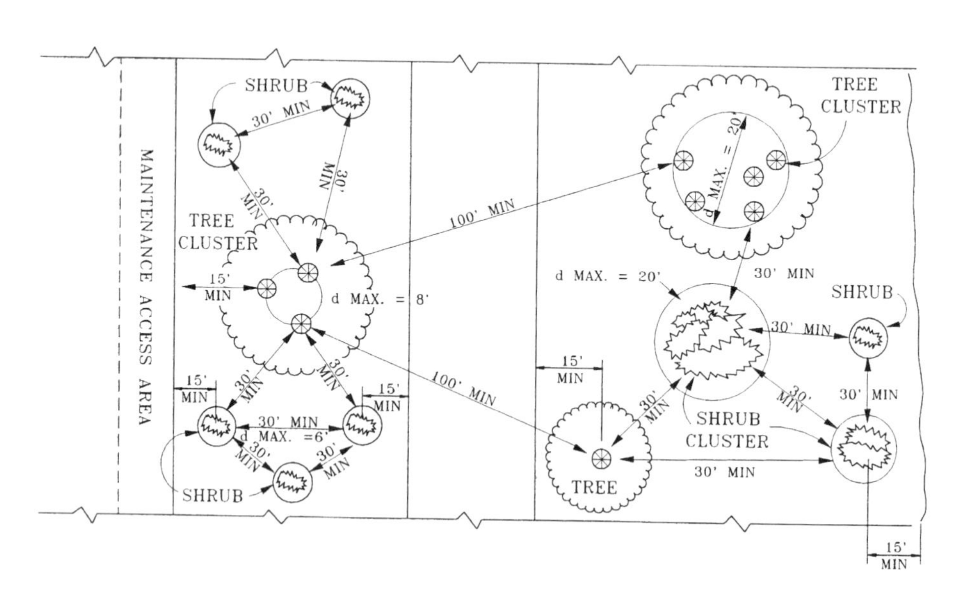

MINIMUM SPACING REQUIREMENTS FOR TREES AND SHRUBS
(PLAN VIEW)

Figure 3-29. Schematic illustration of planting arrangement on a levee slope to create relatively clear lanes for inspection and access purposes. (From Water Resources Control Board, 1987).

a continuous series of concave and convex forms interspersed with swales and berms that grade into the profiles. Revegetation in conjunction with landform grading entails planting vegetation in patterns that occur in nature, as opposed to specifying either uniform or random coverage. Trees and large shrubs tend to require more moisture, and they are also better at stabilizing against shallow slope failures than herbaceous vegetation. Accordingly, trees should be clustered in swales and valleys in a slope where runoff tends to concentrate and evaporation is minimized. Shrubs and trees should also be heavily concentrated along drainage flow lines of each swale. Conversely, seepage and runoff tend to be diverted away from convex-shaped areas. These areas should be planted with more drought-tolerant herbaceous vegetation. Irrigation needs are thus reduced by careful control of drainage pattern on a slope and selection and placement of appropriate plantings for different areas. Landform grading and associated planting patterns for vegetation are discussed in greater detail in Chapter 4.

3.7.4 Coppicing

An interesting approach for mitigating the adverse effects of vegetation on slope stability is the practice of coppicing. Coppicing is a timber harvesting or pruning method that involves the production of new trees from the old stumps. This procedure leaves the root system intact while generating smaller, multiple stems near the cut area, as shown in Figure 3-30. Many tree species that have the ability to regenerate or sprout from dormant buds along their stems lend themselves to coppicing, especially northern hardwoods that have dormant buds on the lowest parts of their trunks. Examples include willows and most maples and locust trees. Some species, such as aspen, also produce new sprouts from their roots, which are referred to as root suckers. Thus whole new forests can be generated from stump sprouts and root suckers.

Best results with coppicing are obtained if the stumps are cut after leaf drop in the late fall or winter (Ecabert, 1993). Red maples, silver maples, and black locust sprouts can grow more than 6 feet the first season. As the stump sprouts grow, they can be thinned and pruned to the desired height and number of trees per stump. Coppicing mitigates two main adverse effects from the legend in Figure 3-2, namely, *surcharge* (#7) and *windthrowing* (#8), while retaining beneficial effects. There may be some initial loss of beneficial influence *interception* (#1), but this is only temporary and greatly outweighed by the attendant benefits. Coppicing allows one to enjoy a view (a frequent reason for tree removal on slopes), use smaller trees, and retain all the hydromechanical benefits provided by a tree's living root system.

3.7.5 Planting and Management Strategies

Several different planting and/or management strategies can be employed to enhance desired characteristics of vegetation at a particular site. More vigorous

Figure 3-30. Example of coppicing showing multiple shoots growing from stump that was cut at approximately breast height.

and deeper rooting can be accomplished in a variety of ways, namely by:

- Watering for longer times at less frequent intervals
- De-compacting or ripping a soil before planting
- Avoiding the use of overly rich topsoil dressings
- Weeding to minimize competition from unwanted plants

Fire is often used as management tool—sometimes with unanticipated consequences. Levees are frequently fired to rid them of woody vegetation. Burning, however, promotes explosive growth of fire adapted species, which may not necessarily be the vegetation of choice for soil erosion control and other purposes.

Figure 3-31. Boardwalk or dune walkover structure protecting dune vegetation against destructive effects of trampling and pedestrian traffic on vegetation.

Another simple yet effective management technique is to control pedestrian and vehicular traffic in critical areas that are protected by vegetation. Coastal dunes are a good case in point. Foredunes play a critical role in a shoreline defense system. Dune vegetation is very effective at trapping drifting sand and helping to build and accrete dunes, but this same vegetation is very vulnerable to trampling and traffic. The use of boardwalks and walkover structures in beach dune areas (see Figure 3-31) is an effective way of protecting vegetation so that it can fulfill its own protective role.

3.8 CONCLUSIONS

Vegetation improves the resistance of slopes to both surficial erosion and mass wasting. Conversely, the removal of slope vegetation tends to accelerate or increase slope failures. Specific hydromechanical mechanisms can be identified through which vegetation affects stability in both beneficial and detrimental ways. Woody vegetation improves shallow mass stability, mainly by increasing the shear strength of the soil via root reinforcement and by a buttressing effect from well-anchored stems. Vegetation also modifies the hydrologic regime by intercepting rainfall in the foliage and by extracting and transpiring soil moisture via the roots. The most effective restraint is provided where roots penetrate across the soil mantle into fractures or fissures in the underlying bedrock or where roots penetrate into a residual soil or transition zone whose density and shear strength increase with depth.

The mechanical or reinforcing effect of plant roots on the stability of slopes can be described and accounted for in a systematic manner. Root fibers reinforce a soil by transfer of shear stress in the soil matrix to tensile resistance in the fiber inclusions. Simple force-equilibrium models are useful for identifying parameters that affect root reinforcement and predicting the amount of strength increase from the presence of fibers in a soil. Shear strength increases from fiber reinforcement can be incorporated into standard slope stability analyses. The presence of roots and their reinforcing effect can improve the factor of safety against sliding significantly at shallow depths.

Although woody vegetation generally has a beneficial influence on stability, it can have adverse effects as well under certain conditions. These effects include surcharge, windthrowing, and scour from large, rigid stems growing on streambanks. Other adverse or undesirable influences unrelated to stability, such as view obstruction and access hindrance, have also been cited. Many of these adverse influences can be mitigated or minimized without compromising the beneficial influence of vegetation by such strategies as selection of appropriate species, proper placement, coppicing, and appropriate management procedures.

3.9 REFERENCES CITED

Amaranthus, M. P., R. M. Rice, N. R. Barr, and R. Ziemer (1985). Logging and forest roads related to increase debris slides in southwestern Oregon. *Journal of Forestry* **83**(4): 229–233.

Biddle, P. G. (1983). Patterns of soil drying and moisture deficit in the vicinity of trees on clay soils. *Geotechnique* **33**(2): 107–126.

Bishop, D. M. and M. E. Stevens (1964). Landslides on Logged Areas in Southeast Alaska. *USDA Forest Service Research Paper* NOR-1, 18 pp.

Bohm, W. (1979). Methods of Studying Root Systems. Ecological Services No. 33, Berlin: Springer-Verlag.

Burroughs, E. R. and B. R. Thomas (1977). Declining root strength in Douglas fir after felling as a factor in slope stability. *Research Paper INT-190*, Intermountain Forest and Range Experiment Station, US Forest Service, Ogden, UT, 27 pp.

Brenner, R. P. (1973). A Hydrologic Model Study of a Forested and a Cutover Slope. *Bulletin Hydrologic Sciences* **18**(26): 125–143.

Coppin, N. J. and I. Richards (1990). Use of Vegetation in Civil Engineering. Sevenoaks, Kent (England): Butterworths.

Ecabert, R. M. (1993). Coppicing: A management program for trees on hillsides that block views. Cincinnati Urban Landscape Tree Care Specialists, 2 pp.

Endo, T. and T. Tsuruta (1969). The effect of tree roots upon the shearing strength of soil. *Annual Report of the Hokkaido Branch*, Tokyo Forest Experiment Station, Vol. 18, pp. 168–179.

Gray, D. H. and A. T. Leiser (1982). Biotechnical Slope Protection and Erosion Control. New York: VanNostrand Reinhold.

Gray, D. H. and H. Ohashi (1983). Mechanics of fiber reinforcement in sands. *Journal of Geotechnical Engineering* (ASCE) **109**(3): 335–353.

Gray, D. H. and W. F. Megahan (1980). Forest vegetation removal and slope stability in the Idaho Batholith. *USDA Research Paper INT-271*, Ogden, UT, 23 pp.

Greenway, D. R. (1987). Vegetation and slope stability, In: *Slope Stability*, edited by M. F. Anderson and K. S. Richards. New York: Wiley.

Hathaway, R. L. and D. Penny (1975). Root strength in some *Populus* and *Salix* clones. *New Zealand Journal of Botany* **13:** 333–344.

Henderson, R. et al. (1983). Morphology of the structural root system of Sitka spruce 1: analysis and quantitative description. *Forestry* **56**(2): 122–135.

Keller, E. A. and A. Brookes (1984). Considerations of meandering in channelization projects: Selected observations and judgements. *Proceedings*, Conference on Rivers, 1983, pp. 384–397.

Kozlowski, T. T. (1971). Growth and Development of Trees, Vol. 2. New York: Academic Press, 520 pp.

Maher, M. and D. H. Gray, Static response of sands reinforced with randomly distributed fibers. *Journal of Geotechnical Engineering* (ASCE) **116**(11): 1661–77.

Megahan, W. F. and W. J. Kidd (1972). Effect of logging and logging roads on erosion and sediment deposition from steep terrain. *Journal of Forestry* **70:** 136–141.

Nilaweera, N. S. (1994). Effects of tree roots on slope stability: The case of Khao Luang Mountain area, So. Thailand. Dissertation No. GT-93-2. Thesis submitted in partial fulfillment of requirements for degree of Doctor of Technical Science, Asian Institute of Technology, Bangkok, Thailand.

Nolan, M. F. (1984). Vegetation on Corps of Engineering project levees in the Sacramento-San Joaquin Valley, California. In: *California Riparian Systems*, edited by R. E. Warner and K. M. Hendrix. Berkeley: University of California Press, pp. 538–547.

O'Loughlin, C. L. (1974). The effects of timber removal on the stability of forest soils. *Journal Hydrology (NZ)* **13:** 121–134.

O'Loughlin, C. L. and A. Watson (1979). Root-wood strength deterioration in Radiata Pine after clearfelling. *New Zealand Journal of Forestry Science* **39**(3): 284–293.

Patric, J. H. et al. (1965). Soil water absorption by mountain and piedmont forests. *Soil Science Society of America Proceedings* **29:** 303–308.

Riestenberg, M. M. and S. Sovonick-Dunford (1983). The role of woody vegetation on stabilizing slopes in the Cincinnati area. *Geologic Society of America Bulletin* **94:** 504–518.

Riestenberg, M. M. (1994). Anchoring of thin colluvium by roots of sugar maple and white ash on hillslopes in Cincinnati. U.S. Geological Survey Bulletin 2059-E, U.S. Government Printing Office, Washington, DC.

Schiechtl, H. M. (1980). Bioengineering for Land Reclamation and Conservation. Edmonton, Canada: University of Alberta Press, 404 pp.

Schor, H. (1980). Landform grading: Building nature's slopes. *Pacific Coast Builder*, (June): 80–83.

Schor, H. (1992). Hills like nature makes them. *Urban Land* (Mar.): 40–43.

Shewbridge, S. E. and N. Sitar (1989). Deformation characteristics of reinforced sand in direct shear. *Journal of Geotechnical Engineering* (ASCE) **115**(GT8): 1134–1147.

Shewbridge, S. E. and N. Sitar. (1990). Deformation based model for reinforced sand in direct shear. *Journal of Geotechnical Engineering* (ASCE) **116**(GT7): 1153–1157.

Shields, F. D. (1991). Woody vegetation and riprap stability along the Sacramento River mile 84.5 to 119. *Water Resources Bulletin* **27**(3): 527–536.

Shields, F. D. and D. H. Gray (1993). Effects of woody vegetation on the structural integrity of sandy levees. *Water Resources Bulletin* **28**(5): 917–931.

Sutton, R. F. (1969). Form and development of conifer root systems. Technical Communication No. 7, Commonwealth Agricultural Bureau, England.

Swanston, D. N. (1974). Slope stability problems associated with timber harvesting in mountainous regions of the western United States. *USDA Forest Service General Technical Report* PNW-21, 14 pp.

Tschantz, B. A. and J. D. Weaver (1988). Tree growth on earthen dams: A survey of state policy and practice. Civil Engineering Department, University of Tennessee, 36 pp.

Tsukamoto, Y. and O. Kusuba (1984). Vegetative influences on debris slide occurrences on steep slopes in Japan. Proceedings, Symposium on Effects of Forest Land Use on Erosion and Slope Stability, Environment Policy Institute, Honolulu, Hawaii.

Turmanina, V. I. (1965). On the strength of tree roots. *Bulletin Moscow Society Naturalists* **70**(5): 36–45.

USDA Soil Conservation Service (1978). Predicting Rainfall Erosion Losses: a Guide to Conservation Planning. USDA Agricultural Handbook #537, Washington, DC.

Waldron, L. J. (1977). The shear resistance of root-permeated homogeneous and stratified soil. *Soil Science Society of America Proceedings* **41:** 843–849.

Waldron, L. J. and S. Dakessian (1981). Soil reinforcement by roots: calculation of increased soil shear resistance from root properties. *Soil Science* **132**(6): 427–35.

Water Resources Control Board (1987). Guide for vegetation on project levees. Final draft. State of California Resources Agency, Sacramento, CA, 22 pp.

Watson, A. J. and C. L. O'Loughlin (1985). Morphology, strength and biomass of Manuka roots and their influence on slope stability. *New Zealand Journal of Forestry Science* **15**(3): 337–348.

Watson, A. J. and C. L. O'Loughlin (1990). Structural root morphology and biomass of three age classes of Pinus radiata. *New Zealand Journal of Forestry Science* **20**(1): 97–110.

Wilde, S. A. (1958). Forest Soils: Their Protection and Relation to Silviculture. New York: Ronald Press, 537 pp.

Wu, T. H. (1976). Investigation of landslides on Prince of Wales Island, Alaska. Geotechnical Engineering Report No. 5, Department of Civil Engineering, Ohio State University, Columbus OH, 94 pp.

Wu, T. H., W. P. McKinell, and D. N. Swanston. (1979). Strength of tree roots and landslides on Prince of Wales Island, Alaska. *Canadian Geotechnical Journal* **16**(1): 19–33.

Wu, T. H., R. M. Macomber, R. T. Erb, and P. E. Beal (1988a). Study of soil-root interactions. *Journal of Geotechnical Engineering* (ASCE) **114**(GT12): 1351–1375.

Wu, T. H., P. E. Beal and C. Lan (1988b). In-situ shear test of soil-root systems. *Journal of Geotechnical Engineering* (ASCE)**114**(GT12): 1376–1394.

Ziemer, R. (1981). Roots and shallow stability of forested slopes. *International Association of Hydrological Sciences,* Publication No. 132, pp. 343–361.

Ziemer, R. and D. N. Swanston (1977). Root strength changes after logging in southeast Alaska. *Research Note PNW-306,* Pacific Northwest Forest and Range Experiment Station, U.S. Forest Service, Portland, OR, 9 pp.

4 Principles of Biotechnical and Soil Bioengineering Stabilization

4.0 INTRODUCTION

All slopes are subject to soil erosion and mass wasting. Various approaches can be used to slow down, if not completely prevent, this degradation. Erosion losses can be minimized by adherence to basic erosion control principles (see Chapter 2) such as preserving native vegetation whenever possible, clearing the land in workable increments, installing facilities to handle increased runoff, and installing erosion control measures as early as possible. Erosion losses can also be minimized by adoption of landform grading practices (see Section 4.3) that mimic natural slopes. Landform graded slopes are characterized by a variety of shapes, including concave-convex forms and landscaping patterns that are adjusted to natural slope hydrogeology.

Many new manufactured products have been introduced during the past few decades for erosion control and slope protection. These products increase our options and enhance our ability to deal with erosion. They also pose a possible danger, namely, that we come to rely exclusively on the products and lose sight of the impact of site grading procedures, the need to handle site runoff, and the importance of observing basic erosion control principles. The proper handling and disposal of surface and subsurface water probably has a greater impact on both surficial and mass stability than any other single factor. In addition, we must always remember the importance of preserving and utilizing natural slope vegetation whenever possible, and remember to take advantage of soil bioengineering methods that capitalize on the advantages of vegetation through the embedment of live cuttings in the ground in various arrays and combinations.

This chapter addresses these issues and concerns by presenting a taxonomy of different methods for slope protection and erosion control and by describing the salient characteristics of soil bioengineering and biotechnical stabilization. Recent developments and potential areas for soil bioengineering application are also included.

4.1 CLASSIFICATION OF DIFFERENT APPROACHES

Slope protection and erosion control methods can be classified in different ways. The methods listed in Table 4-1 fall into three major categories: *live construc-*

TABLE 4-1. Classification of Slope Protection and Erosion Control Measures

Category	Examples
Live Construction	
Conventional planting	Grass seeding Sodding Transplants
Mixed Construction	
Woody plants used as reinforcements and barriers to soil movement	Live staking Live fascines Brushlayering Branchpacking
Plant/structure associations	Breast walls with slope face plantings Revetments with slope face plantings Tiered structure with bench plantings
Woody plants grown in the frontal openings or interstices of retaining structures	Live crib walls Vegetated rock gabions Vegetated geogrid walls Vegetated breast walls
Woody plants grown in the frontal openings or interstices of porous revetments and ground covers	Joint plantings Staked gabion mattresses Vegetated concrete block revetments Vegetated cellular grids
Inert Construction	
Conventional Structures	Concrete gravity walls Cylinder pile walls Tie-back walls

tion (the traditional use of grass and other live plants, primarily for erosion control), *mixed construction* (the use of soil bioengineering and biotechnical methods), and *inert construction* (the use of inert structural or mechanical systems). This classification system is useful and is referred to frequently in this book. It is applicable to both erosion control and the prevention of mass slope failures. Ground cover systems that can be used for erosion control and protection of temporary waterways can be classified in a similar manner (see Chapter 9).

4.2 CONVENTIONAL APPROACHES TO SLOPE PROTECTION AND EROSION CONTROL

4.2.1 Inert Construction

Engineers tend to rely primarily on inert systems for slope stabilization and erosion control. A large array of products and techniques fall into this category. Reasons for their widespread use include availability, ease of installation, familiarity, advertising and promotion, existence of standards, and acceptance by specifiers. Inert materials are presumed to have predictable and invariant properties. In fact, inert materials such as steel, concrete, and synthetic polymers slowly degrade, decompose, and/or decay with time.

Examples of the range of technology and products that fall within this category are summarized below:

Retaining Structures

- Gravity walls (gabions, crib, and bin walls)
- Rock breast walls
- Articulated block walls (Loffelblock®, Keystone®, Dynablock®, etc.)
- Reinforced earth structures (with metal strips, geogrids, or geotextiles)
- Cellular confinement systems (stacked and backfilled three-dimensional webs)

Revetment Systems

- Riprap (quarry stone, rubble, natural rock)
- Gabion mattresses
- Concrete facings (gunnite, concrete filled mattresses)
- Cellular confinement systems (three-dimensional webs that cover the surface and are backfilled with aggregate)
- Articulated block systems (concrete blocks linked by cables or other methods)

Ground Covers

- Artificial mulches (fiberglass roving and cellulose fibers)
- Blankets, mats, and nettings (slope coverings that protect the surface and promote/enhance the growth of vegetation)
- Cellular confinement systems (three-dimensional honeycomb webs that cover the surface and are backfilled with soil or aggregate)

It is interesting to note that many of these systems or products lend themselves to integrated or combined use with vegetation. Basically vegetation can be incorporated into any retaining structure, revetment, or inert ground cover that is porous or that has openings (interstices) at its front face. However, for plant survival, moisture and sunlight must be available. Other requirements for the establishment and maintenance of vegetation are discussed in Chapter 6. Design considerations and stability requirements for structural retaining systems and revetments are treated in Chapter 5.

4.2.2 Live Construction

Although the value of vegetation in preventing soil erosion is now well established, its role in stabilizing slopes against mass wasting is less certain and not as well understood. However, the value of woody vegetation in particular for protecting slopes against shallow sliding and mass wasting has gained considerable recognition in recent years (Greenway, 1987; Coppin and Richards 1990). The role and function of vegetation in protecting slopes against both surficial erosion and shallow mass movement is explored in Chapter 3.

Live construction uses conventional plantings, primarily for erosion control. A dense, tight ground cover of vegetation greatly enhances the erosion resistance of soils. Various types of grasses and herbaceous vegetation are best for this purpose; vegetation can be established from cuttings, transplants, sodding, or direct seeding. Procedures and requirements for the selection and establishment of vegetation on slopes are detailed in Chapter 6. In many instances conventional planting techniques offer the most cost-effective protection against surficial erosion of slopes. Examples of conventional uses of herbaceous plantings for control of surficial erosion are shown in Figure 4-1.

Although grasses and herbaceous vegetation are very effective for erosion control, they are sometimes difficult to establish on slopes because of adverse

(*a*)

Figure 4-1. Use of herbaceous plantings to control surficial erosion along highway rights-of-way. (*a*) Foxgloves, poppies, and grasses protecting cut slope, Highway 1, California. (*b*) Crown vetch ground cover on cut slope, Interstate 94, Michigan.

(*b*)

Figure 4-1. (*Continued*)

conditions such as steepness, droughty conditions, high velocity runoff, wave action, and so on. A large variety of products are available to enhance and promote the establishment of vegetation for erosion control purposes. These consist of hydraulic mulches, tackifiers, nets, erosion control blankets, turf reinforcement mats, and geocellular containment systems.

A different approach is also required for the protection of steep slopes against shallow mass movement. As explained in Chapter 3 woody vegetation is generally more suitable for preventing mass movement. But planting of trees and shrubs alone may not suffice where slopes are very steep, where both surficial and mass stability are a problem, or where site conditions make vegetative establishment very difficult. Under these conditions it is necessary to use a mixed construction, or biotechnical, approach. Soil bioengineering methods that are characterized by the judicious placement and embedment of live cuttings in the ground in various arrays and configuration fall under this category of slope protection and erosion control techniques.

4.3 MIXED CONSTRUCTION SYSTEMS

4.3.1 General Description

Biotechnical stabilization and soil bioengineering methods fall under this category. The distinction between the two is noted below:

Biotechnical Stabilization: This term refers to the use of natural inclusions, living or inert, to reinforce soil and stabilize slopes. The term "biotechnical slope protection" describes the integrated or combined use of living vegetation and inert structural or mechanical components. The inert components include concrete, wood, stone, and geofabrics. The term "geofabric" refers to woven or nonwoven geotextiles and geogrids made either from synthetic polymers or from natural materials such as jute and coir (Yamanouchi, 1986).

Soil Bioengineering: This is a more specific term that refers primarily to the use of live plants and plant parts alone. Live cuttings and stems are purposely imbedded and arranged in the ground or in earthen structures, where they serve as soil reinforcements, hydraulic drains, barriers to earth movement, and hydraulic pumps or wicks. Techniques such as live staking, live fascines, brush-layering, and so on, fall into this category. Live plant parts, that is, stems and branches, may also be used in conjunction with geofabrics. Soil bioengineering techniques are described in detail in Chapter 7.

Vegetation can be used with inert structural or mechanical elements in a variety of ways, as summarized in Table 4-1. Plants can either be introduced on the benches of stepped or tiered retaining wall systems (see Figure 4-2) or they can be introduced in the openings or interstices of porous retaining structures and revetments. Open-front crib walls and gabion walls lend themselves well to the latter approach (see Figure 4-3). Geogrid retaining structures also can accom-

Figure 4-2. Highway cut stabilized by stepped-back retaining wall. Woody shrubs were planted on the horizontal benches of this tiered wall.

Figure 4-3. Highway fill stabilized by open-front, concrete crib wall. Volunteer plants have become established in the openings between headers.

modate vegetative treatment in a variety of ways. Grasses and plants can be established in the soil behind (or beneath) the grid. The shoots and stems grow through the netting and serve multiple functions, namely, shading, filtering, and visual screening. A sequence of photos showing the construction and vegetative treatment of a geogrid-reinforced earth buttress is shown in Figures 4-4 and 4-5. Vegetation can be introduced into virtually any type of porous revetment system. Joint planting, for example, which is a variant of a *live staking* (a soil bioengineering technique), consists of tamping or inserting live cuttings between bank armor stone or riprap.

Reinforced Grass: This term designates a grassed surface that is augmented with either hard armor (e.g., articulated blocks) or soft armor. A typical soft-armor system consists of very porous, synthetic, three-dimensional mats (*turf reinforcement mats*) that are placed on the ground, filled with soil, and seeded. These mats improve the performance of the vegetation by tying (entangling) plant roots together and bridging over weak spots. Grassed surfaces reinforced in this manner resist the tractive force of high-velocity water flow better than a grassed surface by itself. Reinforced grass and other types of biotechnical ground covers are described in detail in Chapter 9.

4.3.2 Soil Bioengineering Attributes and Limitations

Biotechnical stabilization differs in significant respects from conventional approaches to slope protection and repair, and provides important advantages.

(*a*)

(*a*)

Figure 4-4. Geogrid reinforced earth buttress construction sequence. (*a*) During construction, showing geogrid being pulled over earth lifts. (*b*) After construction, showing appearance of slope face prior to vegetative treatment. (Photos courtesy of J. David Rogers.)

(*a*)

(*b*)

Figure 4-5. Vegetated, geogrid reinforced earth buttress. (*a*) Appearance of buttress structure three months after hydroseed treatment. (*b*) Close-up view of vegetated surface showing grass growing through geogrid.

These advantages notwithstanding, biotechnical stabilization should not be viewed as a panacea for all slope failure and surface erosion problems. Vegetation is inappropriate, for example, where highly toxic conditions exist or in sites subjected to high water velocities or extreme wave action. Soil bioengineering, likewise, is not appropriate for all sites and situations. In certain cases, a conventional vegetative treatment, for example, grass seeding and hydro mulching, works satisfactorily at less cost. In other cases, the more appropriate and most effective solution is a structural retaining system alone or possibly in combination with soil bioengineering.

The following can be cited as important attributes of soil bioengineering systems:

Labor/Skill Requirements: Soil bioengineering measures tend to be labor/skill-intensive, as opposed to energy/capital-intensive. Soil bioengineering requires the use of hand labor. Successful implementation requires prebid orientation, on-site training of the workforce, and careful supervision and inspection. Certain techniques, such as brushlayering, can be installed using bulldozers and other mechanized equipment. In spite of the relatively high use of labor, soil bioengineering work often costs less than conventional treatments because it is normally performed in the dormant season, a time of year when labor is often more available.

Utilization of Natural and/or Indigenous Materials: Soil bioengineering relies primarily on the use of native materials such as plants and plant stems or branches, rocks, wood, and earth. Live-cut brush and stems, for example, that are harvested in the wild, tied into bundles, and then placed in the ground in shallow trenches are the main construction component of live fascines (see Figure 4-6). Appropriate vegetation can often be obtained from local stands of species such as willow, alder, dogwood, and others. This stock is already well suited to the climate, soil conditions, and available moisture of the area, and is therefore better suited to survival.

Cost Effectiveness: Field studies have shown instances where combined slope protection systems have proven more cost effective than either vegetative treatments or structural solutions alone. The use of indigenous material accounts for some of the cost savings because plant costs are generally limited to labor for harvesting, handling, and transportation to the site. Where construction methods are labor-intensive and labor costs are reasonable, the combined systems may be especially cost effective. Even where labor is scarce or relatively expensive, labor costs can often be offset by avoided or reduced costs in structural materials, grading, heavy equipment, and other costs.

Environmental Compatibility: Soil bioengineering systems generally require minimal access for equipment and workers and cause relatively minor site disturbance during installation. Over time, systems themselves are visually non-

(*a*)

(*b*)

Figure 4-6. Use of local plant material in the fabrication of live fascines. (*a*) A clump of native willows provides a good source of supple stems and whips. (*b*) Fabrication of live fascine bundles from cut willow brush and stems.

intrusive and blend into the natural surroundings, as shown in the sequence of photos in Figure 4-7. These are favorable attributes in environmentally sensitive areas such as parks, woodlands, riparian areas, and scenic corridors, where aesthetic quality, wildlife habitat, ecological restoration, and similar values are important.

Self-Repairing Characteristics: Unlike conventional, inert systems, soil bioengineering systems become stronger with time as the vegetation roots and becomes well established. Vegetation has the ability to regenerate when subject to stress that does not kill the plants. Replanting and infill planting can also be used to repair damaged areas.

Planting Times: Soil bioengineering systems are most effective when they are installed during the dormant season, usually the late fall, winter, and early spring. This ordinarily coincides with a slowdown in other construction work. Constraints on planting times or the availability of the required quantities of suitable plant material during the allowable planting time may limit the usefulness of soil bioengineering methods in certain circumstances. This time constraint can be partly circumvented by placing cuttings or live plant materials in cold storage until ready for use.

(*a*)

Figure 4-7. Successive views of highway embankment slope treated by brushlayering. North Carolina State Route 126. (*a*) View of slope during fill brushlayer construction. (*b*) After two months of stabilizing growth. (*c*) Brushlayer fill one year after construction.

(*b*)

(*c*)

Figure 4-7. (*Continued*)

Difficult Sites: Soil bioengineering is often a useful alternative for highly sensitive or steep sites where the use of machinery is not feasible and hand labor is a necessity. On the other hand, the usefulness of soil bioengineering methods may be limited by a poor medium for plant growth, such as rocky or gravely slopes that lack sufficient fines or moisture, or by extreme acidity or other toxic conditions in the soil. The biotechnical usefulness of vegetation is also limited on slopes that are exposed to periodic, high-velocity flow or constant inundation. The former constraint can be circumvented to some extent, however, by using structural reinforcement or augmentation with biotechnical ground covers (see Chapter 9).

4.3.3 Compatibility Between Plants and Structures

At first glance biotechnical construction methods may appear unworkable because of compatibility problems, that is, engineering requirements imposed for structural stability may clash with biological requirements of the vegetation. The backfill or cribfill behind a retaining structure, for example, should have certain specified mechanical and hydraulic properties if the structure is to perform satisfactorily. Ideally the fill should be a coarse-grained, free-draining material. The presence of excessive amounts of clay, silt, and organic matter is not desirable. The requirement of free drainage—so essential to the stability of earth-retaining structures—is also important to vegetation, which generally can not tolerate waterlogged soil conditions. Establishment of vegetation, on the other hand, usually requires the presence of fines in the soil to provide some moisture and nutrient retention. In many instances these biological requirements can be satisfied without compromising engineering performance. There are procedures and techniques for establishing vegetation in and around structural backfills and in the frontal openings of porous revetments and retaining structures; these techniques address this concern (see Section 8.2).

Another frequently voiced concern about the use of plants in conjunction with structures is the danger of physical damage or disruption by stems and roots of woody vegetation. A prevalent notion is that plant roots will pry, dislodge, or heave the structure, as shown schematically in Figure 4-8. The evidence for such damage is scant and mostly anecdotal. Roots exhibit a property termed "edaphoecotropism" (Vanicek, 1973), or simply stress avoidance. This means that plant roots tend to avoid zones of stress such as areas of moisture deficiency, light, high temperature, mechanical obstacles (rocks), and low porosity (high density). Thus roots will avoid the face of a porous, open retaining structure because of moisture deficits and phototropic stress, as shown in Figure 4-9. Live wood also has a remarkable ability to literally flow around and engulf obstacles without any disruption or deformation to the obstacle itself, as shown in Figures 4-10 and 4-11. Even so, it advisable not to plant seedlings that will mature into large-diameter trees in the frontal openings of a structure.

The main danger from planting trees in and around structures appears to come from soil moisture extraction that results in ground settlement in the vicin-

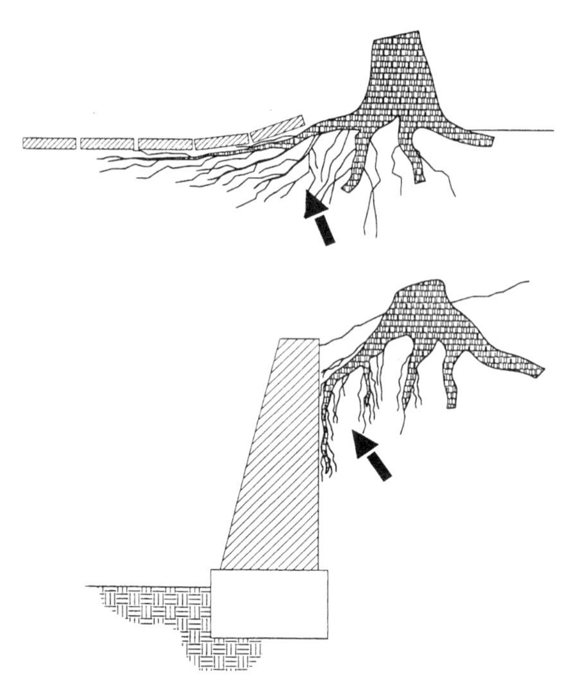

Figure 4-8. Schematic illustration of perceived danger to structures from tree roots.

Figure 4-9. Large mature tree on backfill behind a stone retaining wall. Plant roots avoid growing toward wall. Root reinforcement of backfill compensates for surcharge from weight of tree.

(*a*) (*b*)

Figure 4-10. Photos showing ability of woody stems to flow around and engulf mechanical obstacles. (*a*) Wire fence netting. (*b*) Cast iron fence.

Figure 4-11. Photo showing large mature tree growing on top of stone hedgerow wall with an earthen core. Roots show no evidence of disrupting or compromising the structural integrity of the wall.

ity of the tree. A large, mature pine tree can remove as much as 200 gallons of water from the ground a day by evapotranspiration. The area affected extends approximately 1 to $1\frac{1}{2}$ times the height of the tree. Soil moisture extraction by vegetation poses a serious problem only when trees are planted too close to buildings in soils with poor volume stability, such as plastic clays and silty clays (Bozozuk and Burn, 1960). Soil moisture extraction and other hydrologic effects of woody vegetation are discussed further in Chapter 3.

4.3.4 Applications

Biotechnical and soil bioengineering stabilization methods have been adopted for a variety of applications from the stabilization of both cut and fill slopes along highways to streambank protection (see Table 4-2). Sotir (1991) presented a review of soil bioengineering projects in North America during the past decade. The earliest documented use of soil bioengineering in the United States was the use of willow poles to stabilize eroding streambanks (*Engineering News Record*, 1925). Another often-cited application was reported by Kraebel (1936), who used contour wattling (live fascines) to stabilize steep fill slopes along the Angeles Crest Highway in Southern California. Recent examples of soil bio-engineering solutions for the stabilization of highway cut and fill slopes are described by Gray and Sotir (1992). Kropp (1989) describes the use of contour wattling in combination with subdrains to repair and stabilize a debris flow in a

TABLE 4-2. Applications of Biotechnical Slope Projection

	Engineering Situations												
Applications	Mining and Reclamation	Highways and Railways	Construction Sites	Waste Disposal and Public Health	Airfields and Helipads	Waterways	Land Drainage	Reservoirs and Dams	Coastal and Shoreline Protection	Building Sites	Recreation Areas	Pipeline Right-of-Ways	Site Appraisal
Slope stabilization—embankments and cuttings	■	■	■					■	■				
Slope stabilization—cliffs and rockfaces	■	■							■				
Water erosion control—rainfall and overland flow	■	■	■	■			■	■	■		■	■	
Water erosion control—gully erosion	■	■	■			■	■	■	■		■	■	
Water courses and shoreline protection—continuous flow channels						■	■						
Water courses and shoreline protection—discontinuous (ephemeral) channels	■	■		■		■	■		■	■			
Water courses and shoreline protection—large water bodies (shorelines)								■	■				
Wind erosion control	■		■	■					■		■	■	
Vegitation barriers—shelter	■	■		■					■	■	■		
Vegitation barriers—noise reduction		■		■						■			
Surface protection and trafficability			■		■	■		■	■	■	■	■	
Control of runoff in small catchments	■	■	■	■		■	■				■		
Plants as indicators	■			■		■						■	■

Source: From Coppin and Richards, 1990. Use of Vegetation in Civil Engineering, CIRIA Book 10. Used with permission of Construction Industry Research and Information Association.

shallow, colluvial-filled ravine in Northern California. Other examples and case studies of soil bioengineering stabilization projects are described in Chapter 7.

4.4 IMPACT OF SLOPE GRADING

Both transportation corridors and residential developments in steep terrain require that some excavation and regrading be carried out in order to accommodate roadways or building sites. The manner in which this grading is planned and executed, and the nature of the resulting topography or landforms that are created affect not only the visual or aesthetic impact of the development but also the stability of the slopes and effectiveness of revegetation efforts (Schor and Gray, 1995). Succinct descriptions and comparative definitions of grading designs are presented in Table 4-3. These approaches to grading and revegetation are compared and contrasted in the next few sections.

TABLE 4-3. Comparative Definitions of Grading Designs

A. Conventional Grading

1. Conventionally graded slopes are characterized by essentially linear (in plan) planar slope surfaces with unvarying gradients and angular slope intersections; resultant pad configurations are rectangular
2. Slope drainage devices are usually constructed in a rectilinear configuration in exposed positions
3. Landscaping is applied in random or geometric patterns to produce "uniform coverage"

B. Contour Grading

1. Contour-graded slopes are basically similar to conventionally graded slopes except that the slopes are curvilinear (in plan) rather than linear, the gradients are unvarying, and profiles are planar; transition zones and slope intersections generally have some rounding applied; resultant pad configurations are mildly curvilinear
2. Slope drainage devices are usually constructed in a geometric configuration and in an exposed position on the slope face
3. Landscaping is applied in random or geometric patterns to produce "uniform coverage"

C. Landform Grading

1. Landform grading replicates irregular shapes of natural, stable slopes; landform-graded slopes are characterized by a continuous series of concave and convex forms interspersed with swales and berms that blend into the profiles, nonlinearity in plan view, varying slope gradients, and significant transition zones between man-made and natural slopes; resultant pad configurations are irregular.
2. Slope drainage devices either follow "natural" slope drop lines or are tucked away in special swale and berm combinations to conceal the drains from view; exposed segments in high visibility areas are treated with natural rock
3. Landscaping becomes a "revegetation" process and is applied in patterns that occur in nature: trees and shrubs are concentrated largely in concave areas, whereas drier convex portions are planted mainly with ground covers

Source: Adapted from Schor and Gray, 1995.

4.4.1 Conventional Grading

Conventional grading practice often results in drastically altered slopes and the replacement of natural hillside forms with artificial, sterile, and uniform shapes and patterns. Conventionally graded slopes can be characterized by essentially linear, planar slope surfaces with constant gradients and angular intersections. Slope drainage devices are usually constructed in a rectilinear and exposed fashion; landscaping and plants are applied in random or geometric patterns, as shown in Figures 4-12 and 4-13.

4.4.2 Landform Grading

Landform grading essentially attempts to mimic nature's hills. This approach has been largely developed and pioneered by Schor (1980, 1992), who has successfully applied landform grading in several large hillside developments and planned communities in Southern California. Very few hillsides are found in nature with linear, planar faces. Instead, natural slopes consist of complex landforms covered by vegetation that grows in patterns adjusted to hillside hydrogeology, as shown in Figure 4-14. Shrubs and other woody vegetation growing on natural slopes tend to cluster in valleys and swales where moisture is more abundant.

Figure 4-12. Example of conventional grading with planar slopes and rectilinear drainage ditch in highly visible and exposed location.

Figure 4-13. Conventionally graded and landscaped hill slope with planar face, rectilinear drainage ditch, and uniformly spaced plantings.

Figure 4-14. Natural hill slopes showing vegetation patterns. Woody shrubs and trees tend to cluster in swales and valleys where moisture is more abundant. Grass grows on the convex-shaped interfluve areas.

Landform graded slopes are characterized by a variety of shapes including convex and concave forms, as shown in Figure 4-15. Downslope drain devices either follow natural drop lines in the slope or are tucked away and hidden from view in special concave swale and convex berm combinations, as shown in Figure 4-16. Vegetation patterns that are found in nature are also mimicked. Random patterns or uniform coverage should be avoided. Instead the vegetation is placed where it has a better chance of surviving and where it does a better job of holding soil. Trees and shrubs require more moisture, and they also do a better job of stabilizing a soil mantle against shallow mass wasting. Accordingly, it makes sense to cluster them in the swales and valleys of a slope, where runoff tends to concentrate, as shown schematically in Figure 4-17. Shrubs should also be heavily concentrated along the drainage flow of each swale.

By purposely controlling the drainage pattern on a slope, runoff can be concentrated in concave areas where it is needed or where it can best be handled by woody slope vegetation (see Figure 4-15). Conversely, runoff and seepage will be diverted away from convex-shaped areas. These areas should be planted with grasses or more drought-tolerant herbaceous vegetation. Irrigation needs are thus reduced by careful control of drainage pattern on a slope and selection of appropriate plantings for different areas.

Figure 4-15. Example of landform grading and revegetation with concave and convex slope forms, variable slope gradients, and clustered plantings.

Figure 4-16. Landform grading with curvilinear drainage way that is placed in a special swale-and-berm combination to conceal it from view.

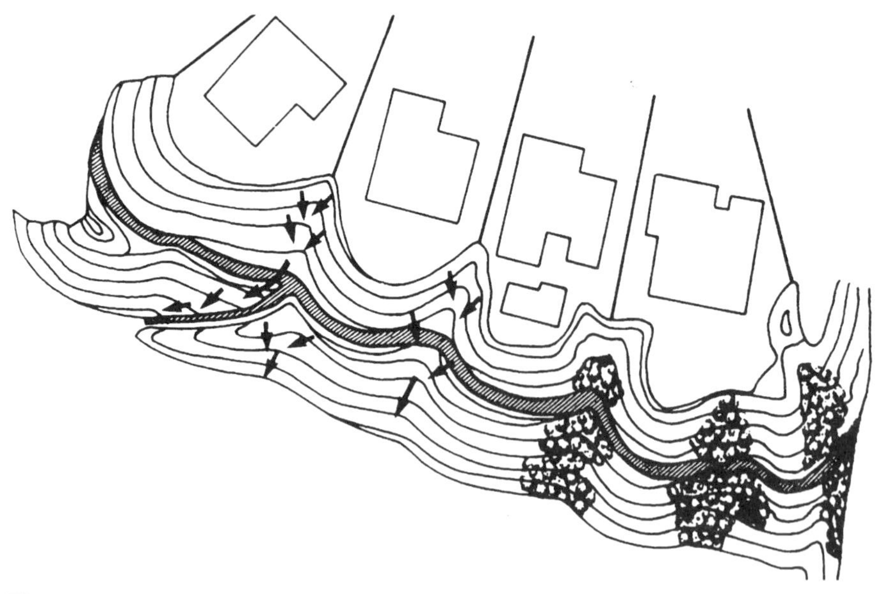

Figure 4-17. Topographic representation of landform configuration showing radial flow of water, foliage placement in swales, and lots that conform with landform grading configuration. (From Schor, 1992. Used with permission of the author.)

4.5 REFERENCES CITED

Bozozuk, M., and K. N. Burn (1960). Vertical ground movements near elm trees. *Geotechnique* **10**: 19–32.

Coppin, N. J., and I. Richards, (1990). *Use of Vegetation in Civil Engineering*. Sevenoaks, Kent (England): Butterworths.

Engineering News Record (1925). Protecting steep banks by planting live willow poles. **94**(20): 822–823.

Gray, D. H., and A. T. Leiser (1982). *Biotechnical Slope Protection and Erosion Control*. New York: Van Nostrand Reinhold.

Gray, D. H., and R. Sotir (1992). Biotechnical stabilization of cut and fill slopes. *Proceedings*, ASCE-GT Specialty Conference on Slopes and Embankments, Berkeley, CA, June, Vol. 2, pp. 1395–1410.

Greenway, D. R. (1987). Vegetation and slope stability, In: *Slope Stability*, edited by M. F. Anderson and K. S. Richards. New York: Wiley.

Kraebel, C. J. (1936). Erosion control on mountain roads. *USDA Circular No. 380*, 43 pp.

Kropp, A. (1989). Biotechnical stabilization of a debris flow scar. *Proceedings*, XX International Erosion Control Association Conference, Vancouver, pp. 413–429.

Schiechtl, H. M. (1980). *Bioengineering for Land Reclamation and Conservation*. Edmonton, Canada: University of Alberta Press, 404 pp.

Schor, H. (1980). Landform grading: Building nature's slopes. *Pacific Coast Builder* (June): 80–83.

Schor, H. (1992). Hills like nature makes them. *Urban Land* (Mar.): 40–43.

Schor, H. and D. H. Gray (1995). Landform grading and slope evolution. *Journal of Geotechnical Engineering* (ASCE) 121 (6T10): 729–734.

Sotir, R. B. (1991). A review of recent soil bioengineering projects in North America. *Proceedings*, NSF Workshop on Biotechnical Stabilization, Ann Arbor, MI, August.

Sotir, R., and D. H. Gray (1989). Fill slope repair using soil bioengineering systems. *Proceedings*, XX International Erosion Control Association Conference, Vancouver, pp. 473–485.

Vanicek, V. (1973). The soil protective role of specially shaped plant roots. *Biological Conservation* **5**(3): 175–180.

USDA Natural Resources Conservation Service (1992). Chapter 18: Soil bioengineering for upland slope protection and erosion reduction. Part 650, 210-EFH, *Engineering Field Handbook*, 53 pp.

Yamanouchi, T. (1986). The use of natural and synthetic geotextiles in Japan. *Proceedings*, IEM-JSSMFE Joint Symposium on Geotechnical Problems, Kaula Lumpur, Malaysia, March, pp. 82–88.

5 Structural-Mechanical Components of Biotechnical Stabilization

5.0 INTRODUCTION

In this chapter we describe various structural measures or components that are appropriate for use with biotechnical slope protection systems. These structural measures include various types of retaining walls and slope revetments. The purpose here is not to catalog and describe all such structural measures, but rather to focus on those that have inherent advantages with respect to economy, ease of construction, appearance, and opportunities for incorporation of vegetation or plantings in the structure. The last named criteria is very important for the purposes of this book.

Retaining structures that meet the above criteria and that are described in this chapter are gravity retaining structures, that is, they resist external forces primarily by means of their weight. We do not provide detailed design and construction specifications in this chapter. This specialized information can be found elsewhere in engineering textbooks and handbooks. Instead emphasis is placed on characterization and selection. Useful information is provided about critical factors that affect the internal and external stability of these structures.

5.1 RETAINING STRUCTURES—GENERAL

5.1.1 Purpose and Function of Structure

Approaches to slope protection and types of retaining structures were discussed briefly in Chapter 4. We elaborate further on the purpose and function of a retaining structure in this section. A structure placed at the foot of a slope helps to stabilize the slope against mass movement and protects the toe against scour and undermining. This can be accomplished by constructing either a toe-wall or toe-bench structure, depending upon right-of-way limitations and other considerations.

Toe-Walls: A low toe-wall placed at the foot of a slope permits local oversteepening of the slope at its base and flattening of the slope above (see Figure 5-1).

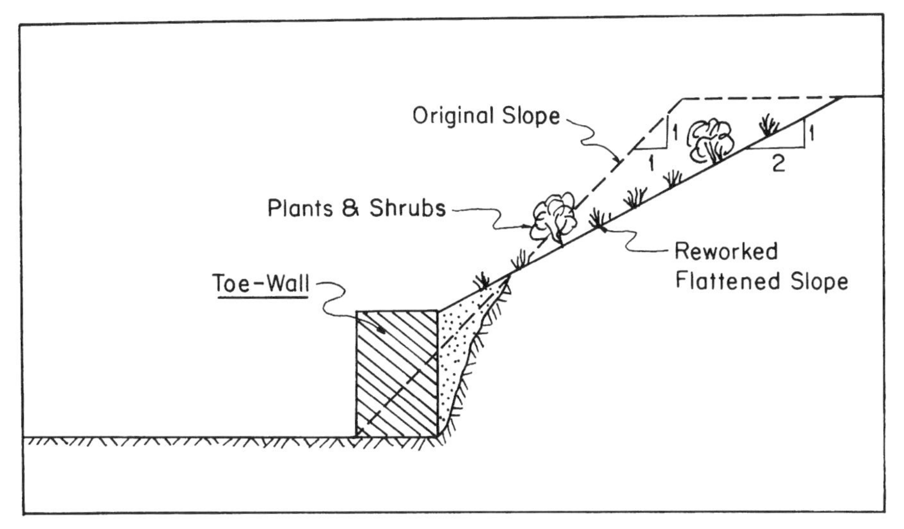

Figure 5-1. Low toe-wall, which permits slope flattening and establishment of vegetation on slope above. Encroachment on adjacent land use or right-of-way at the foot of the slope is minimized.

The latter makes it easier to establish vegetation on the face of the slope and reduces the danger of surficial erosion. Excavation and placement of a retaining structure at the toe reduces the amount of encroachment on an adjacent land use or right-of-way at the base. Unfortunately, excavation at the toe also can cause stability problems in very unstable slopes. This situation may require dry weather excavation or, alternatively, excavation and construction in "slots" or increments. If neither of these alternatives is feasible, then a toe-bench structure that does not entail excavation at the toe can be employed. If the slope above the wall is flattened or graded back, the scaled material can often be used as backfill behind a toe-wall. It may also be possible to use the scaled material within the structure itself (e.g., as cribfill), provided it meets gradation and other requirements for such usage. Ideally it is desirable to balance the cut and fill requirements along a reach of slope to avoid undue soil disposal or borrow problems.

Toe-Bench Structures: A toe-bench structure is similar to a toe-wall and differs primarily in the amount of excavation required at the toe and slope of the backfill (see Figure 5-2). A toe-bench structure: (1) is constructed farther away from the foot of the slope (hence requires little or no excavation at the toe), (2) entails little or no flattening or regrading of the slope above, and (3) incorporates a level or very gently sloping backfill. Toe-benches can be used to provide a fairly level bench at the foot of a slope on which vegetation can be readily established to eventually screen the slope above. They also buttress the base of the slope and catch debris coming off or rolling down the slope.

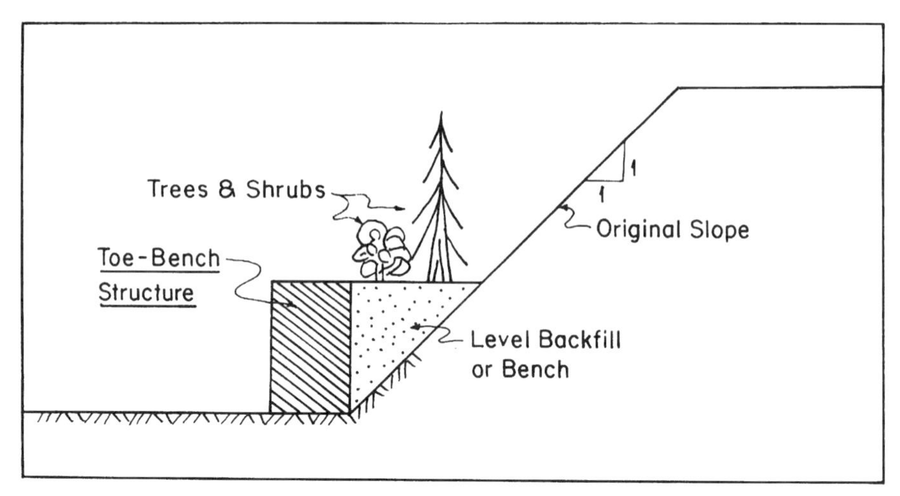

Figure 5-2. Toe-bench structure, which buttresses base of slope and creates level bench on which to establish vegetation. Bench also catches debris coming off the slope above.

Toe-bench structures are suitable for vegetative screening of high, steep rocky slopes that themselves cannot be vegetated because of lack of soil or excessive steepness of the slope (see Figure 5-2). If an open-face or porous structure is employed, the structure itself can be vegetated and screened as well. Toe-bench structures, however, require more clearance at the base of a slope than toe-walls (to avoid excavation at the toe). This may constrain their use somewhat along cut slopes adjacent to a road where the foot of the cut is next to the roadway. If excavation for a toe-wall is not possible and insufficient clearance exists at the base for a toe-bench structure, then a soil bioengineering solution such as a slope grating (see Chapter 7) should be considered.

5.1.2 Basic Types of Retaining Structures

Many different types of retaining structures are available, each with its particular advantages, requirements, and limitations (Palmer, 1987; Randall and Wallace, 1987). Schematic illustrations of basic types of retaining structures are shown in Figure 5-3. In general retaining structures can be classified into two major categories, namely gravity versus nongravity structures. Gravity structures can be subclassified in turn into: buttress structures, coherent gravity walls, articulated block walls, breast walls, and revetments, as noted in Table 5-1.

Gravity structures and retaining walls resist lateral earth forces by their weight or mass. A gravity retaining wall constructed from poured concrete, steel, wire encased stone, or reinforced earth essentially behaves as a coherent, unitary mass. Articulated block walls and rock breast walls are also gravity structures; articulation, interlocking, and/friction between structural units provides coherence and resists local shear or rupture. Gravity walls, like all

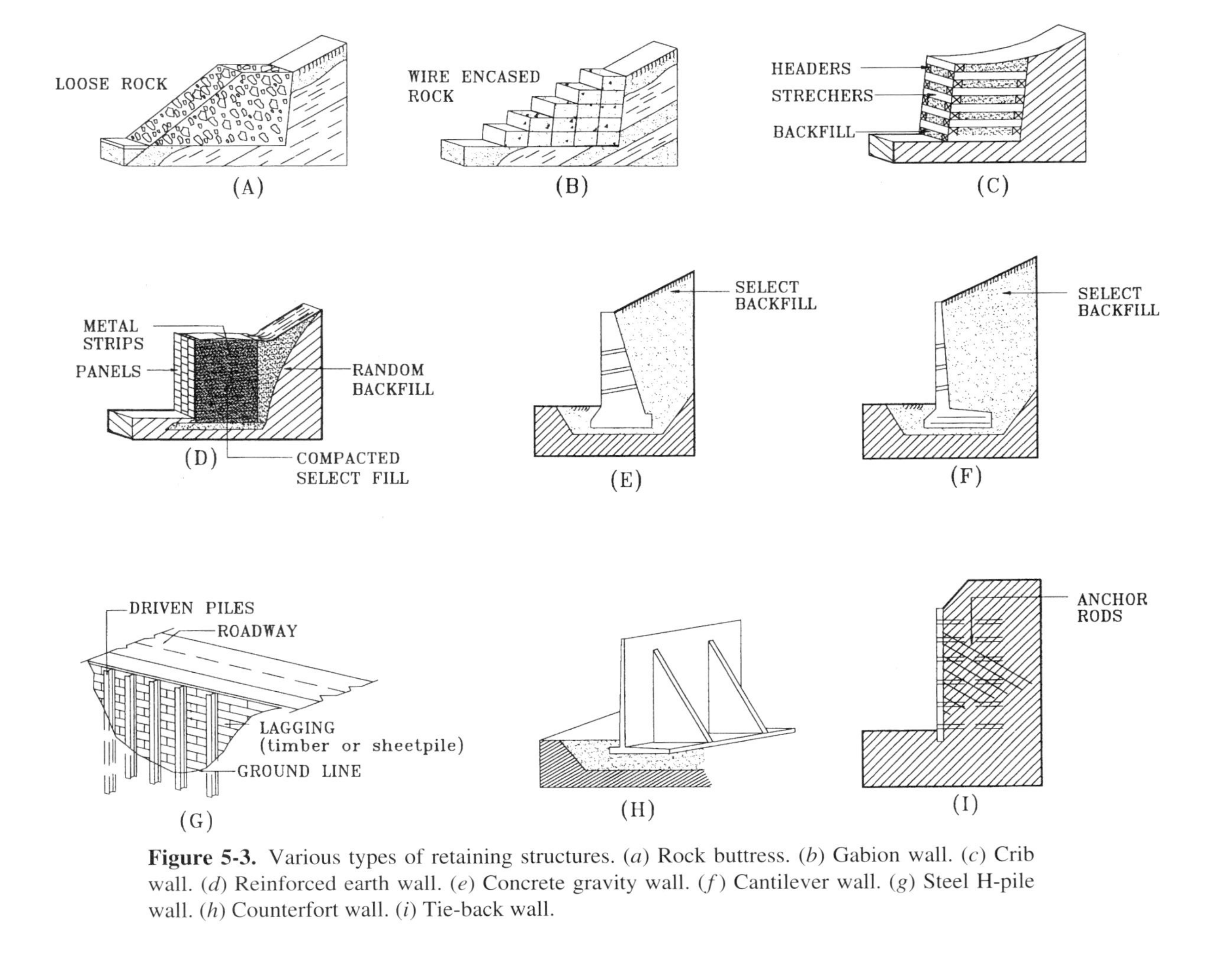

Figure 5-3. Various types of retaining structures. (*a*) Rock buttress. (*b*) Gabion wall. (*c*) Crib wall. (*d*) Reinforced earth wall. (*e*) Concrete gravity wall. (*f*) Cantilever wall. (*g*) Steel H-pile wall. (*h*) Counterfort wall. (*i*) Tie-back wall.

TABLE 5-1. Structural Slope Protection/Retention Systems

Major Types	Examples	Protective/Restraining Mechanism
	Gravity	
Buttress structures (at toe of slope)	Drained rock buttress Earthen berm	Increase in resisting moment due to weight at toe and/or increase in shear resistance by interception of failure surface
Coherent gravity walls	Masonry and reinforced concrete Crib and bin walls Gabion walls Cantilever and counterfort walls Reinforced earth walls	Lateral earth forces, which cause sliding and overturning, are primarily resisted by weight of structure
Stacked, articulated block walls	"Loffelstein" wall "Dyno" blocks "Porcupine" blocks "Keystone" wall	Lateral earth forces are resisted primarily by weight of units; articulation and interlocking between blocks provides coherence and resists local rupture
Breast walls	Rock breast wall Rubble wall	Armoring and partial buttressing action at toe of slope helps to prevent local slumping and erosion
Slope revetments	Gabion mattress Riprap and dumped rock Articulated blocks Concrete mattress	Structural facing armors slope against scour and erosion; armor units maintain position under wave action by their weight and interlocking action
	Nongravity	
Anchored or tied back walls	Sheet pile wall Anchored bulkhead	Lateral earth forces resisted by relatively thin, flexible structure that is tied back to anchor blocks or plates embedded in ground
Embedded, cantilever pile walls	Soldier pile/lagging wall Cylinder pile wall	Lateral earth stresses are resisted by mobilization of passive resistance along embedded length of vertical piles at their bottoms

earth-retaining structures, must be capable of resisting external forces causing overturning and sliding along their bases. These stability requirements are considered in Section 5.1.3.

Cantilever and Counterfort Walls: Cantilever and counterfort walls are constructed from reinforced concrete; they are usually classified as semigravity walls because they do not rely solely on their own weight but also on the weight of the backfill that rests on their bases. Cantilever and counterfort walls can be built to greater heights with a greater economy of material than conventional masonry or concrete gravity walls. Cantilever walls can be used for heights up to 30 feet (9 m), and counterfort walls are commonly used for heights up to 25 feet (8 m). A cantilever wall is reinforced in the vertical direction to withstand bending moments (a maximum at the base of the stem) and in the horizontal direction to prevent cracking. The buttresses behind a counterfort wall are also heavily reinforced to resist tension. Both types of walls are relatively expensive and require careful design and formwork.

Crib Walls: A crib retaining wall consists of a hollow, box-like, interlocking arrangement of logs, timbers, reinforced concrete beams or steel beams filled with soil or rock. A variation on this theme, known as a bin wall, consists of steel boxes or bins that are bolted together in modular units and filled to form a wall. The cribwork can be vertical or tilted backward (battered) for greater external stability. The crib members can be designed to have openings between them at the front face where plants can be established. Crib walls are relatively inexpensive and are usually flexible enough to tolerate some differential settlement. Structurally, cribs and bins are gravity walls and are designed accordingly. In addition, the crib itself must be analyzed for internal stability (i.e., the structural members must be capable of resisting stresses caused by the cribfill and backfill).

Gabion Walls: Gabions are wire baskets made of coarse wire mesh. These baskets are filled with stone or rock and stacked atop one another to form a gravity-type wall. Gabions depend mainly on the shear strength of the rock fill for internal stability, and their mass or weight to resist external, lateral earth forces. Gabions are porous structures that can be vegetated. They are also very flexible, easy to erect, and inexpensive.

Reinforced Earth Walls: The original Reinforced Earth wall (Vidal, 1969) consisted of a granular matrix or fill reinforced with successive layers of metal strips. Shear stresses that develop in the reinforced backfill are transferred via interface friction to tensile resistance in the metal strips. The strips are connected to facing elements—typically thin, precast concrete panels stacked atop one another. Very little lateral earth stress acts on the facing elements at the front of the structure because most of this earth stress is taken up in tensile resistance along the length of the reinforcing strips. The reinforced volume can be regarded and analyzed as a coherent gravity structure (McKittrick, 1978). Internal stability requires in addition that the metal strips or ties be designed to resist breaking in tension or failing by pullout.

Mechanically Stabilized Earth Walls: Many different types of inclusions with various shapes and properties are used to reinforce retaining wall backfills today. These inclusions range from geogrids fabricated from polymeric nets (Koerner, 1990) or welded-wire mesh (Hilfiker, 1978) to continuous filaments fabricated from polyester fiber (LeFlaive, 1982). Backfills that are stabilized or reinforced with such tensile inclusions are referred to as "mechanically stabilized earth." A welded-wire wall is a composite wire and granular soil structure. L-shaped, wire mesh sections are placed and connected between successive lifts of backfill. The wire mesh provides both reinforcement in the backfill and containment at the face of the wall. Welded-wire walls are basically gravity structures that share features of both gabions and reinforced earth walls. Synthetic geogrid walls utilizing tough, flexible, polymeric reinforcements can be constructed by wrapping the geogrid around successive lifts of backfill without need of a separate facing, in a similar manner to welded-wire walls. Both types of walls are relatively low cost, easy to erect, and well adapted to vegetative treatment.

Rock Breast Walls and Articulated Block Walls: Both rock breast walls and articulated block walls can be considered as gravity structures that resist lateral earth forces mainly by their weight. There must be sufficient friction or articulation between the units (rocks or blocks) to resist local shear rupture. These walls must be erected on a firm base and are usually placed against a slope with only a small amount of fill behind them. Neither type of wall is intended nor designed to resist large lateral earth stresses; hence they are limited in height. Their main function is to protect the toe of a slope against undermining and provide some lateral restraint. They are usually constructed with a substantial batter to improve stability and minimize lateral earth forces. Rock breast walls and most articulated block walls are quite porous, which provides opportunities to incorporate plants in the voids and interstices of the wall.

Pile and Tie-Back Walls: Pile and tie-back walls are occasionally used as retaining walls in situations where space constraints or foundation conditions limit use of gravity structures. Pile may consist of a row of bored, cast-in place concrete cylinder piles or, more typically, driven steel H-piles (see Figure 5-3*g*). Driven pile walls have been used to support low-volume roads (Schwarzhoff, 1975) where they traverse steep terrain characterized by weak but shallow residual soils underlain by a zone of weathered rock that increases in competency with depth. The use of driven piles in this situation avoids excessive bench excavations that would be required for a gravity or bearing-type wall. Tie-back walls essentially consist of a relatively thin, flexible facing connected to a network of anchored tie rods (Figure 5-3*i*). The tie rods may be connected to imbedded plates or concrete blocks in the backfill or conversely may be grouted in place.

5.1.3 Selection Criteria

A variety of different systems plus possible modifications results in a wide offering of retaining structures—one or even several to fit nearly any condition. Schwarzhoff (1975) has presented a good review of retaining wall practice and selection procedures for low-volume forest roads. Although any one of several types of retaining walls may do, specified criteria help to select the wall that is best suited for the job. These criteria may include environmental concerns, construction problems, site constraints, esthetics, and cost. Schwarzhoff (1975) discusses each of these criteria and outlines a systematic selection process for choosing among alternatives.

In the case of biotechnical slope protection systems there are certain criteria that are especially important, paramount among them the requirement that the structure blend in harmoniously with its surrounding. Examples of what can be done in this regard include the following:

- Utilizing retaining structures that minimize site disturbance and that preserve existing or nearby vegetation so that it provides screening
- Filling gabions with native stone that draws attention away from the wire to natural, native material
- Using timber and log crib walls in areas such as forests where wooden materials are natural to their surroundings
- Employing structures and/or wall facing elements with shapes, surface textures, and colorations that blend in with the surroundings
- Planting vegetation on the steps of tiered wall systems or interstices of porous structures (e.g., rock breast, articulated block, gabion, open-front crib, and geogrid walls).

The ability to incorporate vegetation into or around a retaining structure is a particularly important attribute for biotechnical slope protection. This characteristic plus the requirements of ease of construction, flexibility, and relatively low cost restrict initial consideration to the following types of structures: (1) rock breast walls; (2) articulated block walls; (3) gabions; (4) crib walls; and (5) geogrid walls. Additional information about the design and performance of these wall systems in provided later in this chapter.

5.2 STABILITY REQUIREMENTS

All the retaining walls listed at the end of Section 5.1.3 can be designed as gravity structures. Gravity walls resist external earth forces chiefly by their weight or mass. In addition to external stability, some of the aforementioned walls must be designed for internal stability as well. Stabilizing elements and structural components must be capable of withstanding various stresses placed upon them. Examples of stabilizing elements and structural components include polymeric and wire mesh reinforcements in geogrid walls and structural members in crib walls (i.e., the stretchers and headers).

External and internal stability requirements are discussed in general in this chapter. Important factors affecting stability, such as wall shape, inclination, and drainage conditions, are also considered. Standard designs have been developed for most of the retaining structures described herein that satisfy external and internal stability requirements. These standard designs are often supplied by manufacturers of various retaining wall systems. Standard designs can be used for initial design and in routine applications, but must be applied with care, particularly at sites with difficult or unusual conditions. Standard designs and their uses are discussed further in Section 5.2.4.

5.2.1 External Stability

Gravity walls require computations to determine the stability of the structure against: (1) overturning, (2) sliding along the base, and (3) bearing capacity failure. In addition, it may be necessary to check against deep seated rotational movement (slope failure) in the case of walls supported on steep hillsides. These four types of failure are depicted schematically in Figure 5-4.

Essential steps in the computation of a gravity wall's stability include determination of the weight of the wall (W) and the lateral force (P) exerted against the wall by the retained earth and surcharge on the backfill. The computations described in this section assume conditions in which no buildup of hydrostatic pressure against the wall can occur. The use of a free-draining, granular backfill in combination with a permeable structure usually precludes

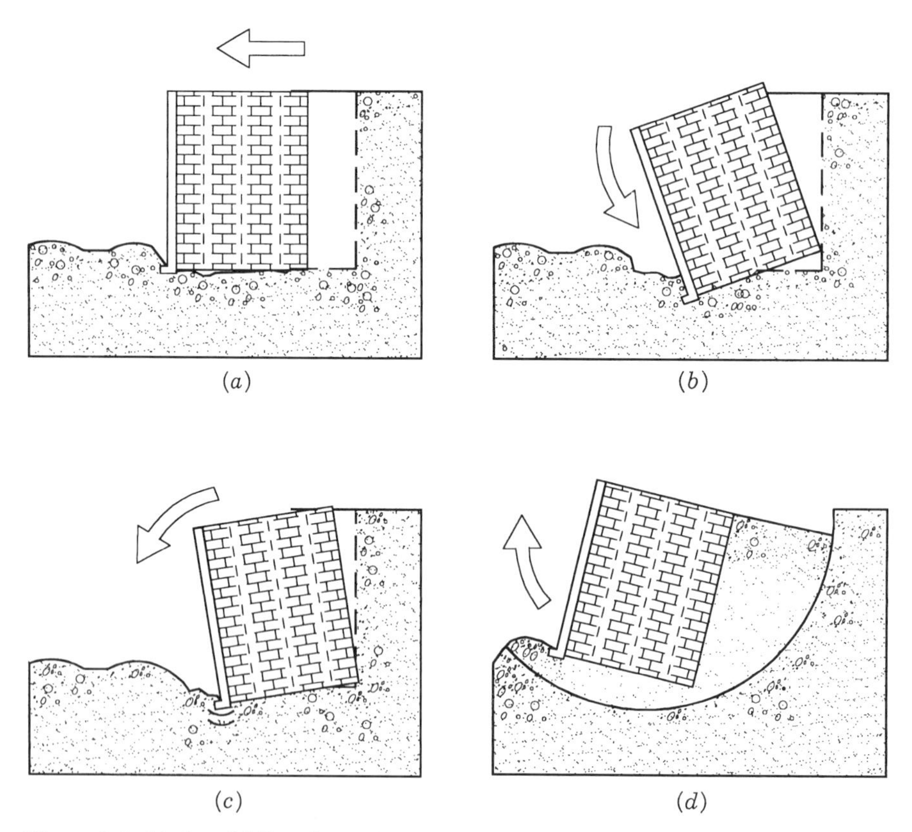

Figure 5-4. Modes of failure for retaining structures. (*a*) Overturning. (*b*) Sliding. (*c*) Bearing capacity failure. (*d*) Deep-seated rotational movement (slope failure).

this possibility—another advantage of the types of porous structures described herein. The influence of water and drainage conditions in the backfill behind the wall are examined in a later section. The presence of water and lack of drainage behind a retaining structure seriously affect stability.

The forces acting on a hypothetical gravity retaining wall are illustrated schematically in Figure 5-5. The magnitude of the lateral earth force (P_A) acting against a gravity retaining wall can be computed by Coulomb's formula:

$$P_A = 0.5\gamma H^2 K_A \tag{5-1}$$

where:
P_A = active earth force per unit length of wall
γ = unit weight of the retained soil
H = height of wall
K_A = coefficient of active earth pressure

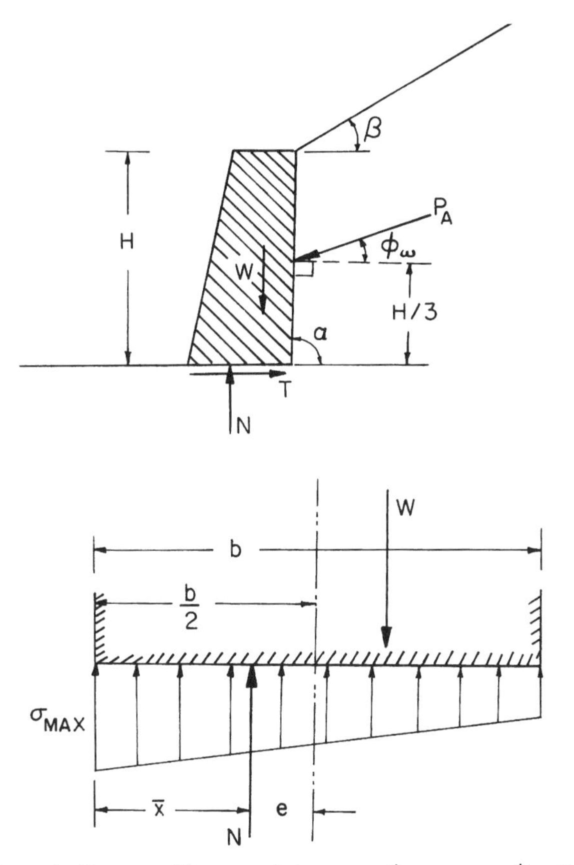

Figure 5-5. Schematic diagram of forces and stresses acting on a gravity retaining wall.

Coulomb's formula is based on so-called active earth pressure conditions; that is, sufficient yielding or movement occurs in the wall to mobilize shear resistance in the backfill itself. This condition is usually met in the case of freestanding gravity walls. The coefficient of active earth pressure (K_A) is a function of a number of wall, slope, and soil parameters:

$$K_A = \left[\frac{csc\,\alpha \sin(\alpha - \phi)}{\sin(\alpha + \phi)^{0.5} + (\sin(\phi + \phi_w)\sin(\phi - \beta)/\sin(\alpha - \beta))^{0.5}} \right]^2 \tag{5-2}$$

where:

β = angle of inclination of backfill or slope
α = batter angle of wall
ϕ = angle of internal friction of retained soil
ϕ_w = angle of wall friction

Values of K_A are tabulated in Table 5-2 for various values of angle of internal friction, wall batter angle, and inclination of backfill for the special case of zero wall friction ($\phi_w = 0$).

The assumption of zero wall friction yields slightly conservative estimates of the coefficient of active earth pressure; that is, lateral earth pressures may

TABLE 5-2. Coefficients of Active Earth Pressure as a Function of Wall and Backfill Inclination[a]

Slope Inclination (β) =		−30°	−12°	±0°	+12° 1 : 4.7	+30° 1 : 1.7
$\phi = 20°$	$\alpha' = +20°$		0.57	0.65	0.81	
	$\alpha' = +10°$		0.50	0.55	0.68	
	$\alpha' = \pm 0°$		0.44	0.49	0.60	
	$\alpha' = -10°$		0.38	0.42	0.50	
	$\alpha' = -20°$		0.32	0.35	0.40	
$\phi = 30°$	$\alpha' = +20°$	0.34	0.43	0.50	0.59	1.17
	$\alpha' = +10°$	0.30	0.36	0.41	0.48	0.92
	$\alpha' = \pm 0°$	0.26	0.30	0.33	0.38	0.75
	$\alpha' = -10°$	0.22	0.25	0.27	0.31	0.61
	$\alpha' = -20°$	0.18	0.20	0.21	0.24	0.50
$\phi = 40°$	$\alpha' = +20°$	0.27	0.33	0.38	0.43	0.59
	$\alpha' = +10°$	0.22	0.26	0.29	0.32	0.43
	$\alpha' = \pm 0°$	0.18	0.20	0.22	0.24	0.32
	$\alpha' = -10°$	0.13	0.15	0.16	0.17	0.24
	$\alpha' = -20°$	0.10	0.10	0.11	0.12	0.16

Source: From Lambe and Whitman (1969).

[a] $\alpha' = \alpha - 90$; $\phi_w = 0$.

be overestimated up to 8 percent. Equation 5-2 can be used to compute more precise values assuming a wall friction of approximately 0.8 to 0.9 times the angle of internal friction of the retained soil. Approximate estimates of the angle of internal friction of the retained soil may be obtained from the chart (Figure 2-10) in Chapter 2.

Overturning: The safety factor against overturning is calculated from the ratio of the resisting moment (moment due to weight of the wall) divided by the net overturning moment due to the earth forces. The net overturing moment is computed from the difference between the vertical and horizontal components of the lateral earth force. Moments are calculated about the toe of the wall. The safety factor against overturning should equal or exceed 1.5. Mathematically,

$$M_R = W x_1 \tag{5-3}$$

$$M_O = (P_A \cos \phi_w) x_2 - (P_A \sin \phi_w) x_3 \tag{5-4}$$

$$\sum M_R / \sum M_O \geq 1.5 \tag{5-5}$$

where: M_O = the net overturning moment due to the earth forces
M_R = the resisting moment due to the weight of the wall
$x_{1,2,3}$ = moment arms about toe

Sliding: The safety factor against sliding is calculated from the ratio of the sum of the vertical forces times the tangent of the friction angle of the base soil divided by the horizontal component of the lateral earth force from the backfill. This safety factor or ratio should equal or exceed 1.5. Mathematically,

$$\frac{\{W + P_A \sin \phi_w\} \tan \phi_b}{P_A \cos \phi_w} \geq 1.5 \tag{5-6}$$

where: ϕ_b = the angle of internal friction of the base soil beneath the wall

Additional safety against sliding of the wall along its base can be provided by passive earth resistance from imbedment. Equation 5-6 purposely omits resistance due to passive earth pressure from imbedment of the wall at the toe. Mobilization of passive resistance requires considerably more deformation; furthermore, this additional margin of safety will vanish if soil is ever excavated away from toe.

Bearing Capacity: The safety factor against bearing capacity failure is equal to the ratio of the ultimate bearing capacity to the average bearing stress exerted by the wall. This ratio should equal or exceed 2.5. Mathematically,

$$q_{\mathrm{ULT}}/q_{\mathrm{AVE}} \geq 2.5 \tag{5-7}$$

where: q_{ULT} = ultimate bearing capacity of soil or ground beneath wall
q_{AVE} = average bearing stress exerted by wall on soil

The resultant of the weight of the wall and lateral earth thrust should pass through the middle third of the base. This means that the resultant normal force on the base (N) should also act within the middle third; otherwise the base at the toe of the wall will exert very high bearing stresses while the part at the heel exerts no stress at all on the underlying ground. Accordingly,

$$\bar{x} \geq B/3 \tag{5-8}$$

where: B = width of wall at base
$\bar{x}$ = distance normal force (N) acts from the toe

The ultimate bearing capacity of the ground beneath a loaded foundation or wall can be computed from various formulas for strip type footings given in standard textbooks in geotechnical engineering (Bowles, 1977). Bearing capacity is governed by the width of the foundation, friction angle and density of the base or foundation soil, inclination and eccentricity of the load, and slope of the ground surface.

Local building codes frequently list allowable bearing stresses or capacities for different types of soils, which can be used to check a retaining wall design. Alternatively, recommended or allowable bearing capacities from handbooks such as the U.S. Department of Navy (1971) Design Manual can be used for this purpose (Table 5-3). Rigid retaining walls (e.g., crib walls, bin walls, counterfort walls, and cantilever walls) exert an uneven, trapezoidal stress distribution on the base (see Figure 5-5) when the normal resultant N acts within the middle third of the base. The greatest stress intensity occurs at the toe. The stress intensity at the edges (toe and heel) can be calculated from the following simple expression:

$$\sigma = \frac{\sum F_v}{B}\left(1 \pm \frac{6e}{B}\right) \tag{5-9}$$

where: e = eccentricity or distance from center point of application of normal resultant on the base
$\sum F_v$ = summation of vertical forces

In the case of a flexible foundation such as a gabion wall, the normal contact

TABLE 5-3. Allowable Bearing Capacities for Different Types of Soils and Soil Conditions

Type of Bearing Material	Consistency in Place	Recommended Value of Allowable Bearing Capacity (tons/ft^2)
Well-graded mixture of fine and coarse-grained soil: glacial till, hardpan, boulder clay (GW-GC, GC, SC)	Very compact	10
Gravel, gravel-sand mixtures, boulder-gravel mixtures (GW, GP, SW, SP)	Very compact	8
	Medium to compact	6
	Loose	4
Coarse to medium sand, sand with little gravel (SW, SP)	Very compact	4
	Medium to compact	3
	Loose	2
Fine to medium sand, silty or clayey medium to coarse sand (SW, SM, SC)	Very compact	3
	Medium to compact	2.5
	Loose	1.5
Fine sand, silty or clayey material to fine sand (SP, SM, SC)	Very compact	3
	Medium to compact	2
	Loose	1.5
Homogeneous inorganic clay, sandy or silty clay (CL, CH)	Very stiff to hard	4
	Medium to stiff	2
	Soft	0.5
Inorganic silt, sandy or clayey silt, varved silt-clay-fine sand (ML, MH)	Very stiff to hard	3
	Medium to stiff	1.5
	Soft	0.5

Source: Adapted from Department of Navy (1971).

stress at the base will not be distributed in a planar, trapezoidal fashion as shown in Figure 5-5. Instead the stress will decrease from a maximum at the point of application of the resultant at the base to lesser values at the edges. The pressure at the toe of a gabion wall is, therefore, generally less than for a rigid wall. Accordingly, the error of assuming a planar, trapezoidal distribution and using Equation 5-9 will yield a conservative or safer estimate of the critical stress at the toe of a flexible retaining wall, compared to a rigid wall.

Safety against overturning is only assured if the pressure under the toe does not exceed the bearing capacity of the soil. Toe pressures for rigid, standard wall designs should be checked in this regard. Toe pressures for standard designs are often plotted as a function of wall height on the specifications sheets. Standard designs for gabion walls, on the other hand, do not include toe pressures but are generally safe for soils having an allowable bearing capacity of at least 2 tons/ foot^2 (see Table 5-3). If computed pressures exceed the allowable bearing capacity for a foundation soil at a site, either the wall height must be reduced, or the heel or toe or both must be extended.

5.2.2 Key Factors Affecting External Stability

Some notion about how different factors affect external stability can be deduced from inspection of the safety factors for overturning, sliding, and bearing capacity. Certain factors play a key role; we discuss them more explicitly in this section. These factors include the shape and inclination of the structure and the presence of water or drainage conditions in the backfill behind the structure.

A critical structural shape factor is the width to height ratio. It is desirable to make this ratio as small as possible to minimize space requirements and right-of-way encroachment. In order to withstand overturning this ratio must usually be set to at least 0.5—more or less. Slightly smaller ratios are possible when the structure is battered (inclined toward the slope); slightly larger are necessary if the fill above the wall is inclined, as shown schematically in Figure 5-6. Strictly speaking, these guidelines apply only to wall cross sections with a simple, rectangular shape; nevertheless, they are useful for initial estimates of required space.

As noted in Figure 5-6, battered or inclined walls require less volume for the same height than vertical walls, that is, smaller width to height ratios. This reduction is possible for two reasons. First, battering shifts the center of gravity away from the toe, into the slope, and increases the resisting moment against overturning. Secondly battering can significantly reduce lateral earth forces acting on a structure. This reduction can be seen more clearly by examining the lateral earth force coefficients listed in Table 5-2. Consider, for example, a vertical wall ($\alpha' = 0$), retaining a backfill with a horizontal surface ($\beta = 0$) and an angle of internal friction (ϕ) of 40 degrees. The corresponding coefficient of active earth pressure (K_A) is 0.22. Battering or inclining the wall 20 degrees reduces the earth pressure coefficient to 0.11—equivalent to a 50 percent reduction in lateral earth force against the wall, all other factors constant.

The presence of water in the backfill and drainage conditions can have a profound influence on external stability. The influence of drainage and water on stability is shown schematically in Figure 5-7 for a hypothetical, concrete wall retaining a sand backfill. Several different cases are shown, ranging from a dry backfill to backfills with either a vertical or sloping drain, and a backfill with a perched water table. Values of the horizontal force, the component that causes overturning, are tabulated in Table 5-4. The worst case occurs when there is blocked or inadequate drainage and a perched water table develops in the backfill. If the perched water table rises to the top of the backfill, the horizontal thrust against the wall increases threefold over the dry case. The provision of adequate drainage, either a sloping or vertical drain, in combination with drain outlets or weep holes in the wall, on the other hand, greatly decreases the lateral force against the wall. This example illustrates the importance of adequate drainage provisions behind retaining structures. This injunction applies to any type of restraining or retaining system whether a wall or buttress type structure.

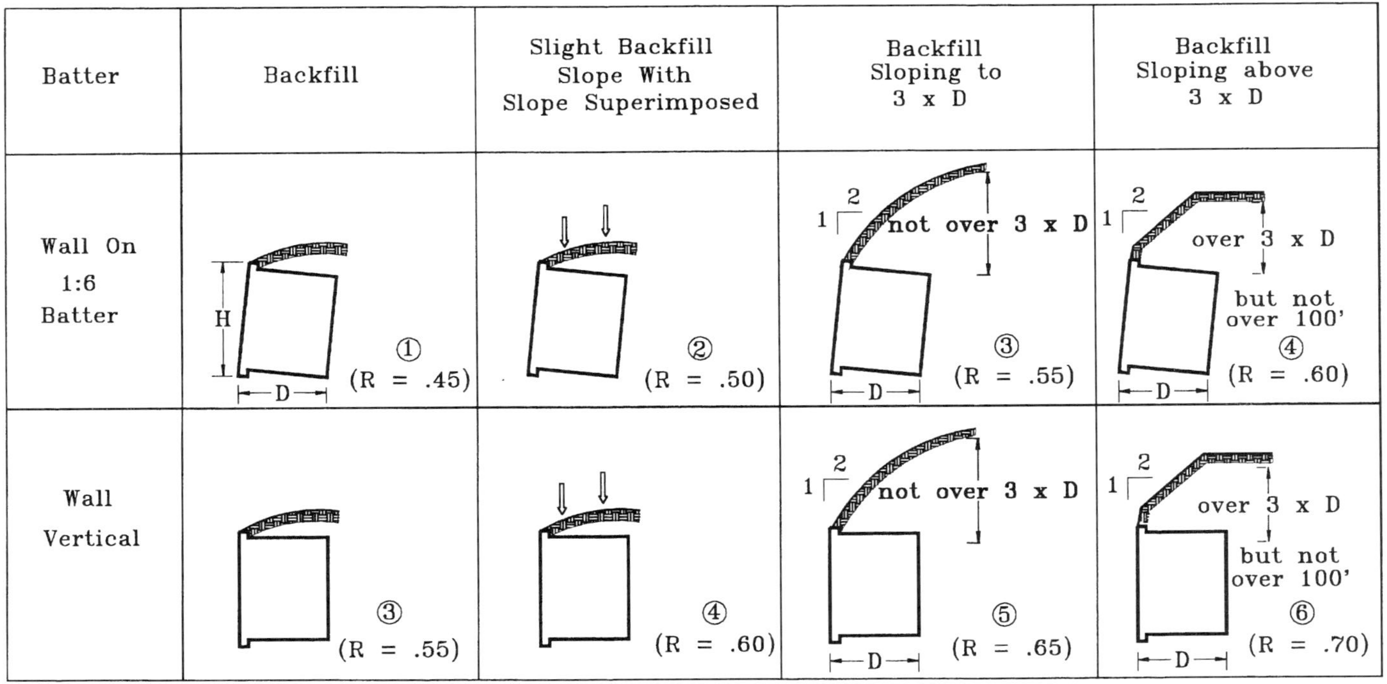

Batter	Backfill	Slight Backfill Slope With Slope Superimposed	Backfill Sloping to 3 x D	Backfill Sloping above 3 x D
Wall On 1:6 Batter	H, D ① (R = .45)	② (R = .50)	1, 2, not over 3 x D, D ③ (R = .55)	1, 2, over 3 x D but not over 100', D ④ (R = .60)
Wall Vertical	③ (R = .55)	④ (R = .60)	1, 2, not over 3 x D, D ⑤ (R = .65)	1, 2, over 3 x D but not over 100', D ⑥ (R = .70)

NOTE: $R = \frac{\text{wall width}}{\text{wall height}} = \frac{D}{H}$

Figure 5-6. Required width to height ratios for gravity structures with simple rectangular cross sections under different wall batter and backfill slope surcharge conditions.

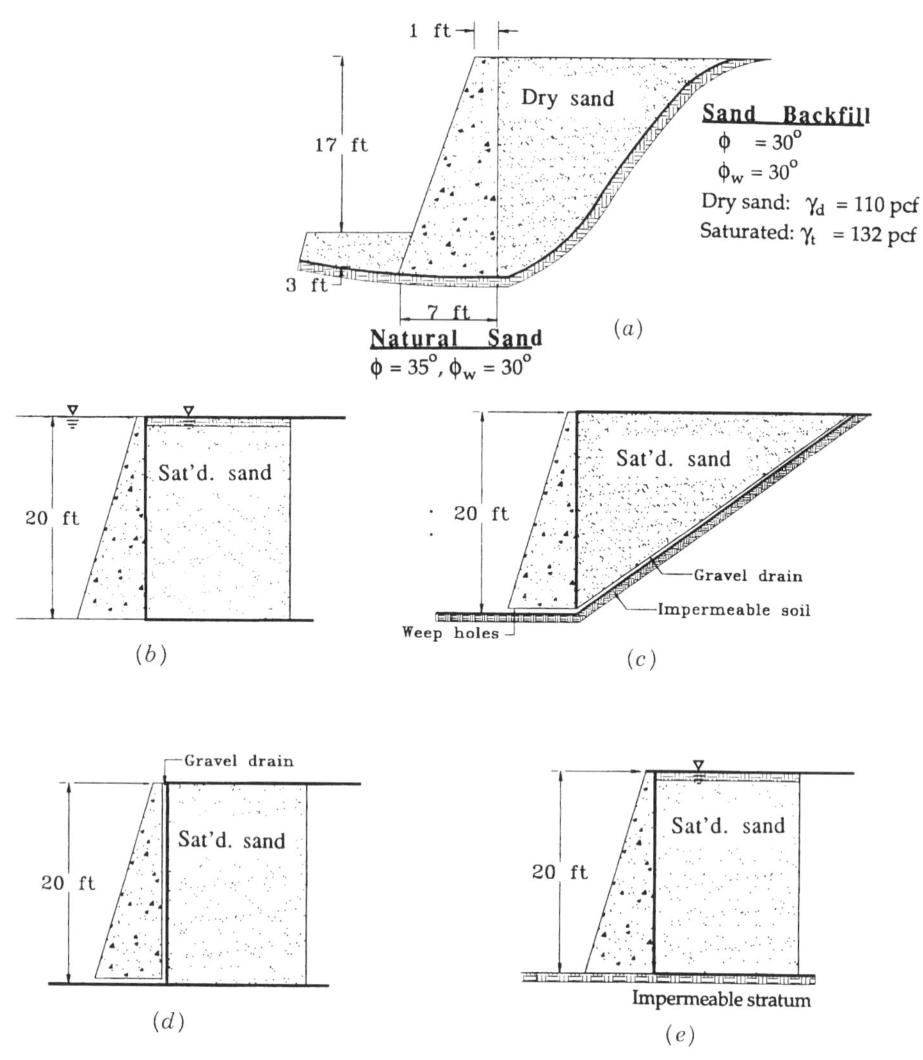

Figure 5-7. Schematic illustrations of different moisture and drainage conditions behind a concrete wall retaining a sand backfill. (*a*) Dry backfill. (*b*) Submerged wall. (*c*) Saturated backfill with a sloping drain. (*d*) Saturated backfill with a vertical drain. (*e*) Backfill with a perched water table. (From: Soil Mechanics by L. W. Lambe and R. V. Whitman. Copyright 1969. Reprinted by permission of John Wiley & Sons, Inc.)

5.2.3 Internal Stability

In addition to withstanding external forces, structural components or members of retaining walls must be capable of safely resisting stresses to which they will be subjected. An example of an internal stability failure of a steel bin wall supporting an interstate highway is shown in Figure 5-8. A vertical support

TABLE 5-4. Horizontal Thrust Against Concrete Retaining Wall (Figure 5-7) for Different Moisture and Drainage Conditions in the Backfill

Backfill Condition	Vertical Force Component (lb/ft of wall)	Horizontal Force Component (lb/ft of wall)	Percent Increase in Horizontal Thrust (relative to dry case)
(*a*) Dry backfill	3,250	5,630	—
(*b*) Fully submerged	2,045	3,540	—
(*c*) Saturated with sloping drain	3,875	6,710	19
(*d*) Saturated with vertical drain	5,100	8,840	57
(*e*) Saturated without drain	2,045	16,020	185

member at the front of the wall has buckled, probably as a result of improper compaction (densification) of the fill within the bin that resulted in settlement that dragged down the sides of the bin. Notice that the external stability of the wall appears unaffected; however, close monitoring of the structure was required to check future performance.

In the case of crib walls the internal stresses consist of: (1) pressures exerted by earth confined in the crib, and (2) stresses in crib members and connectors resulting from earth pressures exerted by both cribfill and backfill. The former can be estimated from theories developed to predict internal pressures exerted by granular materials on the walls of bins or silos (Caughey et al., 1951). In

Figure 5-8. Example of internal stability failure in a steel-bin retaining wall consisting of buckling of front, vertical support member.

the case of member stresses, critical considerations include joint bearing stress, header flexural stress, stretcher flexural stress, and stretcher torsional shearing stress. Schuster et al. (1973) outlined procedures for calculating these stresses. They also analyzed standard designs for timber cribs to detect either overstress of the crib members or potential failure of the entire crib due to external stability problems. In general they found that standard designs for timber cribs were adequate (see Section 5.2.4), the most critical or governing criteria being external stability criteria.

Internal stability requirements are also important in reinforced earth and mechanically stabilized earth walls. The tensile inclusions or reinforcements in these structures must have a sufficient unit tensile strength and/or be placed in sufficient numbers to resist breaking in tension. The inclusions must also be sufficiently long and "frictional" enough to resist failure by pullout. Geogrids tend to have superior pullout resistance compared to strips because of passive resistance mobilized along the transverse elements in a mesh or geogrid. Tie design criteria for reinforced earth walls are discussed by McKittrick (1978), whereas design requirements for geogrids and fabric inclusions are described by Leschinsky and Reinschmidt (1985) and Bonaparte et al. (1987).

Protection of structural components or members against weathering, rot, and corrosion is essential to the integrity and internal stability of inert wall systems. Unlike vegetative stabilization systems, inert structures do not tend to become stronger with age. Some control can be exercised by drainage provisions and constraints on the type of cribfill and backfill used. Structural or reinforcing elements can also be made oversize or thicker to provide a margin of safety against deterioration. Alternatively, members can be coated or impregnated to improve their resistance to deterioration.

5.2.4 Standard Designs

Most retaining walls can be constructed for specified slope-loading conditions and heights from standard designs. These designs have evolved over the years on the basis of both theory and practical experience with a given retaining-wall system. Standard designs can be used safely provided the stated conditions on which the designs are based also pertain at the site in question. It is advisable to read carefully the caveats and general notes associated with each design. In general the factors or conditions to consider include the following:

1. The maximum wall height
2. Backfill surcharge conditions
3. Wall inclination or batter
4. Strength and density of backfill
5. Gradation and compaction requirements on cribfill and backfill
6. Conditions of native soil beneath wall
7. Moisture and drainage conditions in the backfill

8. Type of fasteners and connectors
9. Erection and assembly sequence
10. Strength and finish of the structural members

Standard designs have been developed by and are available from a number of sources; these include: (1) manufacturers of retaining wall systems (e.g., gabions, crib walls, welded-wire walls, polymeric geogrid walls, articulated block walls); (2) trade associations (e.g., American Wood Preservers Institute); and (3) state and federal agencies (e.g., U.S. Forest Service, Federal Highway Administration). Standard designs that lend themselves to vegatative treatment and landscaping have been collected for reference in an earlier guidebook by Gray and Leiser (1982).

5.3 POROUS, GRAVITY RETAINING STRUCTURES

In this section we review in greater detail characteristics and attributes of selected gravity retaining structures that have porous facings or frontal openings into or through which cuttings and rooted plants can be inserted. Techniques for vegetating these structures are discussed further in Chapter 8. The issue of compatibility between structure and vegetation was considered previously in Chapter 4.

5.3.1 Rock Breast Walls

A rock breast wall is a low wall (usually 10 feet or less in height) constructed against the base of a slope. The wall is usually built by stacking rocks atop one another in a single, one-rock width course, as illustrated schematically in Figure 5-9. Ideally each rock should rest on two rocks in the tier below with at least three-point bearing. The main purpose of the wall is to defend the toe of the slope and to prevent slope damage by erosion—especially piping and spring sapping as a result of seepage exiting from the face of the slope.

A breast wall will resist lateral earth pressures to some extent and helps to prevent local slumping and slope failure provided it is well constructed, inclined into the slope sufficiently, and does not exceed design height limits. The stability of a breast wall is governed largely by its batter angle and by its height to width ratio. The greater the batter (inclination off vertical toward the slope) the greater will be the permissible height/width ratio for a specified factor of safety. In most cases this ratio should not exceed three. As rocks greater than 2 feet in size are seldom used, this limits the practical height of a rock breast wall to 6 feet or less.

Angular—as opposed to rounded—rocks are preferable. Angular units well keyed at the base provide more resistance to moment (overturning) because the wall is forced to rotate about a point on the outer edge. Rounded rocks, on the

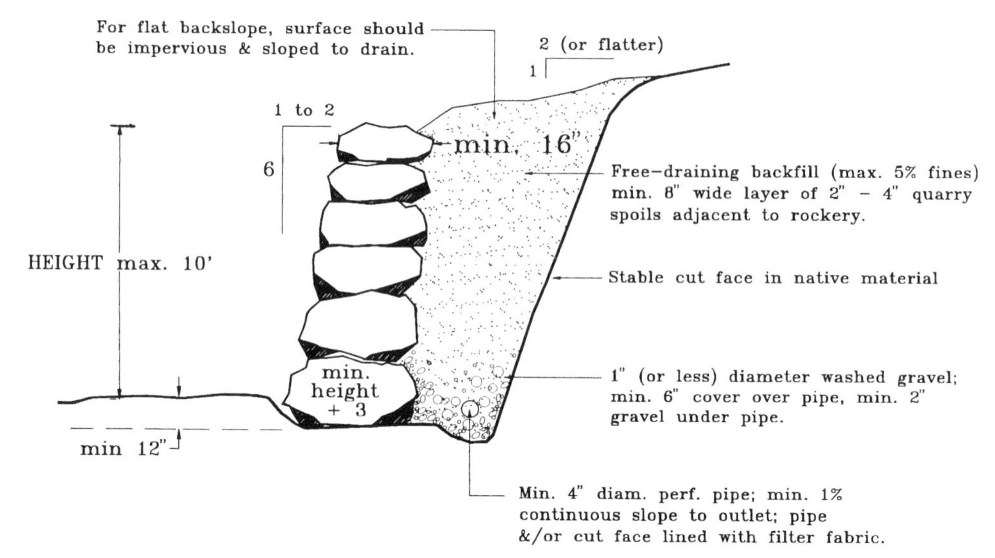

Figure 5-9. Schematic illustration of rock breast wall construction showing placement of rocks atop one another with a slight batter angle. (Adapted from ARC, 1989. Used with permission of the Associated Rockery Contractors.)

other hand, would tend to rotate about a point beneath the center of the basal unit, thus offering less moment resistance. The safety factor against overturning of a rock breast wall constructed from angular shaped rocks can be calculated by the following equation:

$$FS_{OT} = \frac{0.5\gamma_R(H/B)\cos\alpha/\sin^2\alpha + 0.5\gamma_R}{0.5\gamma K_A[0.33(H/B)^2\cos\phi/\sin\alpha - (H/B)\sin\phi]} \tag{5-10}$$

where: H = the total height of the wall
B = the width of the wall (measured at the base)
α = the inclination of the wall (with respect to horizontal)
γ_R = unit weight of rock
γ = unit weight of backfill behind wall
ϕ = angle of internal friction of backfill behind wall
K_A = coefficient of active earth pressure

Permissible values of the height to width ratio (H/B) for different safety factors can be calculated by solving for the roots of Equation 5-10, which is a quadratic equation. The following expression is obtained:

$$\left(\frac{H}{B}\right) = \frac{0.5b \pm \sqrt{b^2 + 0.33(FS)\gamma\gamma_R K_A \cos\phi/\sin\alpha}}{0.33\gamma K_A(FS)\cos\phi/\sin\alpha} \tag{5-11}$$

where: $b = 0.5[\gamma_R \cos\alpha/\sin^2\alpha + \gamma K_A(FS)\sin\phi]$

The critical or maximum height to width ratio $(H/B)_{cr}$ can be found by setting the factor of safety against (FS) equal to unity. Design/analysis curves, based on Equation 5-11, which give the maximum or critical height to width ratio $(H/B)_{cr}$ for a breast wall constructed with angular rocks for different slope conditions are presented in Figure 5-10. These design curves can also be used for articulated block walls. Curves are shown for both a horizontal ($\beta = 0$) and sloping backfill surface for soil backfill and rock with typical properties ($\phi = 30$ degrees $\gamma = 110$ pcf, $\gamma_R = 130$ pcf). These curves, it should be noted, are for the maximum or critical ratio at which the factor of safety is unity (i.e., on the verge of failure). The factor of safety can be increased by reducing the actual (H/B) ratio. Poorly constructed (rounded rock) walls would have lower critical (H/B) ratios compared to well constructed walls. The design curves discount (neglect) the effect of any cohesion in the slope (i.e., $c = 0$), and thus are conservative in this regard. If cohesion is taken into account, then a different set of curves based on an alternative expression for the lateral earth pressure coefficient, K_A, must be used.

Breast walls are also subject to local rupture (bulging) as a result of shear forces acting on the individual rock units. Shear failure is resisted by frictional

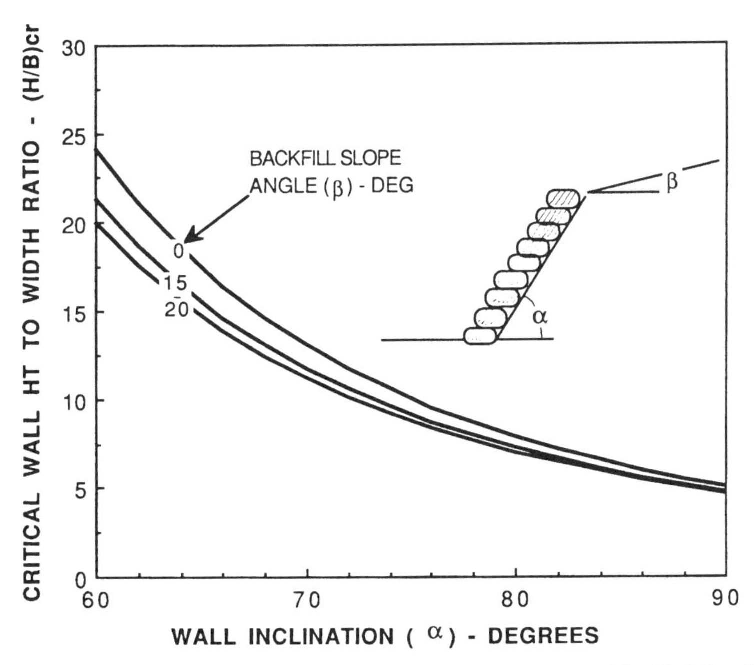

Figure 5-10. Design/analysis curves for estimating maximum or critical height of breast wall constructed with angular rocks or articulated blocks and with the following soil backfill and rock properties: $\phi = 30$ degrees, $\gamma = 110$ pcf, $\gamma_R = 130$ pcf.

forces between the rocks, which in turn depends upon normal contact stresses at bearing points. Analyses show, however, that moment (overturning) is the critical mode of failure, since for any given set of conditions the allowable factor of safety is always higher for the shear criterion than for the moment criterion. This assessment is based, however, upon the wall resting on a firm foundation and the absence of differential settlement. If the latter occurs, arching action in the wall can reduce the contact stress between rocks and permit local shear failure.

Rock breast walls should be constructed on a firm, well tamped earth or foundation; placement of the first course of rocks in an excavated trench (see Figure 5-9) helps in this regard. Piping and concentrated seepage erosion behind a breast wall can be prevented by placing pit run gravel or a filter fabric against the slope face to serve as a filter behind a newly constructed wall. A photo of a rock breast wall is shown in Figure 5-11.

5.3.2 Gabion Walls

Gabions are rectangular containers fabricated from a triple twisted hexagonal mesh of heavily galvanized steel wire. The simplest gabion structure is a 3-foot-high wall using one tier of gabions. A second tier of gabions can be placed on top of the first tier and set back 18 inches (i.e., stepped back slightly). Figure 5-12 schematically depicts typical one-tier and two-tier gabion retaining walls. Gabion walls that are higher than two tiers (6 feet) usually require significant additional design constraints. Higher tiered walls require greater basal widths,

Figure 5-11. Rock breast wall defending toe of low, cut slope along highway.

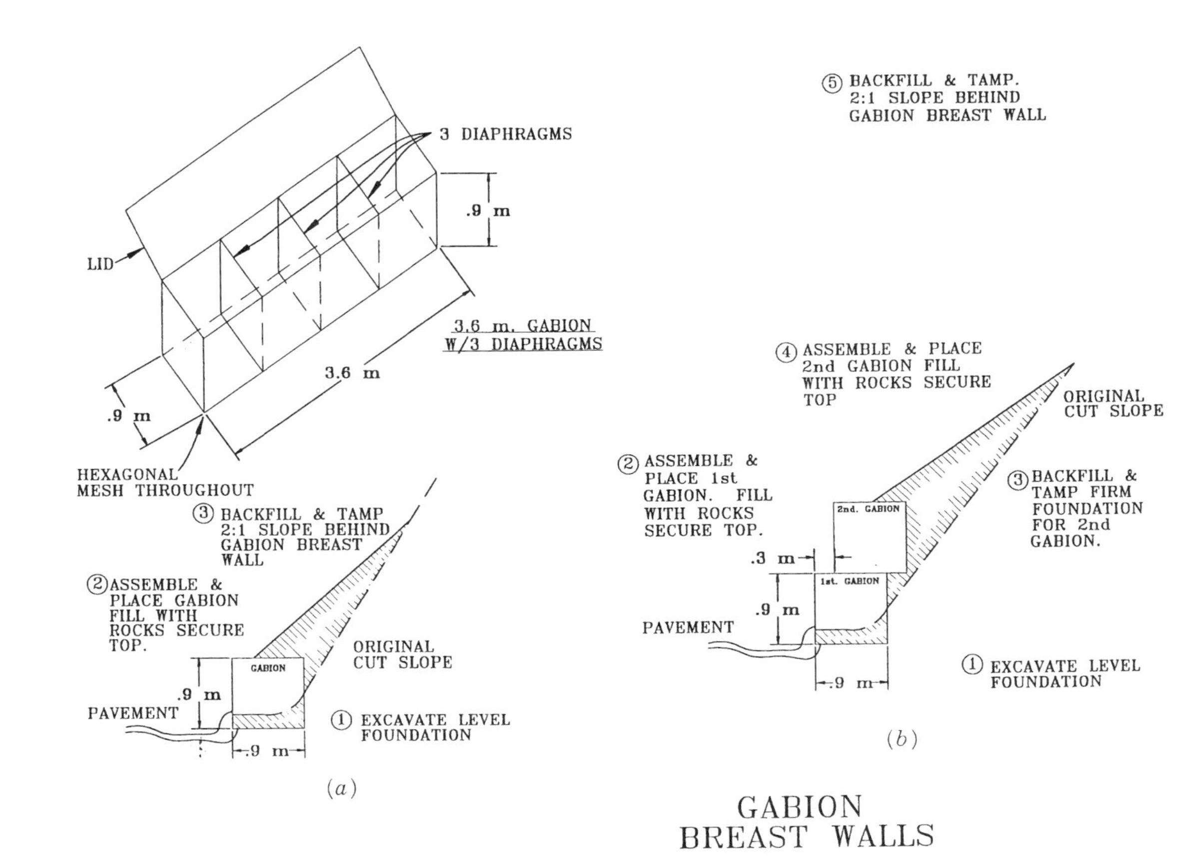

Figure 5-12. Low, gabion toe walls showing the sequence of excavation, assembly, and filling. (*a*) Single tier. (*b*) Double tier. (From White and Franks, 1978.)

and/or the use of counterforts to brace the wall against overturning moments from the backfill behind the wall.

Several different design configurations or stacking arrangements are possible with gabions. They may have either a battered or a stepped-back front. The choice of type depends upon application, although the stepped-back type (see Figure 5-13) is generally easier to build when the wall is more that 10-feet high. The number and arrangement of gabion units also depend upon whether a level or an inclined backfill is used behind the wall. Various standard designs for different wall heights and backfill conditions are supplied by gabion manufacturers.

For easy handling and shipping, gabions are supplied folded into a flat position and bundled together. Each gabion is readily assembled into a rectangular shaped "box" (see Figure 5-12) by unfolding and binding together all vertical edges with lengths of connecting wire stitched around the vertical edges. The empty gabions are placed in position and wired to adjoining gabions. They are then filled with cobblestone-sized rock (10 to 30 cm in size) to one-third their depth. Two connecting wires are then placed in each direction, bracing opposing walls or sides of the baskets. The connecting wires prevent the gabion baskets from bulging as they are filled up, which makes a major difference in the final appearance of the wall.

Low gabion walls and rock breast walls are comparable alternatives. Hand-placed, rock breast walls are more limited in height. Gabions are sometimes criticized as being unsightly; however, use of attractive facing stone toward the front of the wall plus establishment of vegetation in the gabions can amelio-

Figure 5-13. High, stepped-back, gabion retaining wall following construction.

rate this problem. If large rocks are readily available near the site and relatively inexpensive, then their use in construction of a rock breast wall might be preferable. If, on the other hand, rock must be imported from a distance or is only available in small sizes, it is likely that a gabion toe wall would be preferable.

5.3.3 Crib Walls

A crib is basically a box-like structure formed by joining a number of cells together and filling them with soil or rocks to give them strength and weight in order to form a gravity wall. The structural members in most crib walls are assembled "log cabin" fashion to form a cell. The frontal, horizontal members of the cell are termed "stretchers"; the lateral members, "headers" (see Figure 5-14). Forces are transferred between the members at the corner joints. Small blocks called pillow blocks are also placed at critical locations between members to relieve compressive stresses at the joints and reduce bending and torsional stresses in the stretchers and headers. Some newer crib wall designs consist of precast, modular units that are simply stacked one atop the other. These are likewise backfilled with rock or suitable soil. Local shear between these modular units is resisted by some type of articulation, for example, dowels or pins, that connects the modular units to each other.

Timber Cribs: The components and configuration of a typical timber crib retaining wall are shown in Figure 5-14. In modern designs, mechanical con-

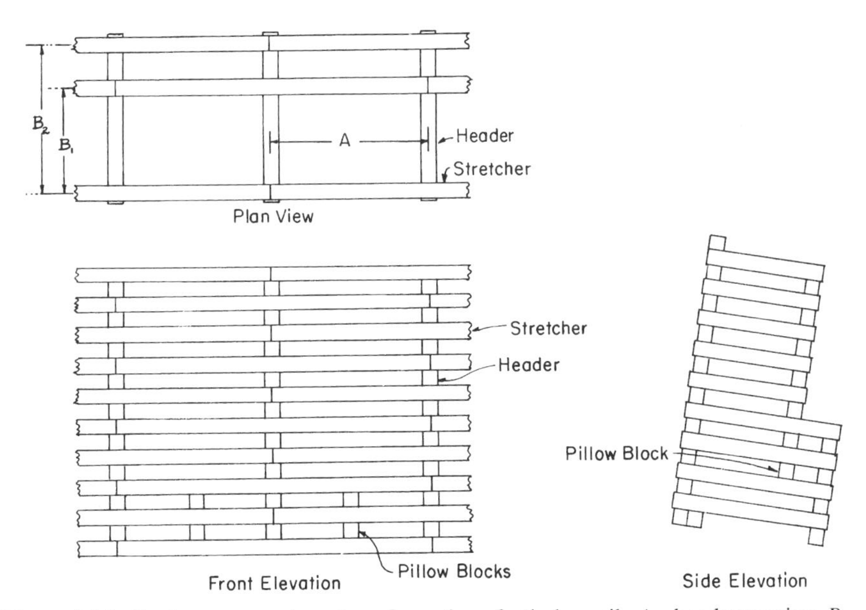

Figure 5-14. Basic components and configuration of a timber crib. A = header spacing, B_1 = stretcher spacing (top), B_2 = stretcher spacing (bottom).

nectors such as drift pins or spit rings are used at the joints; older timber crib walls sometimes used dapped (i.e., notched) joints to transfer forces. Schuster et al. (1973) prepared a comprehensive study and analysis of timber crib retaining walls including an evaluation of then existing walls and a performance comparison of different timber crib designs. A great number of timber walls were built in the early 1990's by the mining industry in the Rocky Mountains. Many of these walls are still standing, some in various states of disrepair. Others, particularly those with well-drained backfill, are 70 to 80 years old and still serviceable. The primary use of timber cribs today is to retain road cuts and embankments, particularly in mountainous areas, and for low-volume haul roads.

A large percentage of timber crib walls that are designed and built today are constructed of dimensioned, treated timber (either 6 × 6 inch or 8 × 8 inch stock). Most commonly the members are rough-cut, structural-grade Douglas fir, although some specifications allow other species. Log crib walls are not subject to exact engineering design and analysis; nevertheless, many have been built. Both the Washington State Department of Highways (1973) and the Federal Highway Administration (1974) have published design standards for log cribbing.

A number of standard designs for timber crib retaining walls are commonly used in the United States. A complete discussion, design details, and evaluation of each of these standard designs have been provided by Schuster et al. (1973). The authors concluded that for level backfills, crib walls formed of treated, dimensional timber and constructed in accord with any of the standard designs can be safely built to the design heights. They noted, however, that in a few cases additional blocking may be necessary at critical points. Timber cribs are also moderately flexible and can withstand differential settlements without a significant effect on the retaining action or gross stability of the wall, even with noticeable crushing of individual member or weakening of the joints.

The American Wood Preservers Institute (AWPI) has developed a number of standard, treated cribbing designs, as shown in Figure 5-15. Details of both these, and also log crib designs can be found in Gray and Leiser (1982). The AWPI (1981) designs all use vertical members in a header-stretcher joint that is different from those used in other designs. The horizontal stretcher forces are transmitted to the header by vertical members bolted to the sides of the projecting headers. These vertical members or posts are also designed to act as a column in resisting the cumulative vertical reactions from the headers. By using various combinations of base widths and either vertical or battered crib sections (see AWPI Basic Designs A through H in Figure 5-15), the AWPI wall is suitable to a maximum height of 30 feet for a variety of surcharge conditions. Although the battered front face configuration is slightly more difficult to construct, it offers several advantages, namely:

- Less structure (timbers and cribfill) is required than for a vertical wall the same height.

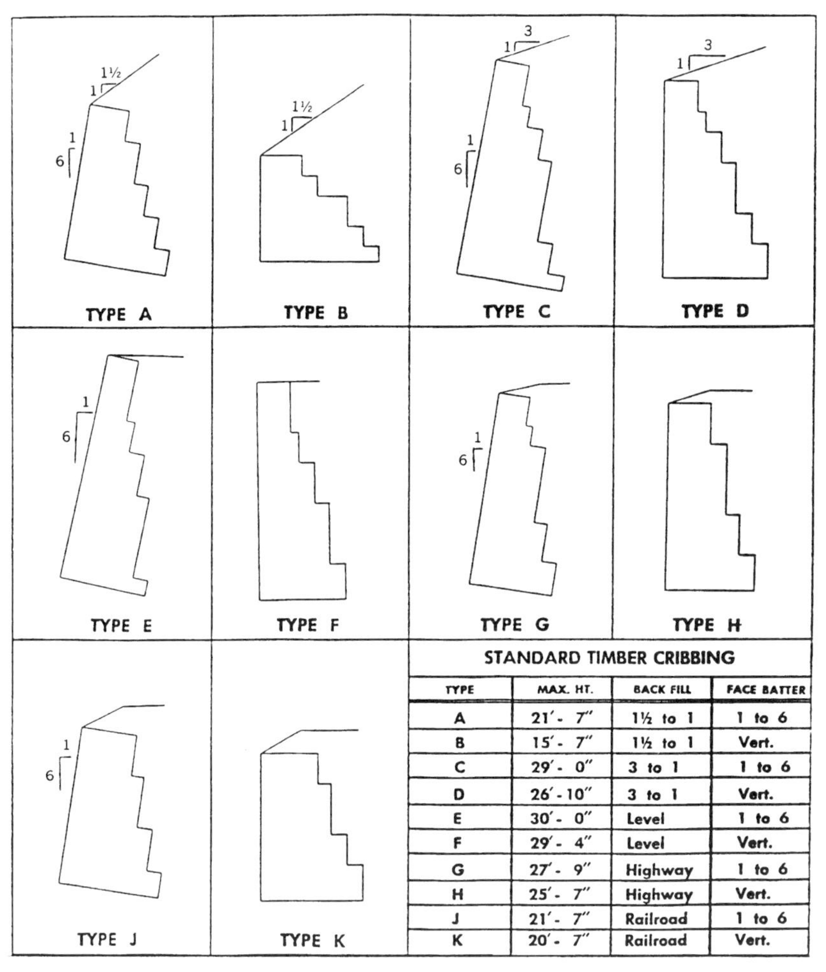

STANDARD TIMBER CRIBBING

TYPE	MAX. HT.	BACK FILL	FACE BATTER
A	21′- 7″	1½ to 1	1 to 6
B	15′- 7″	1½ to 1	Vert.
C	29′- 0″	3 to 1	1 to 6
D	26′-10″	3 to 1	Vert.
E	30′- 0″	Level	1 to 6
F	29′- 4″	Level	Vert.
G	27′- 9″	Highway	1 to 6
H	25′- 7″	Highway	Vert.
J	21′- 7″	Railroad	1 to 6
K	20′- 7″	Railroad	Vert.

Figure 5-15. Standard, treated timber cribbing, basic design or cross section index. (From AWPI, 1981. Used with permission of the American Wood Preservers Institute.)

- The cribfill is less likely to ravel out the front face of the structure.
- Vegetation is more easily planted and established in the interstices between the stretchers.

Drainage characteristics of the backfill and cribfill have a noticeable effect on crib performance and longevity. This is particularly important in mountainous, high-precipitation areas where most timber walls are built. Use of free-draining, granular materials generally ensures that there will not be a buildup of seepage pressures in the fill. These materials also help to keep the crib members relatively dry, thereby prolonging the life of the wood. Timber treatment or preservation is also important in this regard. Most modern timber crib members are pressure-treated with preservatives such as pentachlorophenol in liquid petroleum gas or mineral spirits, or water-borne salts such as copper arsenate. Pressure treatment is commonly performed at a central plant where crib members are individually cut and drilled ready for assembly. Timbers also may be wholly or partly treated in the field by dipping, swabbing, or brushing using these same preservatives. These methods are not as effective as pressure treatment, and cannot be expected to provide the same degree of preservation.

Concrete Cribs: Crib walls can also be constructed from precast, reinforced concrete members. In fact most crib walls in use today are constructed in this fashion. Concrete crib walls are more durable and can be built to greater heights than timber cribs. The basic principle of construction remains the same, namley, cribbing is formed of interlocking structural members or beams, the internal spaces of which are filled with suitable free-draining fill that adds to the stability of the wall.

There are several concrete crib wall systems or designs available. These systems differ in the shape of the structural members (stretchers and headers), wall configuration, structural connections, and erection procedure. A typical design uses "dog-bone"-shaped headers that help to keep the stretchers in place (Figure 5-16). A schematic diagram of this type of crib wall system showing different arrangements and configurations (Hilfiker, 1972) of structural members for different wall heights is presented in Figure 5-17. These crib walls are usually

Figure 5-16. View of open-front concrete crib wall utilizing "dog-bone" header design.

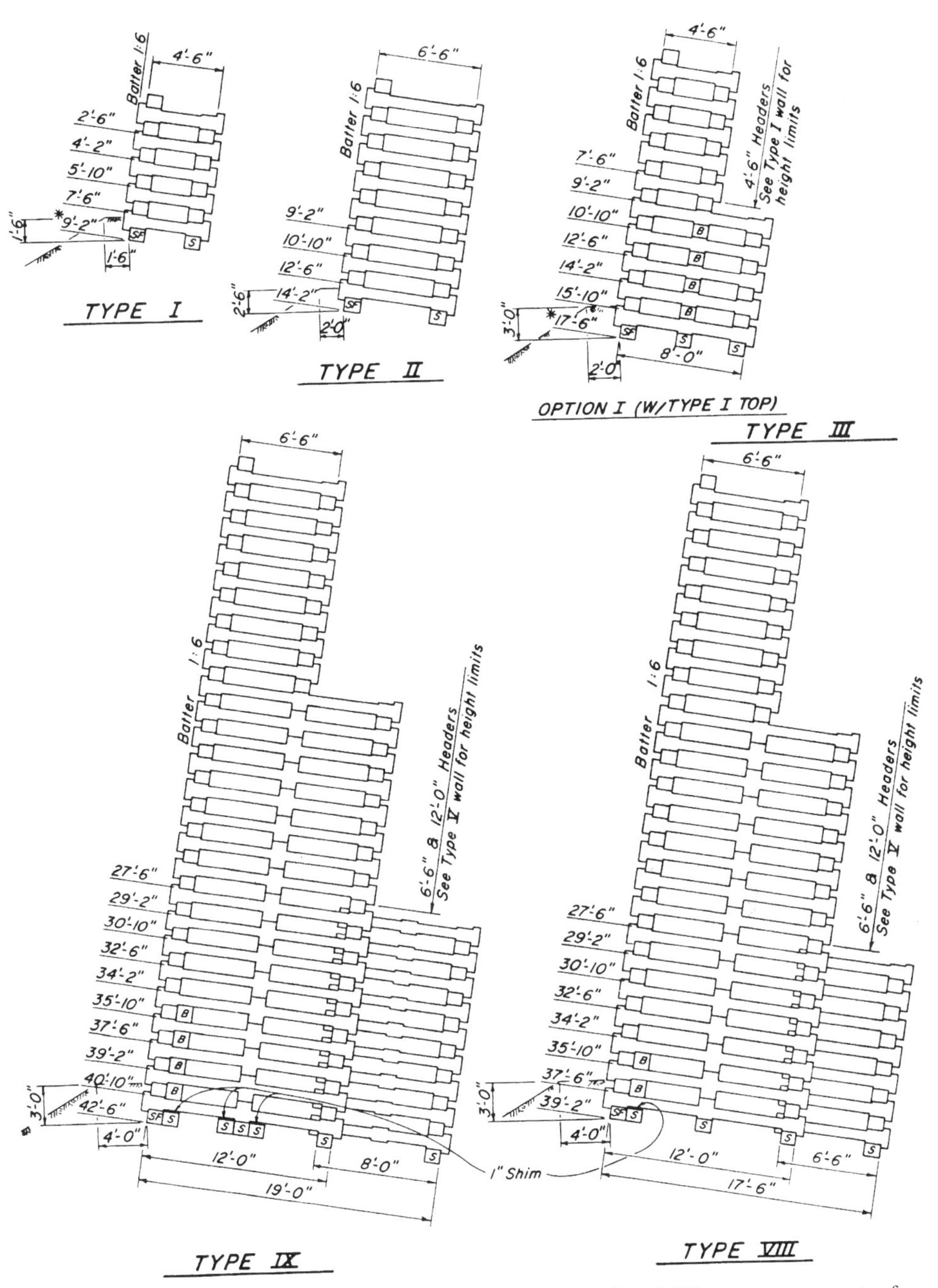

Figure 5-17. Reinforced concrete crib walls showing examples of different arrangements of structural members for different wall heights (From Hilfiker, 1972. Used with permission of the Hilfiker Company.)

constructed with a batter of 1 : 6 to improve stability, although vertical walls can also be specified. An open-front face design facilitates drainage, prevents the buildup of hydrostatic pressure behind the wall, and eliminates the need for weep holes. It also permits the establishment of suitable plants and shrubs in the face of the wall and opportunities for creative landscaping.

A concrete cribwall known as the Evergreen® wall (Jaecklin, 1983; Randall and Wallace, 1987) was developed for the specific purpose of accommodating vegetation in its frontal openings or interstices. This wall utilizes precast, modular units resembling bottomless troughs (Figure 5-18) that are stacked atop one another in successive courses. Cribfill that is suitable for sustaining vegetation is placed in each course after it has been placed, as shown in Figure 5-19. The final appearance of the wall after construction, before planting, is shown in Figure 5-20. Less than 50 percent of the exposed face is concrete.

5.3.4 Geogrid Walls

Many different types of tensile inclusions are used to internally reinforce earthen embankments or, alternatively, to strengthen and reinforce the backfill behind retaining walls. These inclusions range from galvanized, ribbed metal strips to polymeric geotextiles and geogrids. The reinforcing mechanism and effectiveness vary slightly from one type of inclusion to the other. Metal strips reinforce a soil mass by the transferring shear stresses that develop in the soil

Figure 5-18. Precast, trough-like modular unit that is assembled in various stacking arrangements to form an Evergreen crib wall. (Photo courtesy of Evergreen Systems, AG.)

Figure 5-19. Filling a course of Evergreen crib wall units with a soil that is capable of supporting vegetation. (Photo courtesy of Evergreen Systems, AG.)

Figure 5-20. Example of Evergreen crib wall system after construction and before planting vegetation in the frontal bays or openings. (Photo courtesy of Evergreen Systems, AG.)

to tensile resistance in the metal strips via friction at the soil-strip interface. Geogrids function in a similar fashion, but also provide distributed anchorages or interlocking connections to the soil particles. Geogrids tend to have superior pullout resistance (at equal surface areas) compared to strips because of passive resistance developed by transverse members. A comprehensive analysis of alternative earth reinforcement systems and their design considerations has been compiled by Mitchell and Villet (1987).

Geogrid reinforcement systems are of particular interest not only because of their superior pullout resistance but also because of other attributes that lend themselves to biotechnical treatment. A geogrid often doubles at both the reinforcing element in the backfill and the facing element at the front of a retaining structure. This is achieved by simply "wrapping" the geogrid around successive lifts of backfill at the front face. The apertures or openings in the facing portion of the geogrid also provide a locus for the establishment of vegetation in the reinforced backfill.

Welded-Wire Walls: The welded-wire wall (Hilfiker, 1978) is a composite wire and granular soil structure. The wall is constructed from 9-gauge, welded steel wire fabric. The wire fabric or matting is placed between successive lifts of granular fill. The L-shaped form of the mats is designed to both reinforce the granular fill and contain the face of the structure. Exposed vertical ends of the facing portion of each mat are bent over the horizontal wire of the mat above to form a connection between mats at the face. A backing mat and screen (optional) are also inserted to prevent raveling of the fill at the face.

The wire mats are lightweight and easy to transport to and handle at the job site. Typically these mats are fabricated from 2 × 6-inch welded-wire mesh that is 4 to 8 feet wide and sufficiently long to provide stability to the fill, which is sandwiched between the wire. The wire mats are folded in L-shaped-fashion to provide a face that is at right angles to the floor of the mat, as illustrated in Figure 5-21. The face of each section is typically about 16 inches high, and the floor length is equal to approximately 80 percent of the composite height of the wall.

Welded-wire walls have fewer constraints on the material or soil that can be used in the structural or reinforced volume than other wall systems. Free-draining, coarse granular soils are preferable, but any backfill material is classified as GW to SC in conformance with ASTM designation D2487 may be used in standard designs (see Gray and Leiser, 1982). The structural fill should be compacted to 90 percent relative compaction in conformance with ASTM method D698. Both placement and compaction of the fill can be done mechanically with motorized equipment.

Views of completed, welded-wire walls are shown in Figure 5-22 and 5-23. In addition to their low cost, flexibility, and ease of construction, welded- wire walls can be planted with vegetation that will grow through the wire mesh. Vegetation also tends to establish naturally in the face of the wall with time as a result of seeds that are washed or blown into the frontal apertures or openings. This allows the wall to blend harmoniously into the landscape.

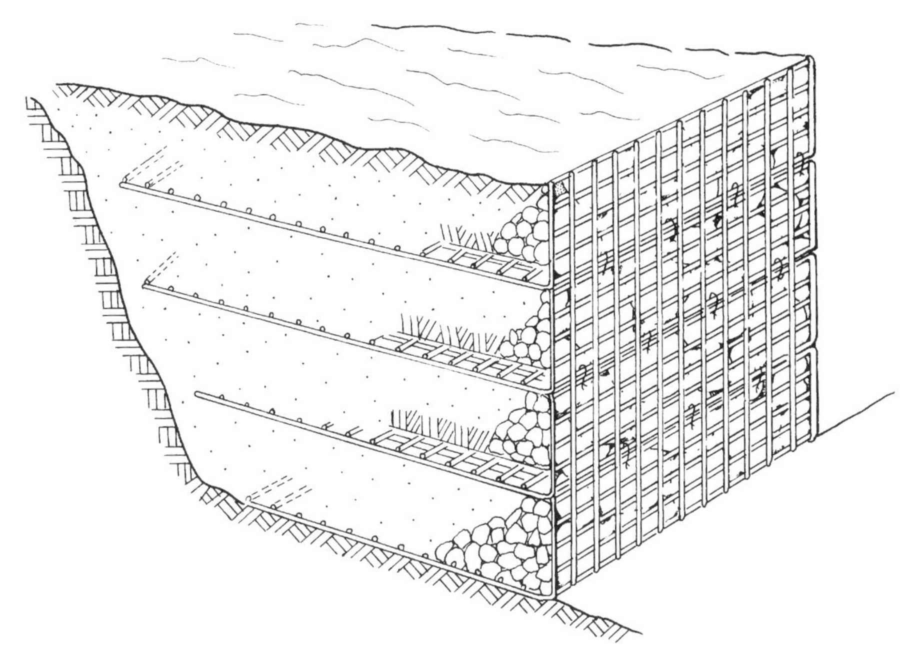

Figure 5-21. Schematic illustration of welded-wire wall. Wire mesh mats are placed between successive lifts of fill and connected together at the front face.

Figure 5-22. View of completed welded-wire wall showing porous front face that permits vegetation to grow through the wire.

Figure 5-23. Vegetation has established naturally in front face of a welded-wire retaining wall that supports a haul road.

Synthetic (Polymeric) Geogrids: Synthetic geogrids fabricated from high-tensile strength polymeric materials are widely used in reinforced earth embankments and retaining walls. They can be used either in a wrap-around fashion to provide both backfill reinforcement and containment at the front face (Figure 5-24) or, alternatively, can be attached to facing units as reinforcements that extend into the backfill. The former approach permits the establishment of vegetation in the frontal openings or apertures, as shown in Figure 5-25. Live cuttings can also be inserted between successive lifts or layers of "wrapped" backfill, as shown in Figure 5-26. Earthen embankment fills or buttress fills constructed with flexible geogrids and live cuttings are referred to as "vegetated geogrids," a soil bioengineering technique described in Chapter 7.

The main considerations in the design of geogrid reinforced earth walls and embankment fills is the required vertical spacing (d) and total length (L) of the reinforcing layers. The total length (L) is comprised of a length or distance required to reach the expected failure surface in the backfill and an additional length, the effective or imbedment length (L_E), extending beyond the failure surface required to prevent pullout. The various soil/slope/reinforcement parameters that affect these two design variables are shown schematically in Figure 5-27. A general expression for the required spacing (d) and effective (or imbedment) length (L_E) for a *vertical*, reinforced structure is given by the following equations:

$$d = \frac{T_{\text{allow}}}{(FS)K_A(\gamma z + q)} \qquad (5\text{-}12)$$

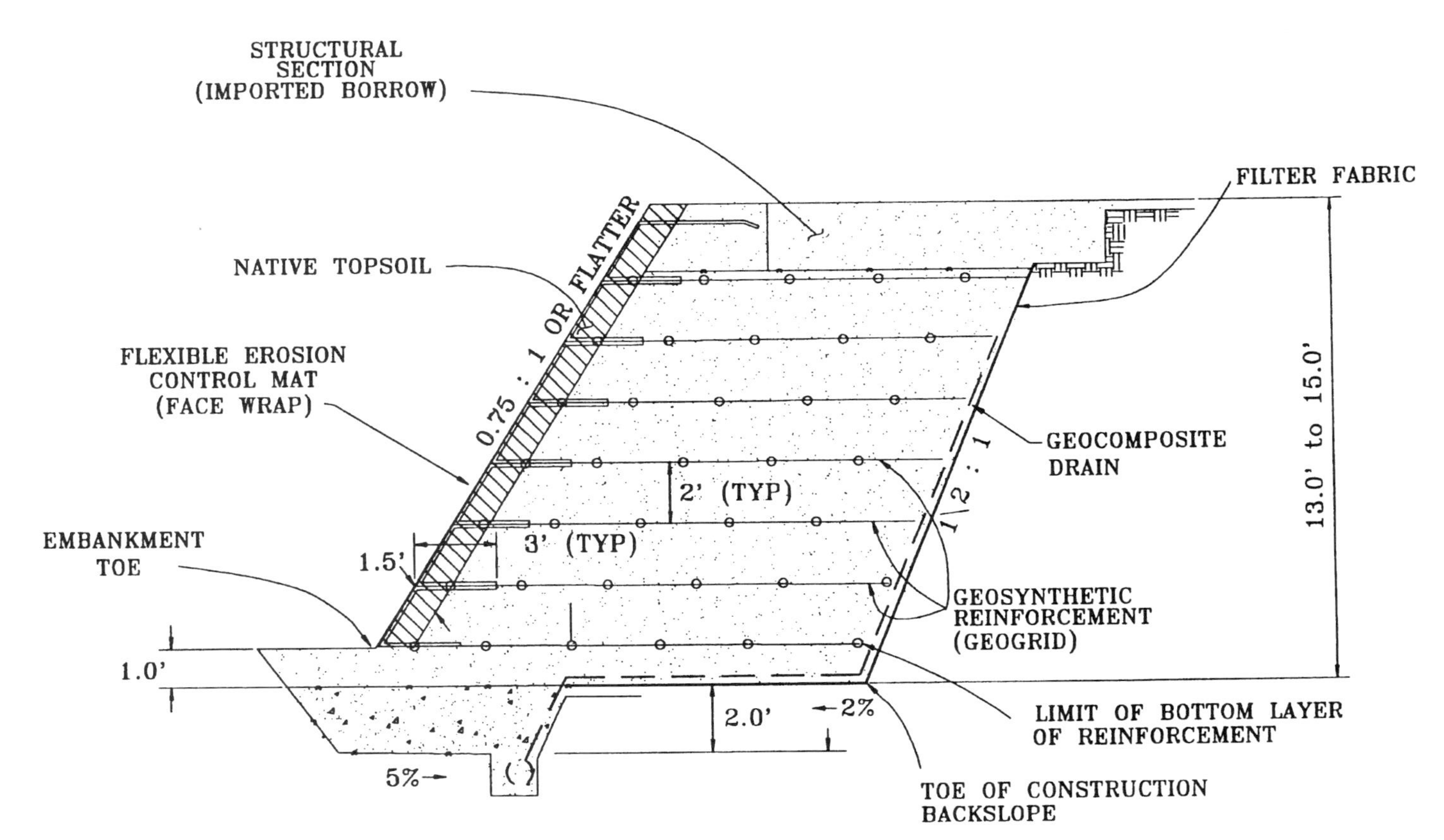

Figure 5-24. Schematic illustration of geogrid reinforced retaining structure showing typical placement and spacing of flexible, polymeric geogrids.

Figure 5-25. Geogrid reinforced retaining structure showing establishment of native plants in the frontal apertures and horizontal benches.

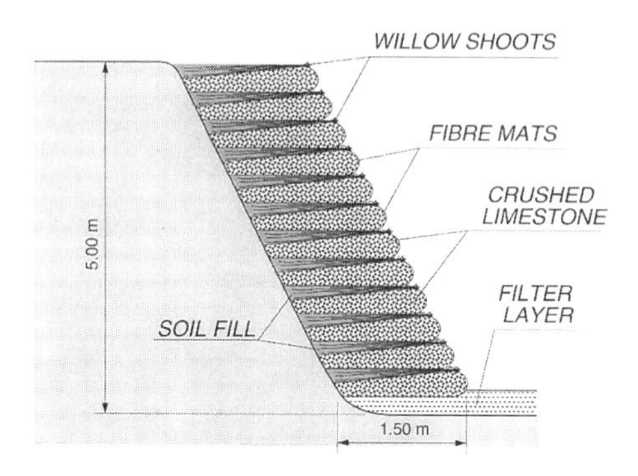

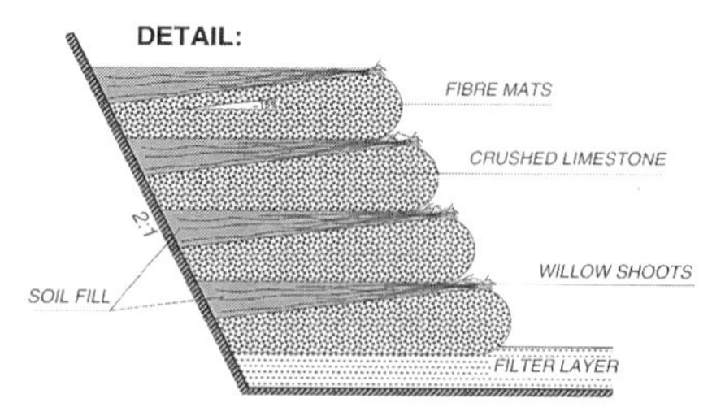

Figure 5-26. Retaining wall constructed from successive lifts of crushed limestone "wrapped" and reinforced with a geotextile. Live cuttings and a thin layer of soil were inserted between successive lifts. (From Smoltczyk and Malcharek, 1982.) Used with permission of Industrial Fabrics Association International.

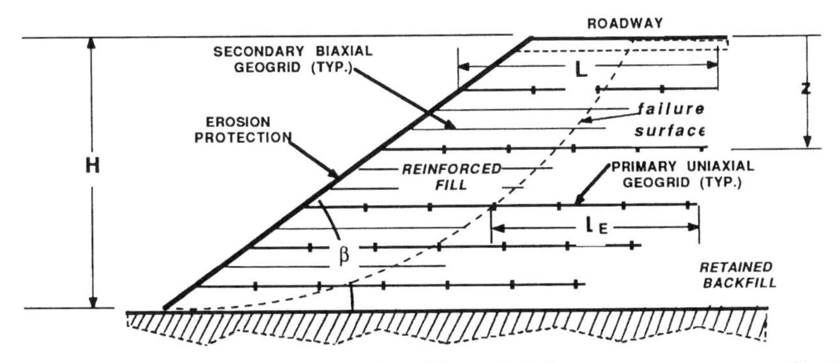

Figure 5-27. Schematic illustration of soil/slope/reinforcement parameters affecting required spacing and length of geogrid reinforcements.

$$L_E = \frac{(FS)T_{\text{allow}}}{2(\gamma z + q)\mu \tan \phi} \tag{5-13}$$

where: FS = desired factor of safety
T_{allow} = the allowable unit tensile load in the geogrids
K_A = the coefficient of active earth pressure
z = the depth to the reinforcing layer
q = uniform (or equivalent) vertical surcharge
γ = the unit weight of the backfill
ϕ = the angle of internal friction of the backfill
μ = an interface friction coefficient between soil and reinforcement (approximately unity for geogrids)

Both the required vertical spacing and the imbedment length vary inversely with the depth of the reinforcing layer, that is, closer spacings are required at greater depths where shear stresses are highest and shorter imbedment lengths are required at greater depths because of increased normal stresses. A conservative design procedure would be to calculate a fixed vertical spacing based on the maximum placement depth (i.e., height of the retaining structure) and fixed length based on minimum placement depth (i.e., at the ground surface where the failure surface is farthest away from the face of the wall). Alternatively, the backfill could be subdivided into zones and required spacing and length based on the midpoint depth of each zone.

5.4 REVETMENTS

5.4.1 Purpose and Function of Revetments

A revetment is a facing placed on a slope to armor it and prevent scour and erosion. Revetments are typically used along stream channels, waterways, inland lakes, and seashores to prevent or control bank erosion. A revetment offers some resistance (or buttressing) against mass movement; however, its primary purpose is to prevent loss of bank material by wave action, fluvial scour, and exfiltration. Revetments are normally placed on slopes no steeper than $1\frac{1}{2}:1$ (30 degrees or 57 percent). Steeper slopes require a retaining wall or other structure.

One of the main uses of revetments is along streambanks for protection against fluvial erosion. Bank armoring with structural coverage is required for erosion control where bank materials are weak and water velocities high, or in areas subject to wave action, ice scour, or emergent seepage. Streambank protection systems lie outside the scope of this book. A good review of streambank protection systems, including the measures cited above, is contained in a U.S. Army Corps of Engineers (1978) Interim Report on its streambank erosion control evaluation and demonstration program. The main focus of this chapter, however, is on the use of revetments to protect sandy coastal landforms (beaches, berms, bluffs) against wave action.

5.4.2 Types of Revetments

In the past, revetments have typically been constructed from hand-placed, dumped, or derrick-laced rock (riprap). Today many types of flexible structural facings are in use; these include:

- Riprap
- Quarry stone
- Concrete rubble
- Gabion mattresses
- Rubber tire networks
- Sand-cement sacks (or mattresses)
- Articulated, precast concrete blocks

The objective herein is not to describe all revetment systems but rather to concentrate on revetments that can be vegetated. Natural vegetation will often invade and establish itself in riprap and other porous structural facings (Figure 5-28). Alternatively, vegetation can be introduced by the insertion of live cuttings, a technique known as "joint planting" (see Section 8.3). Vegetation can also be established in gabion mattresses and articulated, precast block systems.

The stability of revetments or resistance to movement is governed largely by the size (or weight) of the individual armor units, and expected (or design) water velocities and/or wave heights. These stability requirements are discussed in the next section. Vegetated, "hard armor" revetments also fall under the purview of biotechnical or composite ground covers. Vegetated, hard armor systems are used where flow velocities and duration are high (see Chapter 9).

Figure 5-28. Stone armoring or revetment protecting cut slope along a highway. Vegetation has naturally invaded and established itself in this porous revetment.

Rock Riprap: A carefully placed layer of stones and boulders, generally known as riprap, is one of the most common and effective methods of bank protection and armoring. Riprap can settle and conform to the final slope contour if some scour should occur, and vegetation can become established in areas above the waterline. Factors limiting the use of riprap are: availability of suitable-size rocks; difficulty and expense of quarrying, transporting, and placing stone; and the large amount of material needed where streams or offshore waters are deep. The size or weight of the rock that is required depends upon the expected or design velocity and/or wave height, as discussed in the next section.

The main elements of a quarry stone or rock revetment used to protect a coastal slope or bluff are shown in Figure 5-29. Views of typical rock revetments are shown in Figure 5-30. The bank should be graded to a $1\frac{1}{2}:1$ side slope or flatter. The bottom of the revetment should be keyed into the base slightly below the anticipated scour line. Failure to observe this precaution compromises the effectiveness of a revetment. The method of placing riprap is important. Riprap may be either hand-placed, end-dumped, or placed by derrick. For most applications graded rock is placed by end dumping. When a bank must be protected against large wave forces, derrick-placed quarry stone is often used. The riprap must be placed on a filter blanket that is suitably sized to prevent the washout of fines through the armor layer. Either a graded aggregate or filter fabric can be used for this purpose. Care must be exercised to ensure that the blanket is not ruptured or displaced.

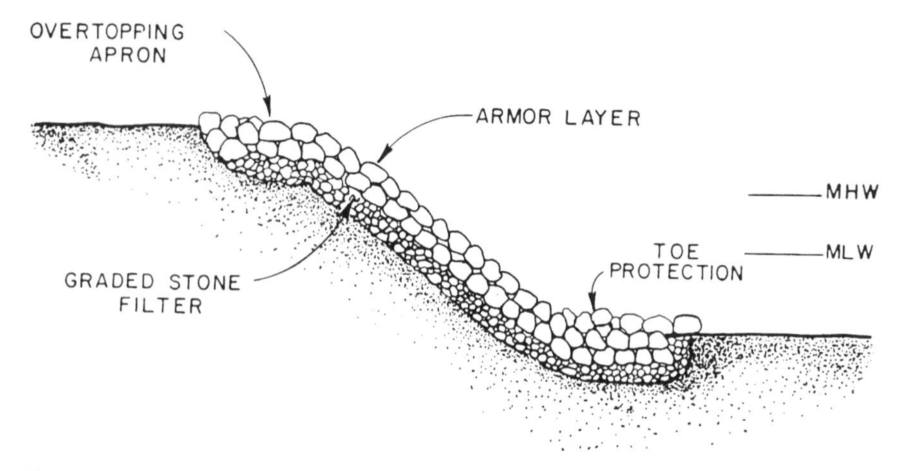

Figure 5-29. Components of typical quarry stone revetment used to protect a coastal slope. MHW = mean high water; MLW = mean low water.

Gabions: Gabions consist of wire baskets that are wired together and filled with rock in place on a slope. When used as revetments (Figure 5-31) the gabions are much thinner or flatter than gabions used in retaining walls and resemble mattresses, which they are sometimes called. Gabions may be considered as alternatives to riprap where suitably sized rock is not available or too costly. When gabion revetments are used to protect river banks or coastal bluffs, a flexible apron is normally included at the base, or alternatively, the apron can be replaced by a single-course gabion toe-wall (Figure 5-31), which not only helps to support the revetment but also prevents undermining. The portion of the gabion above the waterline can be vegetated (see Section 8.4).

The foundation or bed for a gabion revetment should be smoothed, and filter material properly placed under the gabion to prevent washout of fines. The baskets should be assembled in accordance with manufacturer's recommendations. All construction proceeds from the bottom to the top of the installation. The baskets should be stretched slightly while fitting to ensure tight packing and to maintain desired alignment. Views of completed gabion revetments are shown in Figure 5-32.

Articulated Blocks: There has been a rapid development of articulated, precast concrete block revetments during the past decade. The blocks that have been designed and patented for this purpose differ in shape and method of articulation, but share certain common features. These features include flexibility, rapid installation, and provision for installation of vegetation within the revetment. Some block systems are prebonded to a filter cloth to facilitate placement and to eliminate the extra step of placing a filter course beneath the revetment. Examples of articulated blocks are illustrated in Figure 5-33.

(*a*)

(*b*)

Figure 5-30. Views of rock revetments. (*a*) Riprap protecting river bank. (*b*) Quarry stone protecting coastal foredune.

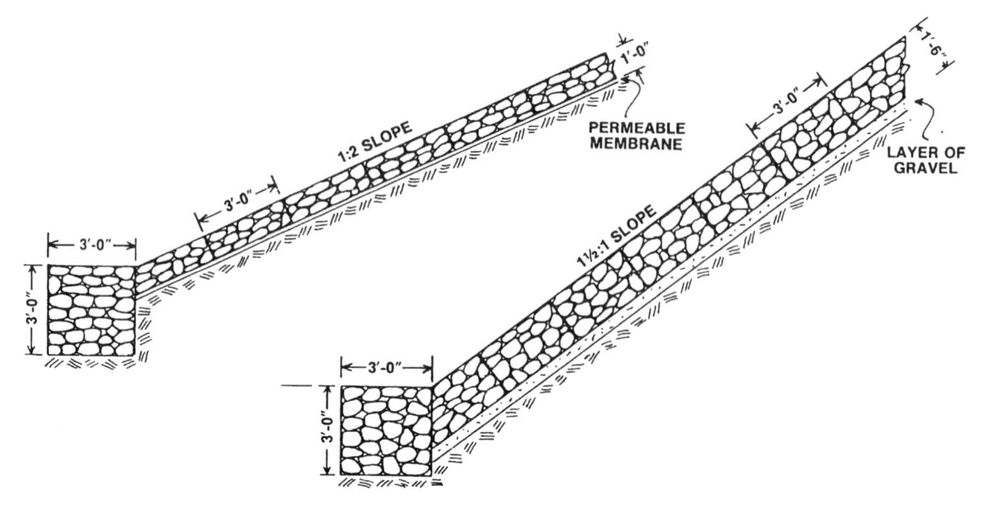

Figure 5-31. Schematic illustration of gabion revetment with toe-wall in lieu of apron.

5.4.3 Stability Requirements

A revetment must be capable of armoring or protecting a slope and resisting displacement from tractive stresses exerted by high-velocity water flows and/or wave action. It must allow for the exfiltration of water from the bank without permitting the movement of fines through the armor layer. Furthermore, a revetment should be imbedded sufficiently at its lower end to defend against scour and undermining. Finally, a revetment must be extended high enough up the slope to prevent overtopping from wave runup or conversely be equipped with an apron to mitigate the effects of such runup.

The following design guidelines (U.S. Army Corps of Engineers, 1981a) apply to quarry stone or rock riprap revetments (Figure 5-29).

Armor Weight: As noted previously the main focus of this section is on the use of revetments to protect sandy coastal landforms (beaches, berms, bluffs) against wave action. Revetments protect coastal slopes and resist movement largely because of the weight and interlocking characteristics of individual armor units. The required weights are very sensitive to wave height, varying with the cubic power of the wave height. Other important parameters include the bank slope, and density/roughness properties of the armor units. The following equation can be used to estimate the required weight (W) of individual armor stones in relationship to the design wave height:

$$W = \frac{w_R H^3}{K_d (s_R - 1)^3 \cot \theta} \tag{5-14}$$

(*a*)

(*b*)

Figure 5-32. Views of gabion revetments. (*a*) Gabion mattress protecting river bank. (*b*) Gabion revetment system protecting highway cut slope.

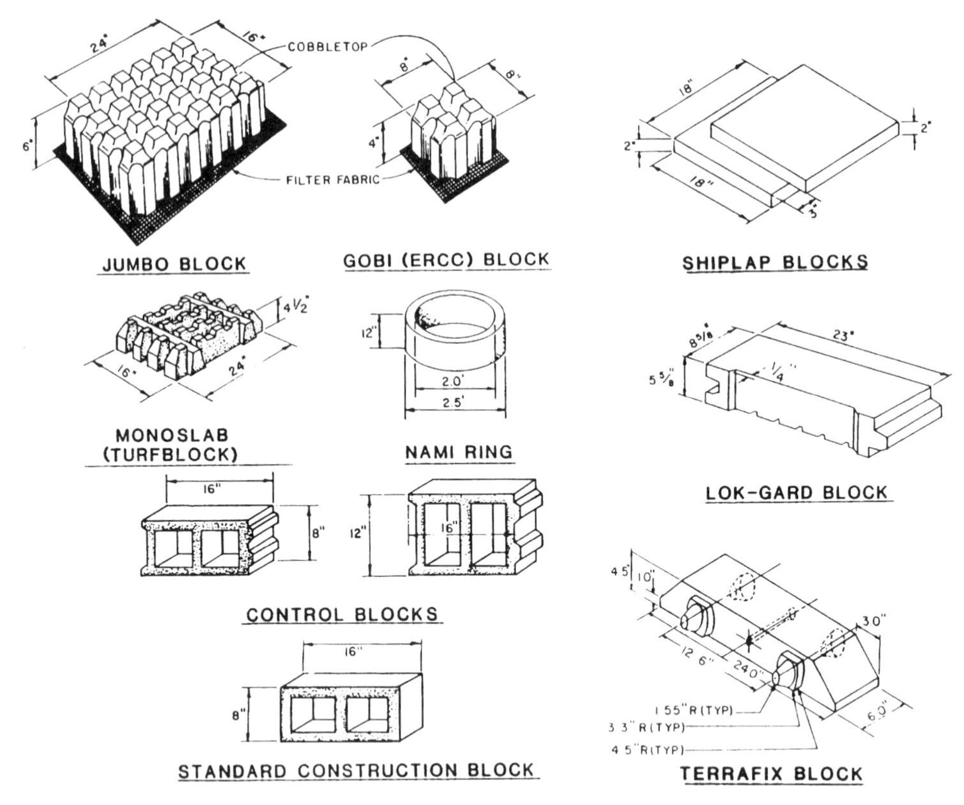

Figure 5-33. Examples of precast concrete block units for revetment. (From U.S. Army Corps of Engineers, 1981b.)

where: W = weight of individual armor stones (pounds)
w_R = unit weight (saturated, surface dry) of rock (pcf)
H = design wave height (feet)
s_R = specific gravity of armor stone
$\cot\theta$ = slope of bank ($H:V$ ratio)
K_D = stability or roughness coefficient (see Table 5-5)

The underlayer stones (Figure 5-29) are used to provide a transition from the armor layer to the filter course. They should be sized at one tenth of the calculated armor stone weight.

Range of Allowable Stone Weights: The recommended range for both the armor and underlayer is $0.75W$ to $1.25W$ with 75 percent of the stones weighing more than W. All stones should be sized so that no side is greater than three times its least dimension.

TABLE 5-5. Roughness Coefficients for Various Types of Armor Stone

Armor Unit	K_d
Quarry Stone	
Smooth, rounded	2.1
Rough, angular	3.5
Graded riprap	2.2

TABLE 5-6. Wave Runup Heights for Revetments and Seawalls

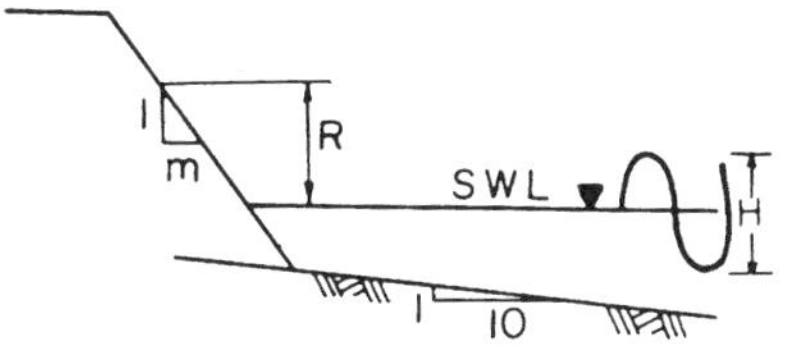

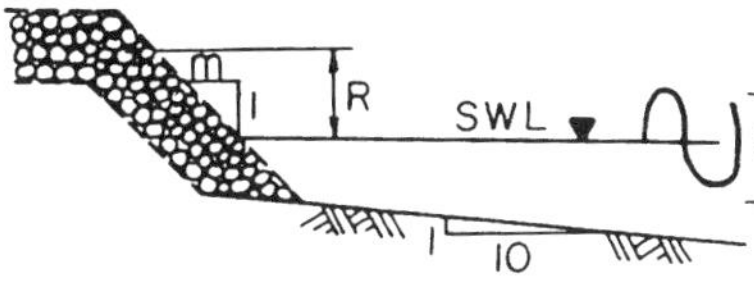

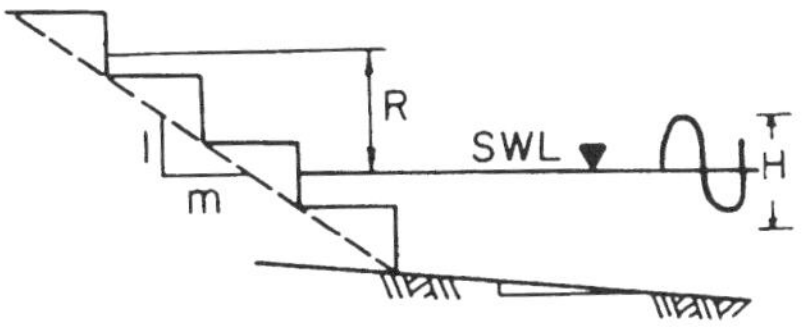

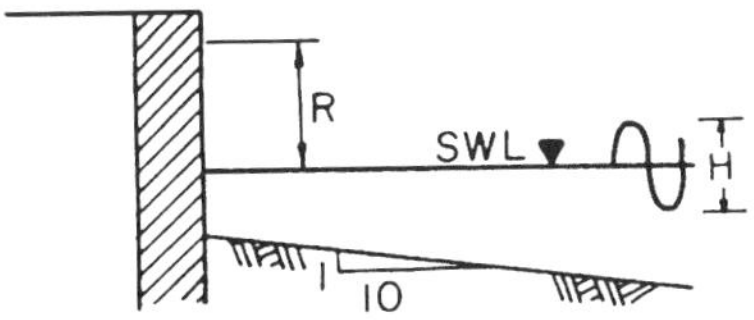

	m	R
Smooth Face (SWL = still water level)	1.5 2.5 4.0	2.25H 1.75H 1.50H
Rough Face	1.5 2.5 4.0	1.25H 1.00H 0.75H
Stepped Face	1.5	2.00H
Vertical Face	—	2.00H

Source: From U.S. Army Corps of Engineers (1981a).

Toe and Flank Protection: The toe should be buried (i.e., extended beneath the bottom of the slope) at least one design wave height, as shown in Figure 5-29. Add an additional layer of armor stone to thicken the toe section. The revetment should also be tied into the existing bank on either side to prevent flanking.

Runup Calculations: A revetment should be carried high enough up the bank or slope to prevent overtopping. Runup heights can be estimated as a function of design wave height and bank slope for different types of revetments from Table 5-6. If it is not possible to extend the revetment to the calculated runup height, then a splash apron must be used.

5.5 REFERENCES

AWPI (1981). *Standard Designs for Treated Timber Cribbing*. Washington, DC: American Wood Preservers Institute.

Bonaparte, R., R. D. Holtz, and J. P. Giroud (1987). Soil reinforcement design using geotextiles and geogrids: in *Geotextile Testing and the Design Engineer*, ASTM STP 952, edited by J. E. Fluet, Jr., American Society for Testing Materials, Philadelphia, pp. 69–110.

Bowles, J. E. (1977). *Foundation Analysis and Design*. New York: McGraw-Hill.

Caughey, R. A., C. W. Tooles, and A. C. Scheer (1951). Lateral and vertical pressures on granular materials in deep bins. Engineering Experiment Station, Bulletin 172, Iowa State University.

Federal Highway Administration (1974). Standard specifications for construction of roads and bridges on Federal highway projects, FP74. Washington, DC.

Gray, D. H., and A. T. Leiser (1982). *Biotechnical Slope Protection and Erosion Control*. New York: Van Nostrand Reinhold.

Hilfiker, W. K. (1972). Reinforced Concrete Cribbing. U.S. Patent #3,631,682 (January 4, 1972).

Hilfiker, W. K. (1978). Fabric Structures for Earth Retaining Walls. U.S. Patent #4,117,686 (October 3, 1978).

Jaecklin, F. P. (1983). Retaining beautifully. *Architect and Builder* (May).

Koerner, R. M. (1990). *Designing with Geosynthetics*. 2d ed. Englewood Cliffs, NJ: Prentice-Hall.

Lambe, T. W., and R. V. Whitman (1969). *Soil Mechanics*. New York: Wiley.

LeFlaive, E. (1982). The reinforcement of granular materials with continuous fibers. *Proceedings*, 2nd International Conference on Geotextiles, Las Vegas, NV, Vol. 3, pp. 721–726.

Leshchinsky, D., and A. J. Reinschmidt (1985). Stability of membrane reinforced slopes. *Journal of Geotechnical Engineering* (ASCE) **111**(11): 1285–1300.

McKittrick, D. (1978). Reinforced earth: Application of theory and research to practice. *Proceedings*, Symposium on Soil Reinforcement and Stabilizing Techniques in Engineering Practice, The New South Wales Institute of Technology, Sydney, Australia, October 16–18.

Mitchell, J. K., and W. C. B. Villet (1987). Reinforcement of earth slopes and embankments. National Cooperative Highway Research Program Report No. 290, Transportation Research Board, Washington, DC, 320 pp.

Palmer, W. D. (1987). Holding back to the earth. *Concrete International* (Nov.): 26–33.

Randall, F. A., and M. Wallace (1987). The many types of concrete retaining walls. *Concrete Construction* **32:** 589–597.

Schwarzhoff, J. C. (1975). Retaining wall practice and selection for low-volume forest roads. Transportation Research Board Special Report No. 160, pp. 128–140.

Schuster, R. L., W. V. Jones, R. L. Sack, and S. M. Smart (1973). A study and analysis of timber crib retaining walls. USDA Forest Service Final Report No. PB-221-447, 186 pp.

Smoltczyk, U., and K. Malcharek (1982). Living sheets on steep slopes. *Proceedings*, 22nd International Conference on Geotextiles, Las Vegas, NV, Vol. 1, pp. 253–257, published by Industrial Fabrics Association, Minneapolis, MN.

U.S. Army Corps of Engineers (1978). The Streambank Erosion Control Evaluation and Demonstration Act of 1974, Interim Report to Congress, Department of the Army, Washington, DC.

U.S. Army Corps of Engineers (1981a). Low cost shore protection: Manual for engineers and contractors. USACE (DAEN-CWP-F), Section 54 Program, Washington, DC.

U.S. Army Corps of Engineers (1981b). Low cost shore protection: Final report on the shoreline erosion control demonstration program, Office, Chief of Engineers, Washington, DC.

U.S. Department of Navy (1971). Design Manual DM-7 for Soil Mechanics, Foundations and Earth Structures, Naval Facilities Engineering Command, Washington, DC.

Vidal, H. (1969). Reinforced Earth. U.S. Patent #3,421,326 (January 14, 1969).

Washington State Department of Highways (1973). Standard Plans for Bridge and Road Construction—1973, Olympia, WA.

White, C. A., and A. L. Franks (1978). Demonstration of erosion control technology, Lake Tahoe region of California. California State Water Resources Control Board Final Report, Sacramento, CA, 393 pp.

6 Vegetative Components and Considerations

6.0 INTRODUCTION

6.0.1 Basic Requirements for Successful Revegetation

Like most endeavors successful revegetation requires a clear understanding of appropriate design, execution, and follow up. It merits the same level of professional attention and diligence accorded to such engineering activities as designing and building retaining structures. Successful revegetation begins with site analysis, which entails assessing all the important variables that might affect the growth and development of vegetation on a site. Site analysis includes examination of the climate, native vegetation, and microsite parameters such as soils, topography, aspect, and possible toxic conditions.

Revegetation can be accomplished with a variety of plant materials and planting methods. Different selection criteria can be used to decide which type of vegetation—trees, shrubs, or grasses—is best suited for a given situation. Each type has its advantages and limitations. Some consideration must be given at the outset to the use of native versus exotic species and the implications of plant succession. Regulatory and other constraints, such as availability and cost, may also affect the plant selection. Vegetation can be introduced and established in several ways. Depending on site conditions, stabilization objectives, cost, and so on, one can choose from seed mixes, transplants, live cuttings, or a combination.

Site preparation is critical to successful revegetation and includes such activities as grading and drainage control. Grading, or shaping of the surface, plays a very important role in drainage. Landform grading practice is usually preferable over conventional grading. Landform-graded slopes are characterized by a continuous series of concave and convex forms interspersed with swales and berms that grade into the profiles. Revegetation in conjunction with landform grading entails planting vegetation in patterns that occur in nature as opposed to specifying either uniform or random coverage. Vegetation will not thrive in soils that lack sufficient aeration and porosity, nor will it do well in sites that are too wet or too dry. Various techniques can be used to improve soil conditions and control both surface and subsurface drainage in order to enhance plant establishment and growth.

Live cuttings that are gathered locally from nearby sites provide the main source of plant material for soil bioengineering construction. Accordingly, this chapter gives special attention to this topic. Procedures for the establishment of vegetation by conventional methods, that is, seeding and transplanting, have been published elsewhere in detail (see Gray and Leiser, 1982; Coppin and Richards, 1990). Readers interested in more detailed information on conventional methods of vegetation establishment are referred to these or other published sources. With regard to the successful establishment of live cuttings, substantial attention must be given first to the acquisition and handling of indigenous cuttings. This includes procedures for harvesting, transporting, and storing cuttings. Proper fabrication and installation procedures must also be followed if cuttings are to grow and flourish. This requires consideration of timing, soil testing, and preparation of the ground for planting. Successful installation and establishment of live cuttings also requires a well-defined procedure of inspection, monitoring, and maintenance.

Revegetation of structures, for example, revetments, breast walls, and retaining structures, presents special problems and requirements. The same principles of plant selection, quality, planting times, and handling still apply, but additional consideration must also be given to engineering requirements imposed on the earthen backfill in or behind these structures. Several techniques can be adopted to satisfy both requirements.

6.0.2 Constraints on Revegetation

Physical: Physical constraints may limit revegetation in several ways. Terrain in steep mountainous areas may pose problems for creating ideal cut and fill slopes for revegetation. When revegetative/soil bioengineering means are used in combination with structural measures, special design considerations may be required in the planning and execution stages. Physical limits often exist due to land availability, where because of cost or other reasons suitable adjoining land cannot be acquired. Under such circumstances steeper, more difficult land may have to be used and more costly solutions considered. In response to regulatory requirements, slopes are often steepened to avoid encroachment onto other areas such as wetlands.

Drainage may also cause problems and constrain optimum revegetative solutions. Engineering requirements for free-draining backfills behind retaining structures may clash with vegetative requirements for sufficient fines content in the backfill to provide moisture and nutrient holding capacity. Several ways for resolving this potential conflict are discussed in Chapter 8, including modification of backfill specifications and the use of amended zones.

Access and viewing requirements constitute yet another form of physical constraint on revegetation. Homeowners and others often insist on unimpeded views. Levee inspection and flood-fighting operations require unobstructed access and viewing fields. These considerations are discussed further in Chapter 3.

Regulatory: Regulatory agencies may require the exclusive use of certain species, such as native plants, and ban the use of other species that otherwise might be excellent for stabilization purposes and in abundant supply nearby. Height restrictions might be imposed on the vegetation growth where overlooks and vistas are desired. Site distances, especially on inside curves of highways and railroads, typically require that the vegetation be installed higher up the slope face to ensure that adequate site distances are maintained.

The issue of using native versus introduced (or exotic) species is an important regulatory consideration. Many government agencies, such as the national and state parks, require that only native plant species be used. This requirement is becoming more common, especially in riparian corridors. Specific species may also be banned because they serve as hosts to diseases or are considered too invasive or are considered noxious weeds. This is often the case with introduced species that have become naturalized. Some of these introduced plants may have been around for over forty years and are difficult, therefore, to distinguish from native plants. While natives are normally appropriate for rural landscapes, they often are considered too wild and unkempt looking for urban settings. This is especially true in highly manicured city parks and along roadways. Additional discussion is included in this chapter (see Section 6.3.5) on the use of exotics versus native plants. It is important in this regard that the long-term vegetation appear right for its locale and that maintenance requirements be well understood.

Other regulatory issues deserve mention as well. When seeding is considered as a means of vegetation establishment, the seed source may come under scrutiny. Some areas require that seeds come from the same providence. Strictly speaking, this means that the seed must come from sources in the immediate area. The Endangered Species Act may affect project designs. In order to protect habitats of endangered species, the Act specifies that construction noise not occur near such areas during mating, nesting and rearing periods. Finally, fertilizers, herbicides, and pesticides may be restricted in some areas. It will be necessary to check local regulations on their use, especially near lake, river, and other drainage systems.

Economic: The costs of revegetation are strongly affected by the time of year the work is done. Typically, the least expensive and most effective biotechnical and soil bioengineering structures are installed during the dormant season. This is a time of year when plant materials (foliage and shoots) are not growing actively. When of necessity projects must be constructed out of the dormant season, they are usually more costly and sometimes less effective. Examples include large highway projects where cut and fill slopes are being constructed the year round. Alternatives may include installing temporary measures until the dormant season returns, preharvesting and refrigerating live cuttings until needed, redesigning the revegetation plan to used rooted stock, or a combination of conventional revegetation and soil bioengineering. These “out of season” or “split season” alternatives require additional temporary measures that increase costs.

Revegetation costs are also affected by the amount of advance planning.

Costs are minimized when sufficient time is allowed to ensure that the most desirable species and needed quantities will be available. Normally the best and least expensive procedure is to install the vegetative components of a project at the same time or in some cases immediately following the earthwork.

6.1 SITE ANALYSIS

Site conditions affect the performance of any biotechnical or soil bioengineering treatment; these conditions must be investigated and understood at the outset so that an appropriate stabilization treatment can be selected and designed. A site investigation entails a review of existing information about climatic, topographic, botanical, and other factors likely to affect revegetation. A site analysis should include information about microclimate, soils, topography, and surrounding vegetation. Topographic data are best obtained from a recent survey and slope cross sections. Soils data can be obtained from borings and soil samples collected for gradation analyses, pH and nutrient tests, and other soil properties affecting plant growth, such as permeability and density.

6.1.1 Climate

A number of climatic variables affect plant growth and should be evaluated. These variables include the following:

- Air temperature averages and fluctuations (including maxima and minima)
- Maximum ground surface temperature
- Length of growing season (taking altitude, latitude, snow cover period, and frost depth into account)
- Rainfall total and seasonal distribution
- Drought duration and time of year

These data are usually available from the National Weather Service. Other more specific data can often be obtained locally. Microsite climate may differ substantially from regional averages and this information may be critical for the plant species selected and the method of installation or propagation.

6.1.2 Native Vegetation

Live plants and live branch cuttings obtained from the vicinity of the project site are usually the most suitable. In natural or rural areas acceptable native species should be selected in preference to exotics. Native vegetation is influenced by such parameters as slope, aspect, climate, elevation, and soil type; therefore, the closer the parameter match between project and harvest site, the better the chance of establishing native vegetation and of encouraging the natural invasion of the surrounding plant community.

Other factors favoring the choice of native vegetation gathered in the wild include regulatory constraints against the use of exotics or introduced species (see Section 6.0.2), a shortage or absence of nurseries growing the desired stock, and possibly a limited knowledge of propagation methods. Factors limiting the use of native plant matrials include restrictions on their removal, and limited availability of nearby, matched natural stands for cutting.

6.1.3 Microsite Parameters

Microclimate: Temperature, rainfall, and sun exposure can vary dramatically depending upon aspect or which direction a site faces. In the Northern Hemisphere, the temperatures on north- and northeast-facing slopes are normally lower than those on south- and west-facing slopes. Cooler temperatures on the north- and northeast-facing slopes reduce evapotranspiration and soil moisture stress on plants. The converse follows on the hotter south- and west-facing slopes. These temperature differences affect the type of vegetation that is selected and also the method of placement or installation. Hotter and drier southern and southwestern aspects may require drought-tolerant vegetation as well as soil bioengineering methods that are installed relatively deep in the ground. Drier sites also require that available water be retained on the slope. Deeply installed soil bioengineering systems, such as brushlayering constructed on contour, often satisfy these requirements and objectives. On the other hand, an aspect that receives large amounts of precipitation may be better suited to systems that are placed at relatively shallow depths and oriented on the slope to enhance surface drainage, such as live fascines installed off contour.

Wind exposure also affects microclimate conditions. In windy areas evapotranspiration will be increased, thereby increasing soil moisture stress on plants. Shade along the edges is also another important microsite consideration. Shade-intolerant vegetation should not be planted in such areas.

Topography: Valleys and low areas will have different microsite conditions compared to high areas or topographic divides. The lower areas often experience cooler soil temperatures and have more soil moisture. Plantings should be selected with this difference in mind. Normally more drought-tolerant vegetation such as grasses do better in drier interfluves or high up on a slope, while shrubs and trees are better suited for wetter drainage swales and valley areas. These considerations are integral to revegetation associated with landform grading practices (see Chapter 4).

Soils: Both physical and chemical properties of soils should be analyzed to determine the soil type and its ability to support plant growth.

Physical properties include:

- Grain size distribution and percent fines (fraction passing #200 sieve)
- Soil structure or texture
- Density or degree of compaction
- Depth to hardpan or impervious layer
- Water repellency
- Soil moisture content

Chemical properties include:

- Plant nutrients (concentrations and availability)
- Soil pH (alkalinity/acidity)
- Soil water salinity (ionic strength)
- Exchangeable sodium
- Toxins (type and amount)

Soil nutrient tests are required prior to proceeding with revegetation plans. These tests should be performed by a competent analytical testing laboratory. Local testing labs should be familiar with soils in the area and are usually able to offer suggestions with regard to needed soil amendments to take care of nutrient deficiencies or other problems such as unfavorable pH, high exchangeable sodium content, or elevated salinity levels.

Soil types range from loose, unconsolidated, sandy gravely materials to heavy, plastic clays and silty clays. Neither extreme is good for supporting plant growth. The former tends to be deficient in fines, lacks nutrients, and has a poor water holding capacity. The latter is poorly aerated, exhibits high soil moisture suction, and impedes root growth. The best soils for plant growth are loamy soils, including silt loams and sandy loams. These soils have sufficient fines for adequate water and nutrient retention while exhibiting good aeration and water transmission properties. The degree of soil compaction is also important. Dense, overcompacted soils tend to have low hydraulic conductivities, poor aeration, and high resistance to root penetration.

6.2 SEEDS AND PLANTING STOCKS

The availability, quality, and quantity of seeds and planting stocks are very important considerations in revegetation designs. To be assured of high standards and specific plant species, specifications are required. In some cases it may be advisable to visit commercial growers to inspect and evaluate their products. The American Association of Nurserymen has developed standards for nursery crops, including bare-root, balled, and burlapped container-grown plants as well as small liner plants (Gray and Leiser, 1982).

6.2.1 Seeds

When specific seed mixes are required for a project they must be either obtained from a commercial source or collected in the wild. The latter is a difficult undertaking and must be done by knowledgeable people. Natural plants differ widely in the quality and quantity of seed production from year to year. Many variables need to be considered, namely: (1) plants may bear seed only in alternate years, (2) weather variations such a drought or unseasonable frosts can affect seed production, (3) seeds may be eaten by birds and other wildlife or be naturally dispersed by wind before they can be collected.

After collection seeds must be carefully cleaned, handled, and stored. Different seeds have different storage requirements with regard to moisture, temperature, light, and oxygen. Individual species also have specific dormancy periods and different storage lives.

6.2.2 Transplants

Transplants are often used alone or in conjunction with other stabilization systems to revegetate a site. Good quality bare root stock that is properly handled, installed, and maintained is usually very successful. It is important that the quality be carefully specified and that the stock be checked when it arrives at the project site. When storage is necessary the maximum period should not exceed two weeks. Transplants should be healed in, watered, and kept in the shade and out of the wind to prevent drying.

6.2.3 Cuttings

Cuttings comprise the main live materials used in soil bioengineering construction. Cuttings are usually collected from the wild, although some commercial sources do exist. Several harvesting sites are normally located in advance when collecting from the wild. Cuttings are normally taken from stems 8 to 10 inches above the ground and tied in bundles that can be handled by one person.

The elapsed time between collecting and installing the cutting should be as short as possible. The fresher the collected material the better the chances for successful establishment. Cuttings that cannot be used immediately may be stored for a few days in water or moist soil. They should also be kept out of the sun and protected at all times from the wind. Additional information about the handling and installation of cuttings can be found in Chapter 7.

Cuttings used in soil bioengineering projects are normally taken and installed during the dormant season. If installation must be delayed, special refrigerated storage is necessary.

6.2.4 Vetiver Grass

Vetiver grass is mentioned separately because of its widespread use for erosion control and land reclamation in tropical and subtropical regions of the world.

Vetiver grass is usually planted close together in hedge-like arrays, on contour across a slope, where it acts as a barrier to downslope movement of soil and as a terrace builder.

Characteristics: Vetiver (*Vetivaria zizanioides*) is not a lawn or pasture grass, but rather a very coarse, tough bunch grass that grows to about 1 meter wide at its base with a clustered mass of dense stems. It is somewhat similar in appearance to citronella or lemongrass. Certain types of vetiver grass bear infertile seed and produce no spreading stolons or rhizomes, so they remain where planted. This may be an important consideration if restrictions or concerns exist about the introduction and spread of exotic, nonnative flora. Vetiver grass has a deep penetrating root system and good soil holding ability. It can survive on many soil types (including sands, shales, gravels, and even aluminum rich soils) with only minimal limitations with regard to fertility, acidity, alkalinity, or salinity. It is also capable of growing in a wide range of climates that range from 300 to 3,000 mm of rainfall per annum and from slightly below 0° to slightly above 50°C.

Uses and Applications: Vetiver grass has been widely employed in tropical and subtropical regions of the world for erosion control purposes with good results (NAS-NRC, 1993). It is used to control erosion in both agricultural settings and for engineering purposes, for example, for erosion control on cut and fill slopes along highways. The grass works best when planted in hedges or hedgerows on contour with the plants spaced approximately 15 cm apart. The recommended spacing between rows varies with the degree and length of slope. From an ecological perspective vetiver grass develops into a monoculture and does not encourage natural succession or biodiversity.

6.3 SELECTION AND SOURCE OF PLANT MATERIALS

6.3.1 Plant Material Spectrum

An immense variety of plants can be used for revegetation purposes; the trick is to decide which types of plants are compatible with site conditions and well suited for achieving stabilization goals. Plant materials are no different from other materials in the sense that they must be selected with care for their intended purpose. They differ substantially from inert materials, however, in the sense that they are constantly in a state of evolution and change. A set of plantings placed in the ground today may bear little resemblance to that set of plantings a decade later in terms of appearance and species composition. Each generation of plants slightly modifies the site and prepares the ground for a new generation that may include a new species.

Plants are very responsive to soil and site conditions; thus growth rates and ultimate species mix or composition depend very much on microsite parameters such as topography, soils, moisture availability and aspect, as noted previously. Nowhere is this more evident than in the coastal dune and bluff environment, where relative degrees of exposure to wind, salinity, and moisture availability determine the types of plants found growing, as shown in Figure 6-1.

6.3.2 Native versus Introduced Species

An often controversial consideration that must be addressed in revegetation efforts is the use of native versus introduced (or exotic) species. Reasonable arguments can be made in favor of both types of plantings. In some instances any choice is precluded by regulatory constraints (see Section 6.0.2). A good case can be made for the use of exotic species on highly disturbed slopes or man-made slopes, such as cuts and embankments, where infertile or highly modified soils are exposed. There may be few or no native species available that grow readily in such soils and site conditions. On the other hand, it may be possible to find exotic species that are well adapted to these conditions. These also may serve as a pioneer or nurse cover that sufficiently modifies the site so that native species can become established. However, caution must be exercised since exotics may spread uncontrollably, compete detrimentally with native species, and form unnatural and undesirable monocultures in the landscape.

The main advantage of using native species is that they are available locally and should be adapted to climatic and other site conditions. A site inventory will reveal what types of plants are thriving in the area and what microsite parameters affect the species mix and relative growth rates. This information can be used to select a spectrum of native plants for use in slope revegetation work.

6.3.3 Plant Succession

Plant succession refers to the variation or change in species composition on a site with time. This variation proceeds until a so-called climax or end composition becomes established that is then relatively stable. Removal of this climax cover by fire or harvesting triggers a new succession. Interruptions in the succession can also return the process to earlier evolutional stages or limit attainment of the climax stage.

Pioneer plants are the first to invade and establish. These pioneer plants modify the site sufficiently for plants in later successional stages to gain a toehold. Plant materials used in most soil bioengineering construction methods (see Chapter 7) belong in the early or pioneer stages of plant succession. One of the most important modifications from pioneer plants is the addition of organic material to the soil and protection against surficial erosion and shallow sloughing. Native woody vegetation and later successional stages may have great difficulty establishing on highly active slopes.

Pioneer plants ultimately may disappear from a site. It is interesting to revisit

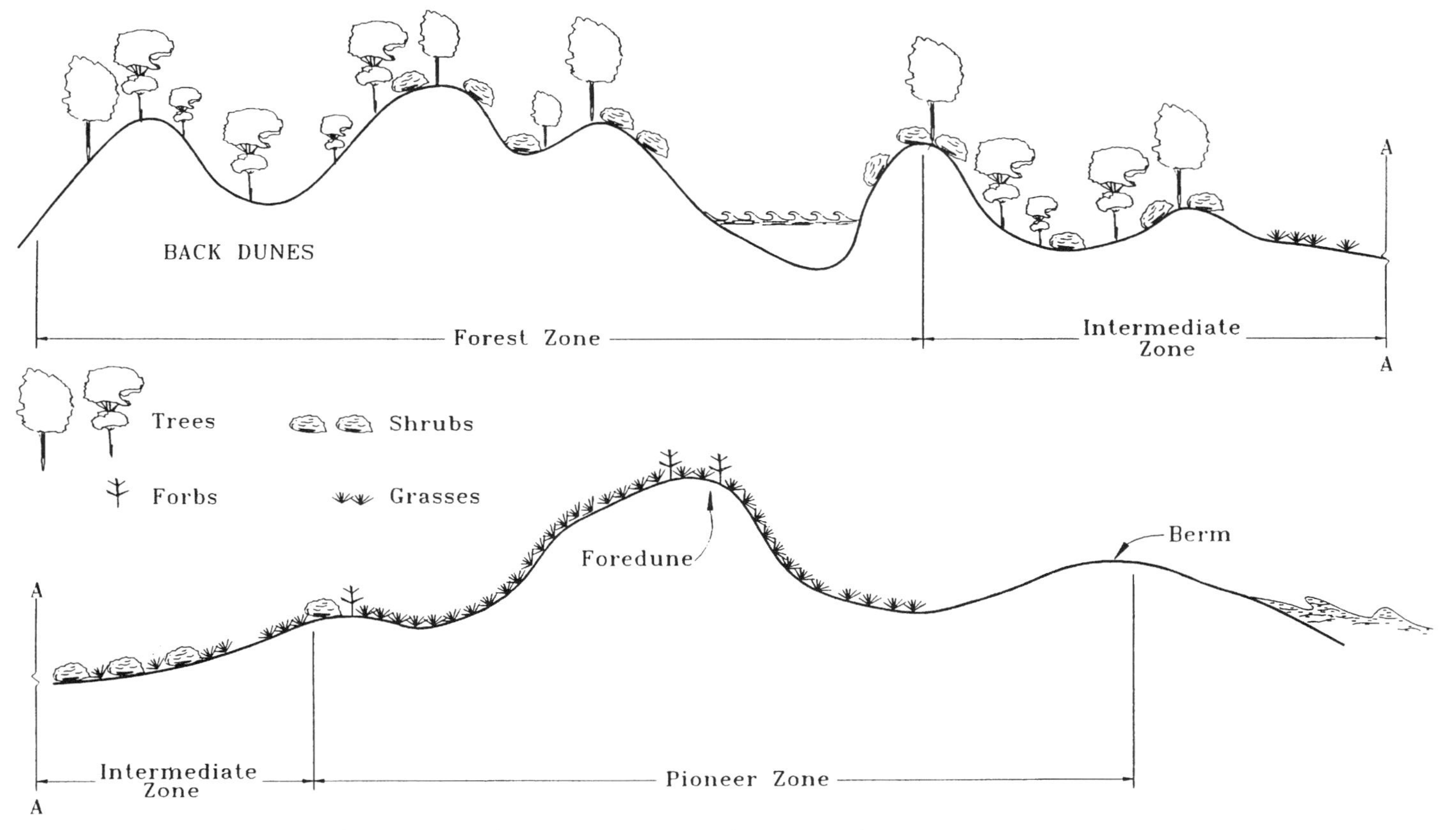

Figure 6-1. Zonation of coastal dune and bluff vegetation according to exposure to wind and salinity, nutrient status, and moisture availability.

a site where soil bioengineering methods have been employed. The site a decade later may bear little resemblance in either appearance or species composition to the initial planting and installation. Willows and alder are commonly used, for example, as the primary plant material in a soil bioengineering project. Few if any willows may remain, however, on the site years later after native vegetation has invaded and taken over. Kraebel's (1936) work on soil bioengineering stabilization of fill slopes along the Angeles Crest Highway in Southern California and Grizzly Peak Boulevard in the Berkeley Hills in Northern California are good examples. So too are coastal bluff sites along Lake Michigan near Benton Harbor that were stabilized by soil bioengineering methods (USDA Soil Conservation Service, 1940). Willow wattles or fascines were used to stabilize the bluffs, as shown in Figure 6-2. Four decades later there is a dense stand of native woody vegetation on the slope, few if any willows, and no evidence of erosion, as shown in Figure 6-3.

6.3.4 Selection Criteria

Site Conditions: Site conditions (see Section 6.1.3) have a profound influence on the choice or selection of plants for revegetation work. Native plants presently growing on the site or surrounding areas are likely good candidates

Figure 6-2. High coastal bluff near Benton Harbor, Michigan, stabilized by willow fascines that have rooted and sprouted in rows on contour along the slope. (From USDA Soil Conservation Service, 1940.)

Figure 6-3. Bluffs in willow fascine stabilization project 40 years later. The bluffs are covered by a dense stand of native woody vegetation. There are few if any remaining willows and no erosion on the site today.

because they are already adapted to these site conditions. Typically the plants used for soil bioengineering purposes are pioneer plants that invade and colonize disturbed land sites. Pioneer plants usually tolerate a wide range of soil conditions; however, soils with very high pH (>7.5) or very low pH (<4.5) require amelioration for plant survival and growth.

Site conditions can be classified in a systematic manner to characterize their suitability for plant growth and to assist in species selection. Table 6-1, adapted from Schiechtl (1980), may be helpful in this regard. The influencing factors and ratings in the table allow a site to be evaluated or classified in terms of its climate, soils, and erosion hazard. This is accomplished by considering each of the influencing factors for a specific classification area. As an example, suppose that a site has been classified as follows: soil (A4) poor, climate (B2) good, and erosion hazard (C3) medium. In this case it would be advisable to choose plant species that are capable of deep rooting, fast growth, and which do not require high fertility.

Stabilization Functions and Objectives: Plants serve various functions and play different roles in stabilizing slopes. These stabilizing roles or functions differ with respect to surficial erosion versus mass stability (see Chapter 3). Stabilizing or ground improvement functions of vegetation and associated desirable plant characteristics are summarized in Table 6-2. In general, "cover and

TABLE 6-1. Classification of Site Conditions for Revegetation

Area of Classification	Range of Classification	Influencing Factors
A. Soil conditions	1 = very good 2 = good 3 = medium 4 = poor (bad) 5 = very poor	Grain size distribution; permeability; texture; moisture and water retention; soil reaction (pH); fertility (especially soil nutrient levels); toxic materials; density (degree of compaction)
B. Climate	1 = very good 2 = good 3 = medium 4 = poor (bad) 5 = very poor	Amount and distribution of precipitation; humidity; evaporation (wind, sun); frequency of dry periods; duration of snow cover; average temperature and fluctuations; light conditions
C. Erosion hazard	1 = very good 2 = good 3 = medium 4 = poor (bad) 5 = very poor	Steepness (slope inclination); soil erodibility; weather (storm frequency and intensity); frost action; groundwater (seeps); geology (orientation of joint surfaces and bedding planes)

Source: Adapted from Schiechtl (1980).

armoring" are more important functions in the case of surficial erosion and "reinforcement and support" in the case of mass stability. Accordingly, a dense, low ground cover with a shallow, fibrous root mat is more effective in preventing surficial erosion, whereas, good rooting depth, strength, and a high root/shoot biomass ratio are more important in the case of mass stability.

The stabilizing functions and requirements of vegetation on streambanks and levees present a special case. These slopes are subjected to substantial scouring and surface tractive stresses. Both surface armoring and soil reinforcement are important functions. Stems with sufficient flexibility to bend over and cover

TABLE 6-2. Stabilizing or Ground Improvement Functions of Vegetation

Function	Desirable Plant Characteristics
Capture and restrain	Strong, multiple, and flexible stems; rapid stem growth; ability to re-sprout after damage; ready propagation from cuttings and root suckers
Cover and armor	Extensive, tight, and low canopy; dense, spreading, surface growth (e.g., grasses, legumes and forbs); fibrous root mat
Reinforce and support	Multiple, strong, deep roots; rapid root development; high root/shoot biomass ratio; good leaf transpiration potential
Improve habitat	Shade and cover to moderate temperature and improve moisture retention; soil humus development from litter; nitrogen fixation potential

the bank and deep enough roots to stabilize and reinforce the soil at depth are desirable characteristics.

Suitability for Slope Stabilization: Both herbaceous and woody vegetation can be used for slope stabilization. The former includes grasses and forbs; the latter shrubs and trees. The suitability of different types of vegetation for slope stabilization is summarized in Table 6-2. Each type has inherent advantages and limitations. In general, grasses and forbs are superior for preventing and controlling surficial erosion, whereas woody vegetation (shrubs and trees) are superior for preventing shallow slope failures or mass erosion.

As noted previously, streambanks are subject to both mass wasting and surficial erosion processes. Both armoring and cover are important functions in addition to deeper seated soil reinforcement in this case. Shrubs are generally best in this regard because they combine some of the desirable attributes of both trees and grasses while minimizing their respective liabilities (see Table 6-2).

Another way of capitalizing on advantages and compensating for liabilities of each vegetation type is by interplanting and species mixing. This is done to a certain extent even within a single vegetation type. Grass seed mixes for erosion control, for example, usually contain annual grasses for rapid establishment and slower growing perennials for long-term protection.

Suitability for Soil Bioengineering Construction: Plant species that are selected must be suitable for their intended use and be well adapted to the site's climate and soil conditions. Both rooted plants and vegetative cuttings are used in soil bioengineering construction. Species that root and propagate readily from cuttings are required for such measures as live fascines, brushlayers, live stakes, and branchpacking (see Chapter 7). Pioneer plants, such as willow, dogwood, and privet, generally work well.

Important attributes of plants under consideration for soil bioengineering use include availability, habitat value, size/form, root type, and ease of propagation from cuttings (*see Appendix 1*). Other important considerations include tolerance of plants to adverse site conditions such as drought, flooding, salinity, and deposition (*see Appendix 2*). These appendices (1 and 2) should be consulted in advance of selecting any plant species for soil bioengineering construction, slope stabilization, and/or site reclamation work.

6.3.5 Sources of Plant Materials

Seeds and Transplants: Both native seeds and rooted plants are available for purchase from a number of commercial growers and seed producers across the country. A large body of information has been published on seed and transplant sources that can be consulted by the reader (see Gray and Leiser, 1982).

Cuttings: Live cut plant materials for soil bioengineering work can be gathered in the wild, i.e., harvested from existing native growing sites. Alternatively, these materials can be obtained from commercial nurseries that stock suitable cultivars. Attributes of cuttings from these two sources are summarized below:

Native Sources: Live cut plant material can be taken from existing, healthy, native growing sites. Such sites may be found within a few feet of the proposed treatment site, or could be located upwards of fifty (50) miles away in some cases. Longer hauling distances require more careful project coordination and impose greater constraints on handling and storage. A different mix of species should be harvested whenever possible.

Nursery Sources: The USDA has released cultivars of dogwood and willow species to commercial growers and nurseries that may be suitable for soil bioengineering stabilization work. Plant materials in question include "Streamco" purpleozier willow (*Salix purpurea L.*), "Bankers" dwarf willow (*Salix X cottetii Kerner*), and "Ruby" redosier dogwood (*Cornus stolonifera Michx.*). These cultivars were initially selected by the Soil Conservation Service for outstanding performance as streambank stabilization plants. They are easy to clone and produce roots readily when cuttings are placed in moist soils.

6.4 SITE PREPARATION

6.4.1 Grading and Shaping

Grading and shaping play key roles in site preparation and successful slope revegetation. The importance of grading can be appreciated in the following checklist of conditions that must be met for successful establishment of vegetation on slopes:

- Grade the slope profile back to a stable angle and shape.
- Protect the toe of the slope against scour and undermining.
- Protect the surface of the slope against raindrop splash and frost action.
- Intercept and divert overland flow away from the top of the slope.
- Intercept and prevent seepage from emerging at the face of the slope.

Grading back to a stable angle for most slopes means a gradient or inclination of no more than $1\frac{1}{2}:1$ $(H:V)$. This inclination can be exceeded if certain soil bioengineering techniques are employed, such as slope gratings or vegetated geogrids (see Chapter 7). A slope angle $2:1$ $(H:V)$ or less is preferable to facilitate planting and establishment of vegetation.

Most engineered slopes are graded with a linear, planar, and unvarying gradient. This is not the most visually pleasing configuration; furthermore, evidence suggests (Schor and Gray, 1995) that it is not the most stable over time. Landform grading practice, on the other hand, is characterized by a continuous series of concave and convex forms interspersed with swales and berms that grade into the profiles (see Chapter 4). This type of grading also leads to a favorable modification of the hydrologic regime in the slope, which can also enhance establishment of vegetation.

Cut Slopes: Both transportation corridors and residential developments in steep terrain require that some excavation and regrading be carried out to accommodate roadways or building sites. Cuts should be rounded at tops and sides to eliminate overhangs, to blend with surroundings, and to provide a better environment for plant establishment. Terracing may be beneficial on long, steep stable cuts to help control surficial erosion. The terraces may be relatively small serrations or relatively large benches, as shown in Figure 6-4. The terraces slow down overland runoff and provide lodgment for seeds and plants. A midslope bench has the additional advantage of providing access or a location for the installation of horizontal drains or surface water collection ditches (see Figure 6-11).

Scaling or removal of boulders, stumps, and other large debris at the tops of slopes may be necessary on old cuts. Overhangs at the crest of a cut and all loose rocks should be removed, working from top to bottom. Large boulders should be broken into smaller pieces to avoid damage to the cut face below.

(*a*)

Figure 6-4. Use of terracing on cut slopes to control surficial erosion and assist plant establishment. (*a*) Serrated slope. (*b*) Benched slope.

(*b*)

Figure 6-4. (*Continued*)

Large rocks that have been scaled can be used to construct breast walls at the base. Other scaled material may be suitable for use as a backfill behind breast walls or gabions.

Smoothing of the slope face to eliminate rills and gullies may be required on old cuts. Rills and small gullies can be filled or obliterated during scaling operations. Larger gullies may need filling and treatment with fascines or branchpacking (see Chapter 7) to keep the gully fill in place.

Fill Slopes: Old fill slopes may require preparations similar to that used on old cut slopes, namely, scaling, smoothing of rills and minor gullies, and filling of larger gullies. A combined smoothing and reduction of gradient may be possible by importing additional fill and placing it in a wedge-shaped prism against the old fill. A frequent problem on old fills is overcompaction of soils near the face. While it is true that erodibility decreases with increased density (or decreased void ratio), overcompaction of soil also makes establishment of vegetation difficult if not impossible. The adverse influence of compaction on plant root development is shown in Figure 6-5.

Overcompaction can be corrected by scarifying the surface with a disk harrow or other mechanical device. This procedure may increase the chances of erosion in the short run, but will greatly improve the establishment of vegetation that provides the best protection in the long run. Furthermore, mulching, mulch blankets, or other temporary ground cover can be used (see Chapter 9) to enhance short-term erosion protection.

6.4.2 Drainage and Water Control

Surface Flow: Runoff or surface water flowing over cuts and fills is a common cause of erosion. Ditches, diversions, and berms can be used to control this water. An interceptor dike or berm (Figure 6-6) is a temporary or permanent ridge of compacted soil constructed at the top of cut or fill slopes. It intercepts and diverts overland flow away from unstabilized, unprotected, or newly constructed slopes. Dike interceptors are normally used as temporary or interim measures, but are also appropriate as permanent installations. They are

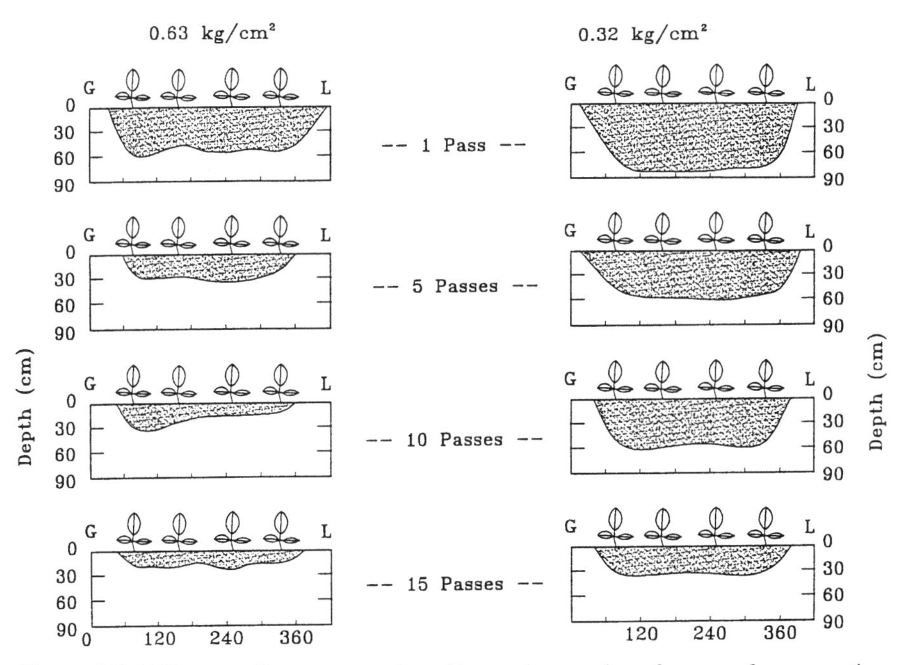

Figure 6-5. Effect on soil root penetration of increasing number of passes of a compaction plant over the soil prior to seeding. (From Cassell, 1983.)

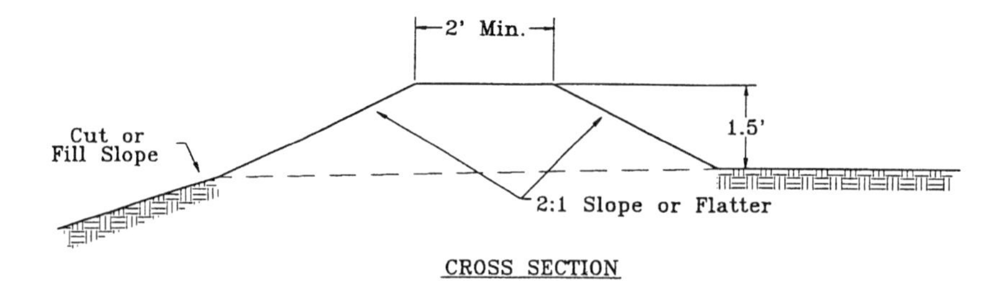

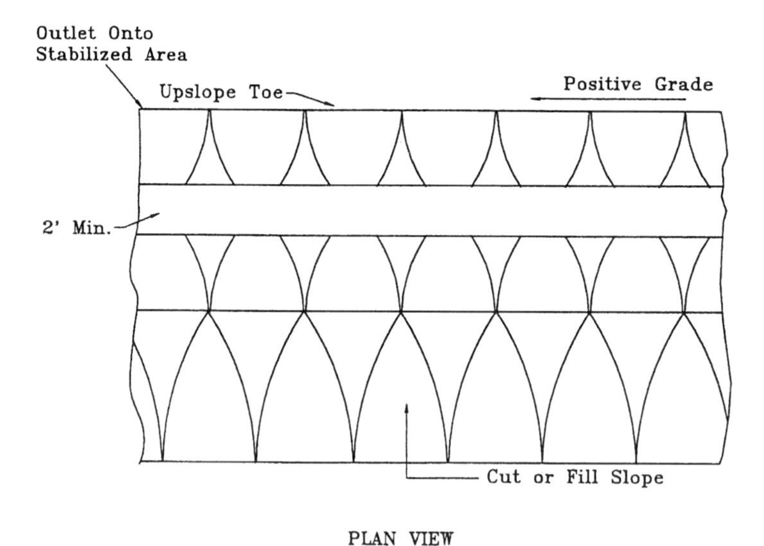

Figure 6-6. Schematic diagram of dike interceptor for protecting cut or fill slopes from surface runoff.

effective when used as roadberms in combination with flexible downdrains to protect the faces of newly constructed embankments from water flowing off the roadway surface, as shown in Figure 6-7.

A diversion is a channel, with or without a supporting ridge on the lower side, constructed across the top of a slope or across any sloping land surface. A diversion is a relatively permanent structure whose purpose is to divert surface runoff away from critical areas and to transfer sediment-laden runoff to sites where it can safely be disposed (e.g., to a sediment basin).

Ditches or channels should be sized to handle maximum expected flows and also be stable against the erosive tractive forces of the flow. Vegetation in the channel (primarily grass) can provide this needed protection if the flow velocity is not excessive nor of long duration. Table 6-3 gives permissible veloc-

Figure 6-7. Use of roadway berms in combination with flexible downdrains for protecting fill embankment from water flowing off roadway surface.

TABLE 6-3. Permissible Velocities for Vegetative Protection in Channels and Temporary Waterways

Cover	Slope Range (%)	Permissible Velocity[a] (fps) Erosion-Resistant Soils	Easily Eroded Soils
Reed canary grass	0–5	7	5
Tall fescue	5–10	6	4
Kentucky bluegrass	Over 10	5	3
Grass-legume[b] mixtures	0–5	5	4
	5–10	4	3
Annuals[c]			
Sudangrass, small grain (rye, oats, barley)	0–5	3.5	2.5
Red fescue	0–5[d]	3.5	2.5

Source: From USDA Soil Conservation Service (1975).

[a]Velocities may exeed 5 fps only where good vegetative cover and proper maintenance can be obtained.
[b]Do not use on slopes steeper than 10 percent, except for side slopes in a combination channel.
[c]Annuals are used as temporary protection until permanent covers are established.
[d]Do not use on slopes steeper than 5 percent, except for side slopes in a combination channel.

ities for channels lined with grass vegetation. Velocity in a ditch or channel can be controlled by limiting the channel gradient and by selecting an appropriate cross section. If velocities are high and/or of long duration, it may be necessary to augment or reinforce the vegetative cover with a "soft" or "hard" armor system. The former includes erosion control revegetation mattresses and turf reinforcement mats; the latter, gabion mattresses and articulated concrete blocks. Descriptions of these augmented ground cover systems and guidelines for their use are given in Chapter 9.

Energy Dissipation: Energy dissipation is required where the velocity of overland flow is excessive or where flow of water is concentrated in erodible areas. The latter is sometimes a problem near the outlet or discharge point of ditches and diversions. Energy dissipation can be accomplished either by spreading the flow out or by increasing the amount of turbulence and friction. The former is achieved by means of level spreaders; the latter by the use of rock layers and baffles. A level spreader (Figure 6-8) is an outlet constructed at zero grade

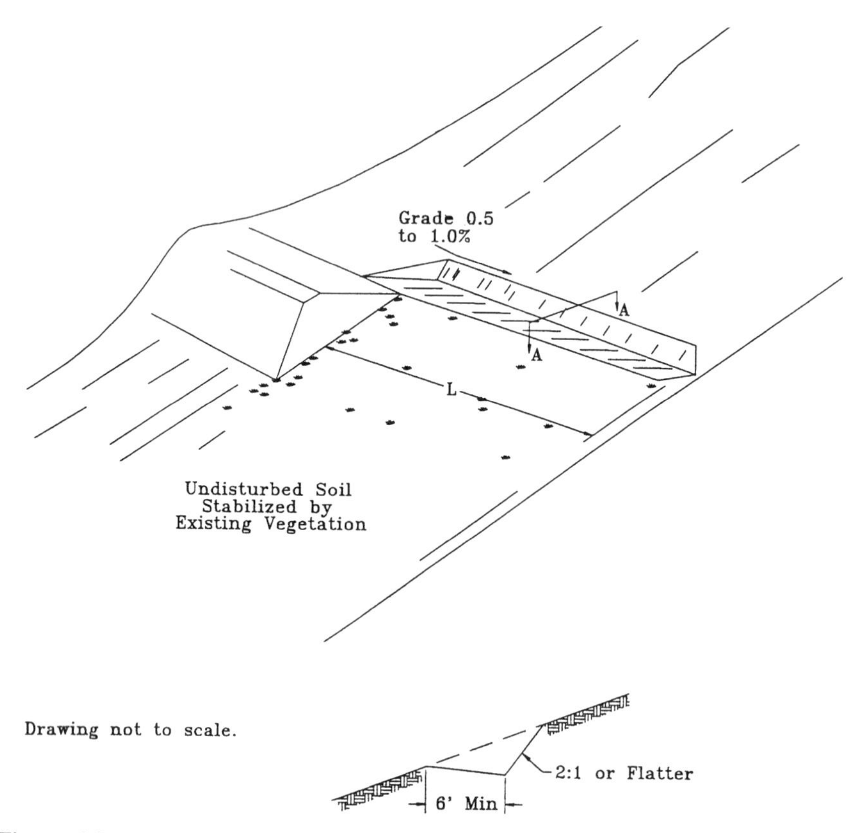

Figure 6-8. Schematic diagram of level spreader used to convert concentrated flow to noneroding sheet flow.

across a slope where concentrated runoff may be spread at noneroding flow velocity over slope areas already stabilized by vegetation.

Subsurface Flow: Seepage on cut slopes may be a major cause of slumping and gullying. The seepage usually emanates from water-bearing strata. Conversely, water infiltrating at the top of the slope may flow vertically downward and then move laterally toward the face of the slope when encountering a soil horizon of lower hydraulic conductivity. Seepage water daylighting at the face can trigger piping and erosion, which can eventually lead to the formation of major gullies, as shown in Figure 6.9.

Shallow seepage can be controlled by using trench drains. A trench drain is typically 2 to 4 feet deep and 18 to 24 inches wide; it intercepts shallow seepage and conducts the water away from critical areas such as the head of steep cut slopes. Several variants of a trench drain design are shown in Figure 6-10. The simplest design is a so-called French drain, which consists of a trench that is backfilled with coarse aggregate. Greater drain efficiency can be achieved by laying a 6-inch perforated collector pipe in the bottom and backfilling with a graded aggregate. Alternatively, a filter fabric and coarse, open-stone combination can also be used.

Deep-seated seepage can be intercepted and diverted by using horizontal drains, slotted or perforated pipes driven or jetted into a slope, as shown in Figure 6-11. Horizontal drains also improve the mass stability of a slope by

Figure 6-9. Major gullies in regraded river bluff formed as a result of seepage erosion at the face of the slope.

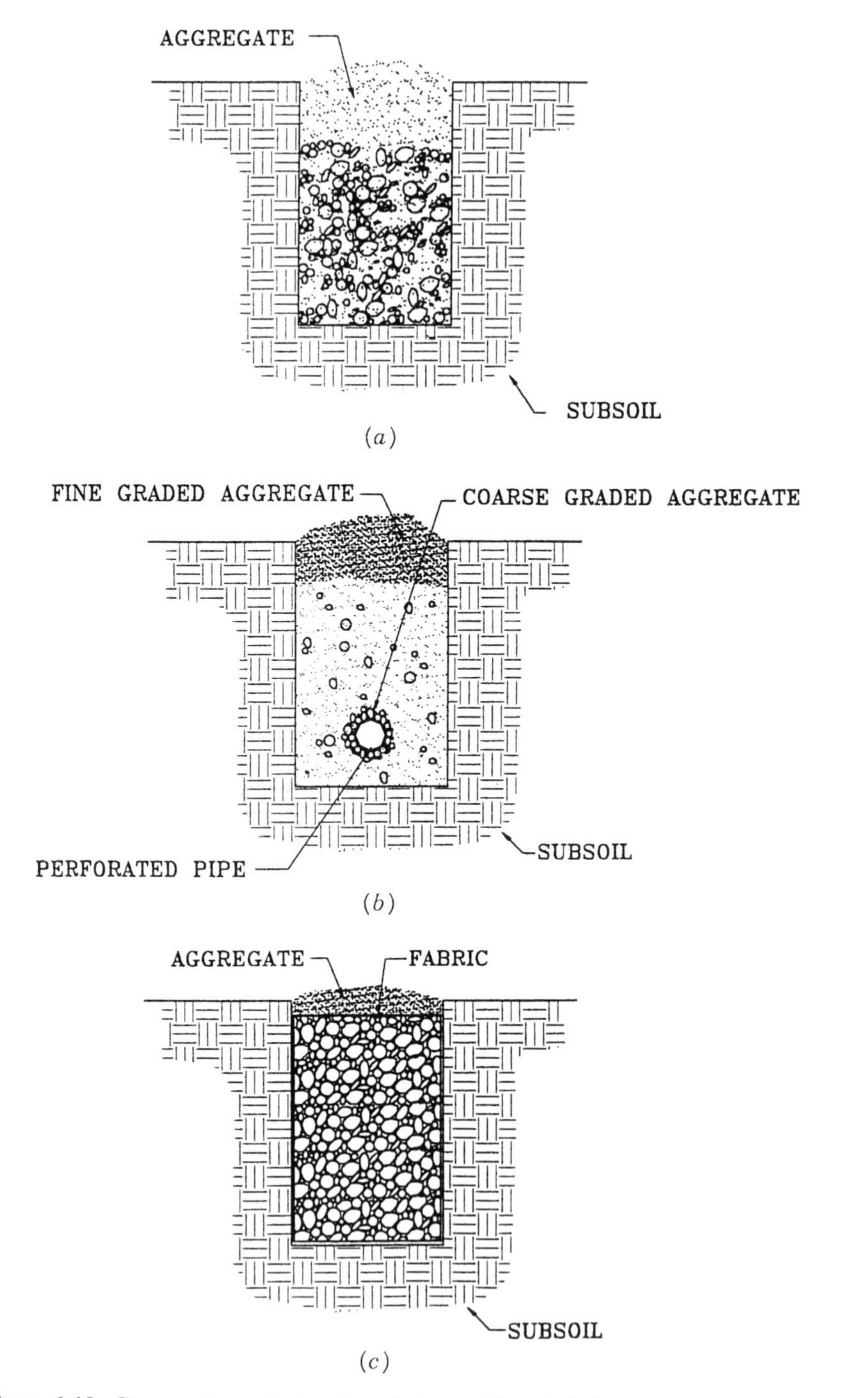

Figure 6-10. Cross sections of subsurface drains. (*a*) French drain. (*b*) Conventional trench drain with graded aggregate backfill. (*c*) Trench drain with filter fabric.

(*a*)

(*b*)

Figure 6-11. Horizontal drain installations. (*a*) At base of highway cut, discharging to ditch. (*b*) On midslope bench, discharging to collector pipe.

relieving pockets of hydrostatic pressure. Outlets of horizontal drains should extend to protected areas such as gutters or culverts. Information on the planning and layout (spacing, depth, grade, and so on) of horizontal drain installations has been published by Smith and Stafford (1957).

6.5 ACQUISITION AND HANDLING OF INDIGENOUS CUTTINGS

6.5.1 Harvesting and Gathering

Chainsaws, bush axes, loppers, and pruners are recommended for cutting living plant material, as shown in Figure 6-12. Safety precautions, such as the use of goggles, gloves, and hardhats, are encouraged when using these tools (see Figure 6-13). The harvesting site plant material should be handled with great care. In some areas, whole sections may be cut and in other sites, cutting must be done selectively. Cuts should be made at a blunt angle, 8 to 10 inches from the ground. This assures that the source sites will regenerate rapidly and in a healthy manner.

6.5.2 Handling

Live branch cuttings should be bound together securely at the collection site, in bundles, for easy loading, handling and for protection during transport (see

(*a*)

Figure 6-12. Tools and methods for harvesting live cuttings. (*a*) Chainsaw. (*b*) Loppers and bush axes.

(*b*)

Figure 6-12. (*Continued*)

Figure 6-14). Live branch cutting are grouped in such a way that they stay together when handled. Side branches and brushy limbs should be kept intact at this time.

Cut plant material should arrive on the job site within 8 hours of cutting. Plants not installed on the day of arrival at the job site should be stored and protected until they are installed. Under normal conditions, all live plant materials should be used within 2 days after cutting. Outside storage locations should be continually shaded and protected from the wind. Live cut plant materials should be heeled into moist soil, or kept in water. They must be protected from drying at all times. When the temperature is 50°F or above, the live cut branches should be installed on the day they are cut, and should not be stored.

When the live branches have been cut and trimmed for live stake installa-

Figure 6-13. Protective gear worn during cutting and harvesting of live plant materials.

(*a*)

Figure 6-14. Tied bundles of live cut branches collected at harvesting site in preparation for transport to project site. (*a*) Tied bundle. (*b*) Hoisting bundle onto truck.

(*b*)

Figure 6-14. (*Continued*)

tion or similar use (e.g., for joint planting), they should be used that day. It is unwise to store trimmed material. Water that may be required for storage of live cuttings during construction should not contain toxic elements that could be harmful to plant growth.

6.5.3 Transportation

During transportation, the live cut branch groups or bundles should be placed on the transport vehicles in an orderly fashion, to prevent damage and facilitate handling and off-loading. Dump trucks, flat bed trucks, and covered van or trailer type vehicles can be used for transportation (see Figure 6-15). Basal ends should be oriented to the rear of dump or other trucks that discharge their loads

(*a*)

(*b*)

Figure 6-15. Transportation of bundled, live cut branches. (*a*) Dump truck. (*b*) Closed van.

by tilting the truck bed. The live cut materials should be covered with a tarpaulin during transportation. Transporting without a cover often causes drying and, therefore, additional stress to the live cuttings.

6.5.4 Storage and Timing

Arrival Time: Cut plant material should arrive on the job site within 8 hours of cutting. Plants not installed on the day of arrival at the job site should be stored and protected until they are placed. Under normal conditions, all live plant materials should be used within 2 days after cutting.

Normal Storage: Outside storage locations should be continually shaded and protected from the wind. Live cut plant materials should be heeled into moist soil, or kept in water, as shown in Figure 6-16. They must be protected from drying at all times. When the temperature is 50°F or above, the live cut branches should be installed on the day they are cut, and should not be stored. When the live branches have been cut and trimmed for live stake installation or similar use (e.g., for insertion through gabion mattress units), they should be used that day.

Refrigerated Storage: Cuttings used in soil bioengineering projects are normally taken and installed during the dormant season. If installation must be

(*a*)

Figure 6-16. On-site storage of live cut branches. (*a*) Depositing freshly cut branches along edge of pond. (*b*) View of basal ends of branches submerged in water.

(*b*)

Figure 6-16. (*Continued*)

delayed, special refrigerated storage is necessary. Hardwood cuttings that are to be stored for future use should be refrigerated at 34°F and 90° humidity. Refrigeration vans are readily available in most areas. These vans typically hold 8,000 to 10,000 stems, 6 to 8 feet long. An example of a refrigerated van for storage of live cuttings is shown in Figure 6-17.

6.6 INSTALLATION AND ESTABLISHMENT OF LIVE CUTTINGS

6.6.1 Installing Cuttings

Installation Time: Installation of live cuttings should begin concurrently with earth moving or grading operations, assuming the latter are being conducted during the dormant season. The best time for installation is during the dormant season, which generally occurs from September to March. This more or less spans the complete time range, from the North of the United States to the South. Each region will have its own dormant season within this range. Yearly variances also should be taken into account. *Live cuttings should not be placed in the ground once they have broken bud.*

Planting Medium: Soil bioengineering projects ideally use on-site stockpiled soil as the planting medium of choice. Soil bioengineering systems should be installed in a planting medium that includes fines and organic material and that

Figure 6-17. Refrigerated van for storage of live cuttings when installation must be delayed.

is capable of supporting plant growth. The soil should be free of any material or substance that could be harmful to plant growth and should be nutrient tested prior to use. Gravel is not a suitable material for use as a fill around live plant materials. Muddy (saturated) soils that otherwise meet these requirements should not be certified as suitable backfill until they have been dried to a workable moisture content. Heavy plastic clays should be mixed with sandy and organic soils to increase porosity.

Soil Tests and Preparation: Agronomic soil samples of the on-site soils should be taken prior to live woody plant installation. Soil samples should also be taken from all fill materials that are brought to the site prior to their use. Nutrient testing by an approved laboratory should include analyses for a full range of nutrient and metal contents: nitrogen, phosphorus, potassium, pH, as well as any toxins present that would be harmful to plant growth. The laboratory reports should also include fertilizer and lime amendment requirements for woody plant material. All fill soil around live plant cuttings should be compacted or tamped, by foot or by machine, to densities approximating the surrounding native soil densities. The soil around the plants should be free of large voids.

6.6.2 Inspection and Quality Control

Maintaining good quality control by means of an adequate inspection schedule will help to insure the success of a soil bioengineering project. The following guidelines are recommended:

Preconstruction

- Select plant species for conformance to requirements.
- Locate and secure source sites for harvesting.
- Define construction work area limits.
- Fence off sites requiring special protection.
- Complete and inspect the following preparations:
  - Layout
  - Excavation
  - Bench size, shape, angle
  - Site preparation, that is, clearing, grading, shaping
  - Disposal of excess gravel, soil, debris
  - Vegetation to be preserved/removed
  - Stockpiling of suitable soil and/or rock

During Construction

- Inspect each system component, at every stage, for the following:
  - Angle of placement and orientation of the live cuttings
  - Backfill material/rock and stone material
  - Fertilizer, type method and quantity applied
  - Lime, type method and quantity applied
  - Preparation of trenches or benches in cut and fill slopes
  - Staking
  - Pruning
  - And so on
- Ensure that proper maintenance occurs during and after installation.
- Inspect daily for quality control:
  - Check all cuttings, remove unacceptable material
  - Inspect the plant material storage area when in use
  - And so on

Establishment Period

- Inspect biweekly for the first 2 months. Inspections should note infestations, soil moisture, and so on.
- Inspect monthly for the next 6 months. Systems not in acceptable growing conditions should be noted, and so on.
- Needed reestablishment work should be performed every 6 months during the initial 2 to 5 year establishment period, and so on.

- Extra inspections should always be made during periods of drought or heavy rains, and so on.

Final Inspection: A final inspection should be held 2 to 5 years after installation is completed. Vigorous, healthy growing conditions should exist by that time. Satisfactory performance can be gauged as follows:

- Healthy growing conditions in all areas refers to overall leaf development and rooted stems defined as follows:
  - Live stakes .... 70–100% growing
  - Live fascines .... 20–50% growing
  - Brushlayers .... 40–70% growing
  - Branchpacking .... 40–70% growing
  - Live cribwall .... 30–60% growing
  - Live slope grating .... 30–60% growing
  - Live gully repair .... 30–50% growing
  - Joint planting .... 50–70% growing
  - Vegetated geogrid .... 40–60% growing
- Growth should be continuous with no open spaces greater than 2 feet in linear systems. Spaces 2 feet or less will fill in without hampering the integrity of the installed living system.

6.6.3 Maintenance and Aftercare

Under normal conditions maintenance requirements should be minor after inspection and acceptance of the established system. In general, maintenance consists of light pruning and removal of undesirable vegetation. Heavy pruning may be required to reduce competition for light or to stimulate new growth.

More intensive maintenance may be required to repair problem areas created by high intensity storms or other unusual conditions, for example, heavy browsing by wildlife. Site washouts should be repaired immediately. Generally reestablishment should take place for a 1- to 2-year period following construction completion and consist of the following practices:

- Replacement of branches in dead, unrooted sections
- Soil refilling, branchpacking, and compacting in rills and gullies
- Insect, disease and weed control
- Wildlife and people control

Gullies, rills, or damaged sections should be repaired through the use of healthy, live branch cutting, preferably installed during the dormant season. The branchpacking system can be used to repair large breaks or depressions, and the live gully repair system for breaks up to 2 feet wide and 2 feet deep. Rooted stock should be considered if the dormant season has passed.

6.7 SPECIAL METHODS FOR STRUCTURES

6.7.1 Crib Walls

Methods for vegetating the open-fronted bays of crib walls will depend somewhat on the nature of the cribfill and the timing of the construction. Detailed guidelines for the vegetative treatment of crib walls is given in Section 8.7. Engineering specifications for the cribfill often make it too coarse and granular for optimal plant growth. When specifications are less stringent and permit a minimum content of fines (clay and silt-size material), planting holes may be dug in the open bays and transplants or cuttings inserted directly. Some open-front crib walls purposely provide this opportunity. If construction coincides with the proper planting season, both planting and cribfilling may be done simultaneously. If not, the plants should be inserted during a more favorable time of year. If the cribfill is patently unsuitable for planting, that is, if it is too granular and droughty to support plant growth, then several alternative approaches may be adopted (see Section 8.2).

6.7.2 Breast Walls

Rock breast walls lend themselves to vegetative treatment, either by insertion of live cuttings between the rocks or by placement of cuttings and rooted stock on the bench above the wall. A similar procedure can be followed in the case of articulated block walls. Breast walls are usually thin enough that the basal end of cuttings inserted through the openings or interstices in the wall can be placed directly against native soil behind the wall. Detailed guidelines for vegetative treatment of breast walls is given in Section 8.6.

6.7.3 Gabion Walls

Gabion walls consist of various stacking arrangements of wire baskets (typically $3 \times 3 \times 6$ feet) that are backfilled with small rock to create a monolithic gravity retaining structure. Opportunities for vegetative treatment exist by insertion of live cuttings between successive gabion courses. The rock in the baskets is far too coarse and droughty to sustain vegetative growth; accordingly, the basal end of the cuttings must extend into the earthen backfill behind the wall. This requirement limits vegetative treatment to gabion walls with a height of no more than 12 feet; such walls typically have a width ranging from 6 to 7 feet at the base, decreasing to 3 feet at the top. Detailed guidelines for vegetative treatment of gabion walls is given in Section 8.5.

6.7.4 Gabion Revetments

Gabion revetments consist of much flatter (and thinner) wire baskets or mattresses (typically 1 × 3 × 9 feet) that are filled with rock and used to armor slopes, streambanks, shorelines, and channels. These mattresses are easier to vegetate with live cuttings (compared to gabion walls) because the cuttings do not have to pass through such a large distance to reach native soil beneath the gabion. Iron bars can be inserted or driven through the gabion mattress first to create a pilot hole for the cuttings. Further information and guidelines for the vegetative treatment of gabion revetments is given in Section 8.4.

6.8 REFERENCES

Cassell, D. K. (1983). Effects of soil characteristics and tillage practices. In: *Crop Reactions to Water and Temperature Stresses in Humid Climates*. Edited by C. D. Raper and P. J. Kramer, Westview Press, Boulder, CO.

Coppin, N. J., and I. Richards (1990). *Use of Vegetation in Civil Engineering*. Sevenoaks, Kent (England): Butterworths.

Gray, D. H., and A. T. Leiser (1982). *Biotechnical Slope Protection and Erosion Control*. New York: Van Nostrand Reinhold.

Kraebel, C. J. (1936). Erosion control on mountain roads. *USDA Circular No. 380*, 43 pp.

NAS-NRC (1993). *Vetiver Grass: A Thin Green Line Against Erosion*. Washington, DC: Board on Science and Technology for International Development, National Research Council, National Academy Press, 171 pp.

Schor, H., and D. H. Gray (1995). Landform grading and slope evolution. *Journal of Geotechnical Engineering* (ASCE) **121**(10): 729–734.

Schiechtl, H. M. (1980). *Bioengineering for Land Reclamation and Conservation*. Edmonton, Canada: University of Alberta Press, 404 pp.

Smith, T. W., and G. V. Stafford (1957). Horizontal drains on California highways. *Journal of Soil Mechanics and Foundations Division* (ASCE) **83**:1301–1326.

USDA Soil Conservation Service (1940). Lake bluff erosion control. Open File Report prepared by USDA Soil Conservation Service, Michigan State Office, Lansing, MI, 81 pp.

USDA Soil Conservation Service (1975). *Engineering Field Manual for Conservation Practices*. USDA Soil Conservation Service Engineering Division, Washington, DC.

7 Soil Bioengineering Stabilization: Techniques and Methods

7.0 INTRODUCTION

7.0.1 Definition

Soil bioengineering measures are a special case of biotechnical stabilization in which plants and plant parts, primarily live cuttings, are imbedded and arranged in the ground in special patterns and configurations. These imbedded cuttings act as: (1) soil reinforcements, (2) barriers to earth movement, (3) moisture wicks, and (4) hydraulic drains. Adventitious rooting along the length of imbedded stems and branches provides secondary reinforcement.

7.0.2 Salient Characteristics

Soil bioengineering systems are in their most vulnerable state when first installed. However, they gain more strength with time as the vegetation roots, foliage leafs out, and plants become well established. The systems are designed to provide sufficient reinforcement and stability to an earthen slope or retention structure at the outset. They can also be regarded as pioneer or precursor systems that provide enough stability to a site so that natural invasion and colonization by native vegetation can occur that will eventually take over the stabilizing role.

7.0.3. Uses and Applications

Soil bioengineering methods can be used to prevent and control surficial erosion and shallow mass wasting. Different methods or combination of methods can be used on: (1) natural hillslopes, (2) cut and fill slopes along roadways, (3) landfill covers, (4) spoil banks, and (5) streambanks. Some methods are better suited than others for particular site conditions and objectives. *Live fascines* (*wattling*), for example, provide good protection against erosion and are relatively easy to install on both cut and fill slopes. *Brushlayering*, on the other hand, provides better reinforcement and protection against shallow mass wasting but is more difficult to install on cut slopes. *Live crib walls* provide additional restraint at the base of slopes and also protect the toe.

Soil bioengineering methods can be used alone or in combination with structural or conventional methods. *Live staking*, for example, can be used to provide ancillary protection around a check dam in an eroding channel, or the stakes can be tamped through openings in a rock revetment or gabion mattress on a stream bank to improve both the performance and appearance of the rock armor (see Chapter 8). Soil bioengineering systems by their nature provide drainage; in wet slopes, however, some ancillary form of drainage, such as surface diversions or subsurface drainage measures (e.g., chimney or trench drains) may be required.

7.1 LIVE STAKING

7.1.1 Description

Live staking involves the insertion and tamping of live, rootable vegetative cuttings into the ground, as shown in Figure 7-1. If correctly prepared and placed, the live stakes will root and leaf out. The procedure is simple, economical, and fast. Please note that the rooted/leafed condition of the plant material shown in Figure 7-1, and successive schematic diagrams, is not representative of conditions at the time of installation.

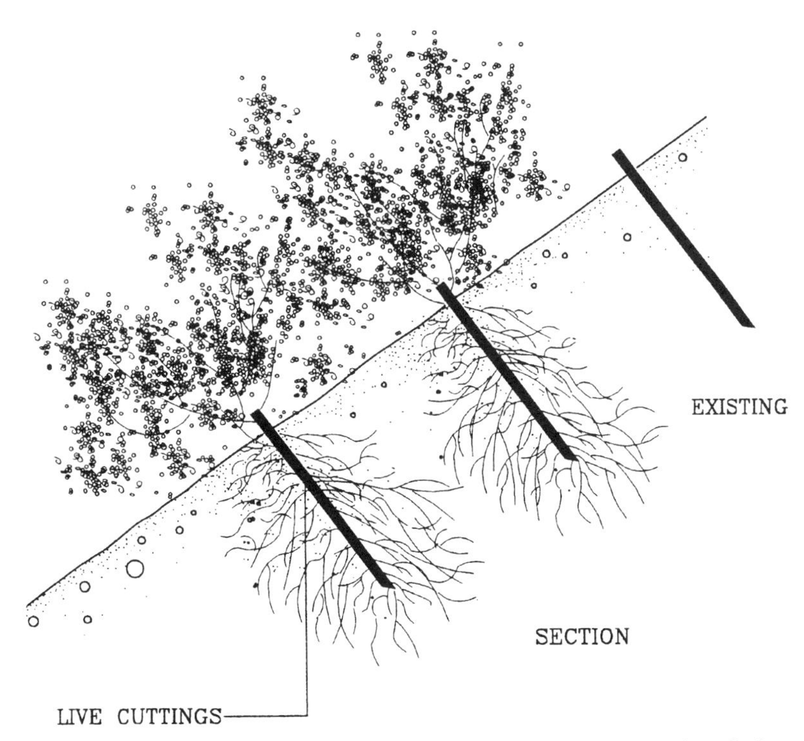

Figure 7-1. Schematic diagram of an established growing live stake installation.

7.1.2 Uses

Live stakes may be used as a primary treatment or in conjunction with live fascines, or other soil bioengineering measures. They can be used alone to repair small earth slips and slumps that are quite wet. Live stakes can be placed in rows across a slope to help control shallow mass movement, as shown in Figure 7-2; they can also be tamped through and used in conjunction with jute or coir netting, as shown in Figure 7-3. Installing of willow stakes in clustered arrays along the sides of gullies may be beneficial in slowing velocity, trapping sediment, and controlling erosion. The clusters are installed in chevron-like rows that point downstream, as shown in Figure 7-4. The rows should extend from the top of the bank in a downstream direction to the toe. The stakes should be set 12 to 18 inches apart in the row clusters, and the rows themselves should be spaced so that the top of the downstream row overlaps the bottom of the upstream row, as shown in Figure 7-4. Live stakes can also be used to provide ancillary protection around check dams and gully-head plugs (see Chapter 8).

7.1.3 Preparation

The cuttings are usually $\frac{1}{2}$ to $1\frac{1}{2}$ inches in diameter and 2 to 3 feet long. Stakes must have side branches cleanly removed and the bark intact. The basal ends should be cut at an angle for easy insertion into the soil. The top should be cut square as shown in Figure 7-1.

Figure 7-2. Live stakes being installed in rows across a slope to help control shallow sloughing or mass movement.

Figure 7-3. Use of live stakes in conjunction with jute netting to protect a slope.

7.2.4 Installation

The following guidelines and procedures should be followed when placing live stakes in the ground:

- Tamp the live stake into the ground at right angles to the slope. Use a *dead blow* hammer to avoid splitting the stakes.
- Install the stakes 2 to 3 feet apart using triangular spacing. The density of the installation should range from 2 to 4 stakes per square yard.
- Four-fifths of the stake should be buried in the ground, and soil firmly packed around it after installation. The buds should be oriented up. A photograph of a healthy, growing live stake is shown in Figure 7-5.

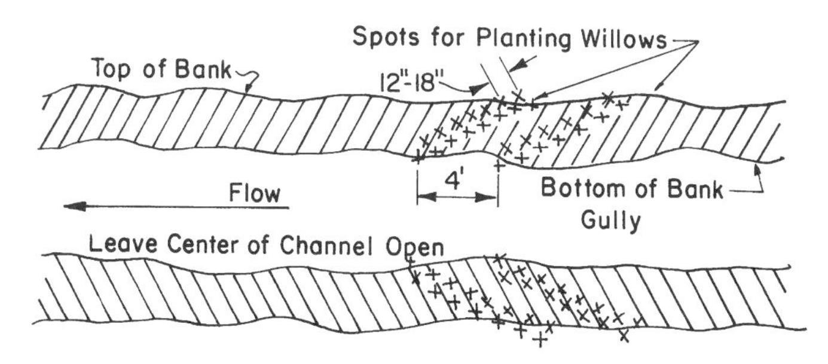

Figure 7-4. Plan of a gully showing planting points for live willow clusters.

Figure 7-5. Healthy, growing live stake in the first year of growth.

7.2 LIVE FASCINES

7.2.1 Description

Stems and branches of rootable plant material (e.g., willow, dogwood, alder) are tied together in long bundles and placed in shallow trenches. The bundles are tied together with twine and anchored in the trench with wooden construction stakes and/or live stakes, as shown in Figure 7-6. The trenches are typically excavated by hand and normally follow the contour of the bank or slope. On very wet slopes the trenches may be excavated off-contour at a slight angle to facilitate drainage. After the live fascines are secured with stakes, the trench is backfilled with soil until just the tops of the live fascine bundles are exposed (see Figure 7-7).

7.2.2 Uses

Live fascines are used for a variety of slope stabilization purposes, such as the stabilization of highway cuts and embankment fills, gullied areas, and other areas where erosion is a problem. The rows of live fascines create a series of benches on a slope that slow runoff and trap sediment, as shown in Figure 7-8. Live fascines are easier to install then brushlayers in cut slopes because they are installed at a shallower depth. Live fascines provide excellent protection against surficial erosion. They are less effective than brushlayers, however, in

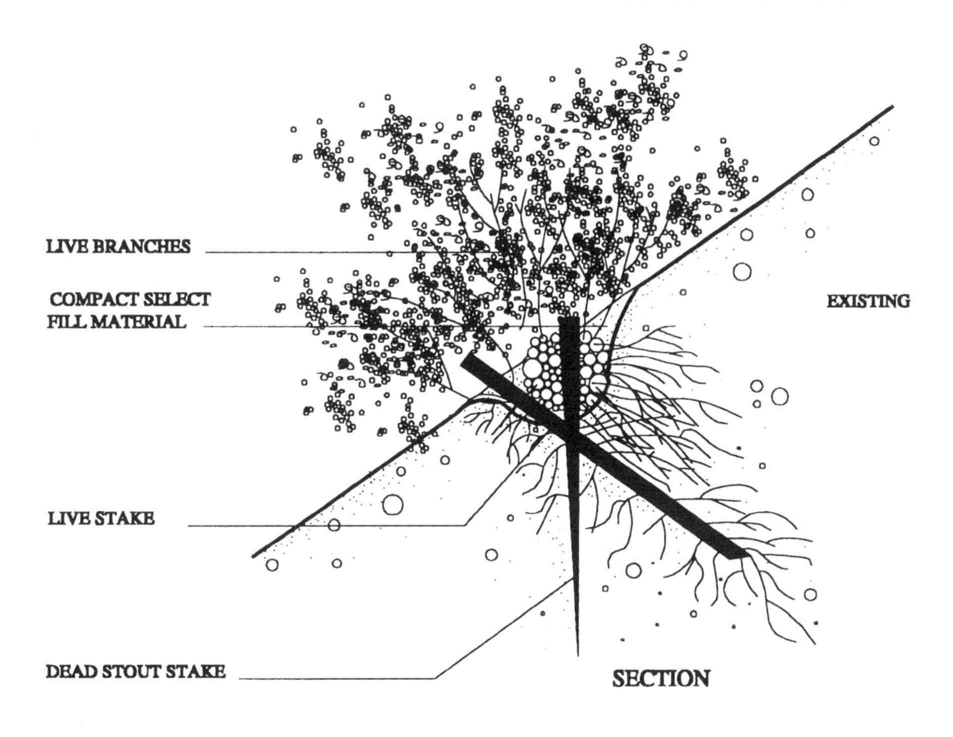

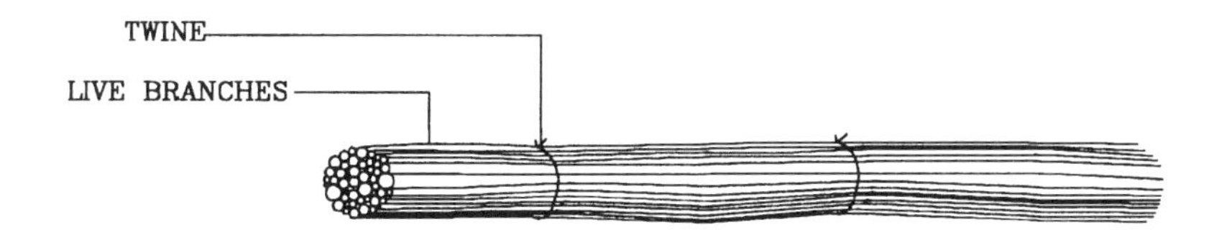

Figure 7-6. Schematic diagram of an established, growing live fascine intallation showing live fascine bundles and method of placement in a slope.

Figure 7-7. Live fascine that has been placed in a shallow trench, staked in place, and backfilled.

Figure 7-8. Photograph taken during the first growing season of live fascines, showing how they create a series of benches and barriers on contour that slow runoff, trap sediment, and provide lodgment for seeds.

preventing shallow mass movements because the reinforcements do not extend as deeply into the slope. Nevertheless, over time roots from the fascines do penetrate into the slope (see Figure 7-9) and provide some protection against shallow sliding.

7.2.3 Preparation

Long, slender, straight branches cut from young willow, alder, and dogwood work best. The cuttings are tied together in bundles (live fascines) that vary in length from 5 to 30 feet, depending on site conditions and handling limitations. The completed bundles (see Figure 7-6) should be 6 to 8 inches in diameter, with all the growing tips oriented in the same direction.

7.2.4 Staking

Inert stakes made of wood are used to secure the live fascine bundles in the trenches. The inert stakes are $2\frac{1}{2}$ feet long in cut slopes and 3 feet long in fill slopes. They are cut from 2×4 lumber by sawing along a diagonal, as shown in Figure 7-10. The inert stakes are driven through the live fascines, as shown in Figure 7-6. Live stakes can also be employed; they are generally installed on the downslope side of the bundles. The tops of the live stakes should protrude 2 to 3 inches above the top of the live fascine after it is placed in the trench.

Figure 7-9. Live fascine that has been wash excavated, exposing sinker roots that extend into and help stabilize the slope.

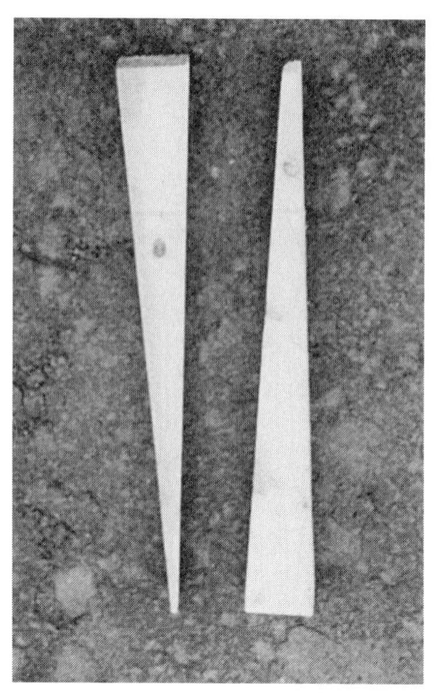

Figure 7-10. Inert construction stake used to secure live fascines; cut from 2 × 4 lumber.

7.2.5 Installation

Live fascine installations begin at the base of the slope and proceed upward. The following guidelines and procedures should be followed when placing live fascines in the ground:

- Beginning at the base of the slope, dig a trench on contour just large enough to contain a fascine bundle. Dig the trench to a depth slightly less than the size of the bundles.
- Place a live fascine into the trench (see Figure 7-11) and drive an inert stake directly through the bundle every 2 to 3 feet along its length, as shown in Figure 7-12. Extra stakes should be used at connections or bundle overlaps.
- Work and tamp moist soil into and along the sides of the bundles. The live fascines should not be completely buried; a few twigs and leaves should protrude above the surface.
- Dig additional trenches at intervals up the slope and repeat the preceding steps to the top of the slope using the spacing interval shown in Table 7-1. When possible place one or two rows over the top of the slope. The slope spacings listed in Table 7-1 are conservative, that is, they are less than the equivalent spacings recommended by Kraebel (1936) based on his work on steep, erodible fill slopes. Kraebel recommended a 3-foot *vertical* spacing between contour trenches. This fixed, 3-foot limit translates into

Figure 7-11. Live fascine placed in shallow trench near base of slope.

Figure 7-12. Driving stakes directly through live fascines that have been placed in a trench. The stakes are driven through the bundles every 2 to 3 feet along their length.

TABLE 7-1. Recommended Spacings for Live Fascines on Slopes

Slope Steepness ($H : V$)	Slope Distance Between Trenches (ft)	
	On Contour	On Angle
1 : 1 to 1.5 : 1	3–4	2–3
1.5 : 1 to 2 : 1	4–5	3–5
2 : 1 to 2.5 : 1	5–6	3–5
2.5 : 1 to 3 : 1	6–8	4–5
3.5 : 1 to 4 : 1	8–9	5–7
4.5 : 1 to 5 : 1	9–10	6–8

variable slope distances that are some 25 percent greater than the spacings recommended in Table 7-1.

- Long straw or similar mulching material should be placed between rows on 1.5 : 1 ($H : V$) or flatter slopes, as shown in Figure 7-13. Jute or coir fabric or netting should be used on steeper slopes. The fabric can be anchored in place by extending it into the trenches and staking the live fascine over the fabric, as shown in Figure 7-14.

Figure 7-13. Use of long straw mulch between rows of live fascines protects exposed soil and further retards erosion.

(*a*)

(*b*)

Figure 7-14. Jute or coir netting used between rows of live fascines. (*a*) Netting extends into trench beneath live fascine bundles. (*b*) Installation of live fascines in fabric lined trench.

7.3 LIVE FASCINES USED IN POLE DRAINS

7.3.1 Description

Rows of live fascines are installed chevron-fashion connecting to a central drain. Typically the side chevron sections are composed of single, live fascine bundles, whereas the central drain, which serves as the primary collector drain, is constructed with three fascine bundles grouped together as shown in Figure 7-15. Views of a completed pole drain system immediately after construction and one year later are shown in Figure 7-16.

7.3.2 Uses

The system is used on wet slopes where there is evidence of subsurface seepage causing destabilization of the slope.

7.3.3 Installation

Live fascines are prepared and installed for the side chevron portions conventionally as previously described.

- The central system and side trenches are constructed first. Beginning a short distance down from the top of the slope, dig the center trench to the bottom of the slope or to a collector system such as a rock drain. This trench should be trapezoidal in shape, 18 to 20 inches deep, 12 to 14 inches wide at the bottom, and 16 to 18 inches wide at the top.
- The side, or lateral trenches, should be constructed starting at or near the top of the slope and angling down 20 to 45 degrees to connect with the center trench. The first lateral should meet the central trench a few feet below the top of the trench. These lateral trenches should be spaced 3 to 8 feet apart, generally parallel to each other, and have a length of 10 to 25 feet.
- The first fascine for the central drain should be 12 to 14 inches in diameter. It is placed on the bottom of the trench and staked in the ground every 3 feet along its length with inert stakes. This fascine bundle is constructed from live materials, but is not expected to grow because of the depth of placement.
- The two live fascine bundles that will be placed on top of the previously described bundle are prepared in usual way. They are placed on top of the first fascine bundle and secured in place by staking through the bundles into the ground below. They should be staked in a staggered fashion every 3 feet along their length.
- The chevron side, or lateral, live fascine bundles are then installed in the

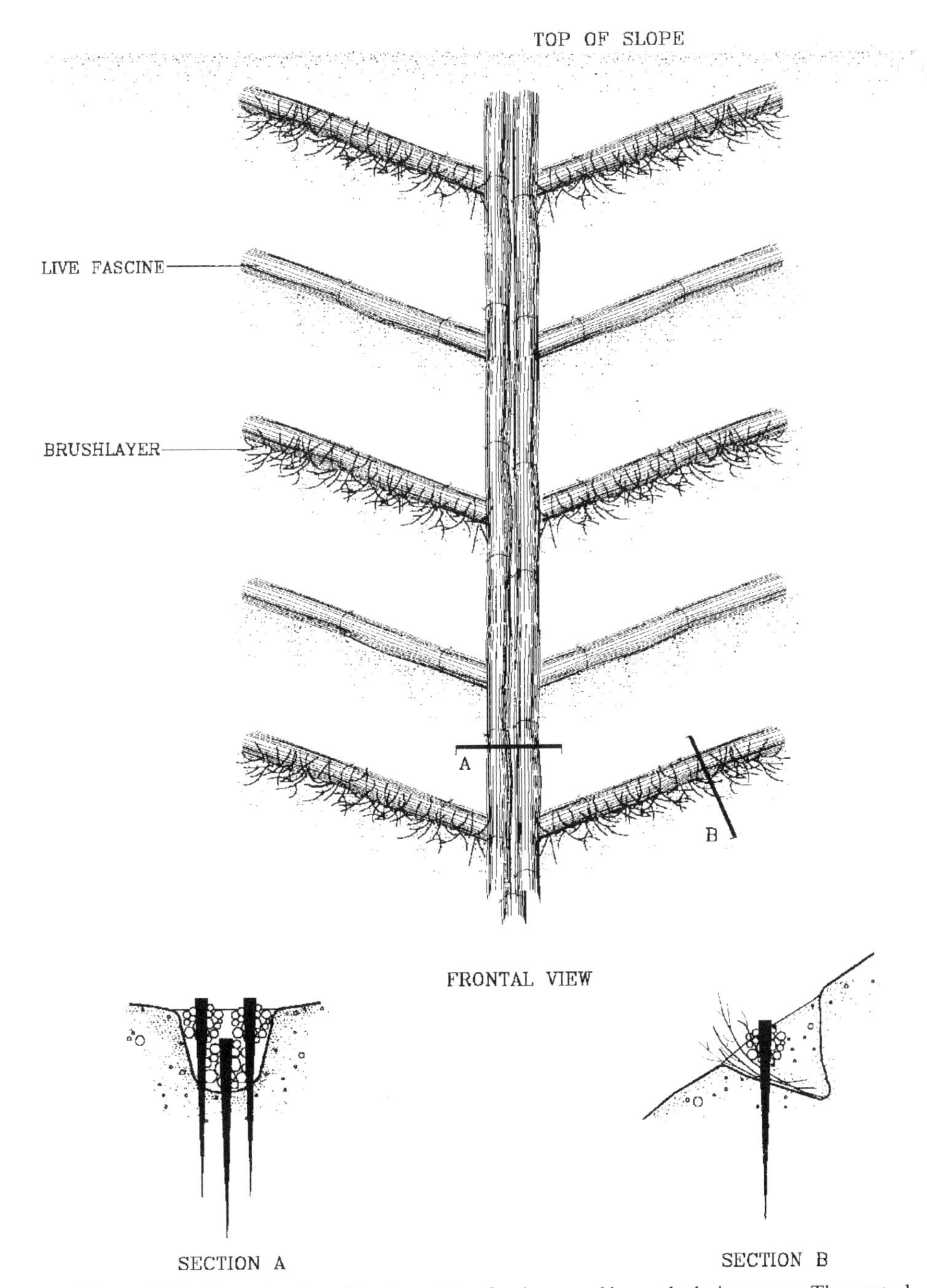

Figure 7-15. Frontal and profile view of live fascines used in a pole drain system. The central fascine system intercepts subsurface seepage and also acts as a collector drain.

(*a*)

(*b*)

Figure 7-16. View of pole drain systems to facilitate drainage on a wet slope. (*a*) Immediately after construction. (*b*) One year later.

previously constructed trenches. They should be long enough to collect the water from the seepage sections on the slope and should join into the central collector drain.

- The bundles are backfilled as previously described.

7.4 FASCINES WITH SUBSURFACE INTERCEPTOR DRAIN

7.4.1 Description

Rows of fascines are installed on contour on a slope in the conventional manner. In addition, a subsurface drain, oriented downslope and perpendicular to the fascines, is placed in an axial trench beneath the rows of fascines to intercept and collect seepage, as illustrated in Figure 7-17.

7.4.2 Uses

This system is used on very wet sites where there is evidence of substantial subsurface seepage that is causing piping and destabilizing the slope. It can be considered for use in filled gully areas, colluvial ravines, or swales where groundwater is likely to collect and concentrate, as shown in Figure 7-18.

7.4.3 Installation

Fascines are prepared and installed in conventional fashion as described previously. The seepage collection trench is excavated first, and a perforated pipe is

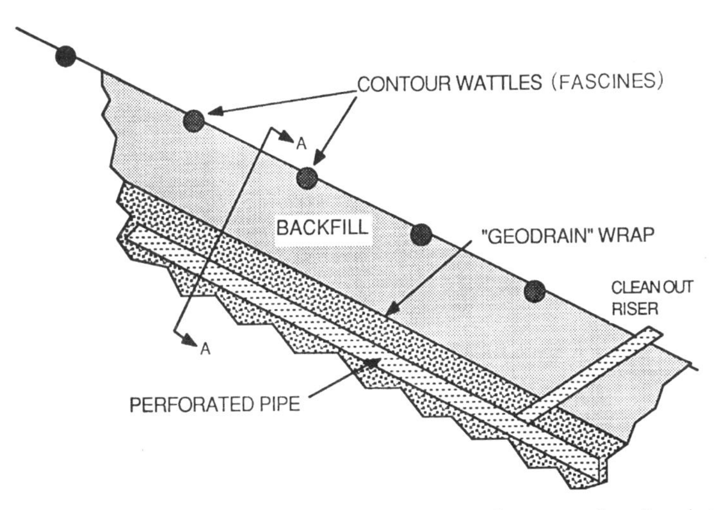

Figure 7-17. Profile view of a fascine system constructed over a subsurface interceptor drain.

Figure 7-18. Fascine system placed in drainage swale. Axial interceptor drain was placed below the fascines to intercept and collect subsurface water.

placed in the bottom of the trench. The drain pipe should be imbedded within crushed rock that is placed in the trench and wrapped with a geotextile filter cloth. Alternatively, a composite geodrain can be wrapped around a perforated pipe, as shown in Figure 7-19. The porous core of the geodrain should face in toward the pipe, with the filter cloth backing facing outward. This design

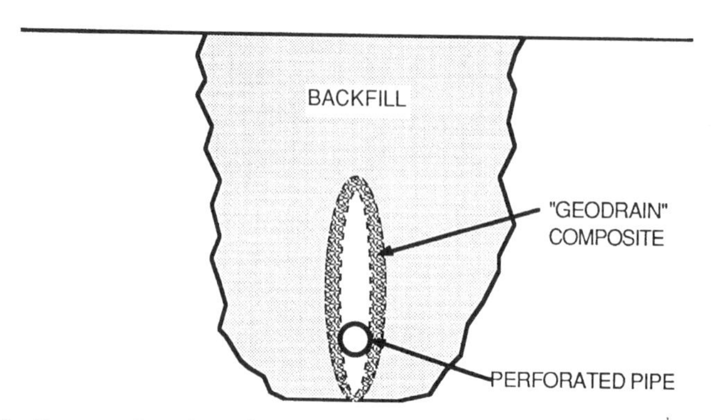

Figure 7-19. Cross section view of a *composite geodrain*/perforated pipe interceptor drain installed beneath a fascine system.

Figure 7-20. Live fascine and interceptor drain system. A clean-out access tube is visible in the photo.

avoids the need to import clean crushed rock to the site. All drains should be constructed and installed with clean-out access tubes, as shown schematically in Figure 7-17. A fascine and interceptor drain system with a clean-out access tube visible in the photo is shown in Figure 7-20.

7.5 BRUSHLAYERING

7.5.1 Description

Brushlayering consists of live cut branches interspersed between layers of soil, as shown in Figures 7-21 and 7-22. The brush is placed in a crisscross, or overlapping pattern, so that the tips of the branches protrude just beyond the face of the fill, where they retard runoff velocity and filter sediment out of the slope runoff. The stems extend back into the slope in much the same manner as conventional, inert reinforcements, for example, geotextiles and geogrids. Unlike conventional reinforcements, however, the brushlayers root along their lengths and also act immediately as horizontal slope drains.

Fill Slopes: Brushlayering works best when done in conjunction with the construction of a conventional fill slope operation. The brushlayers are imbedded between successive lifts or layers of fill at the outside edge, as shown in Figure 7-21. Each layer of brush or live branches is covered with soil and lightly com-

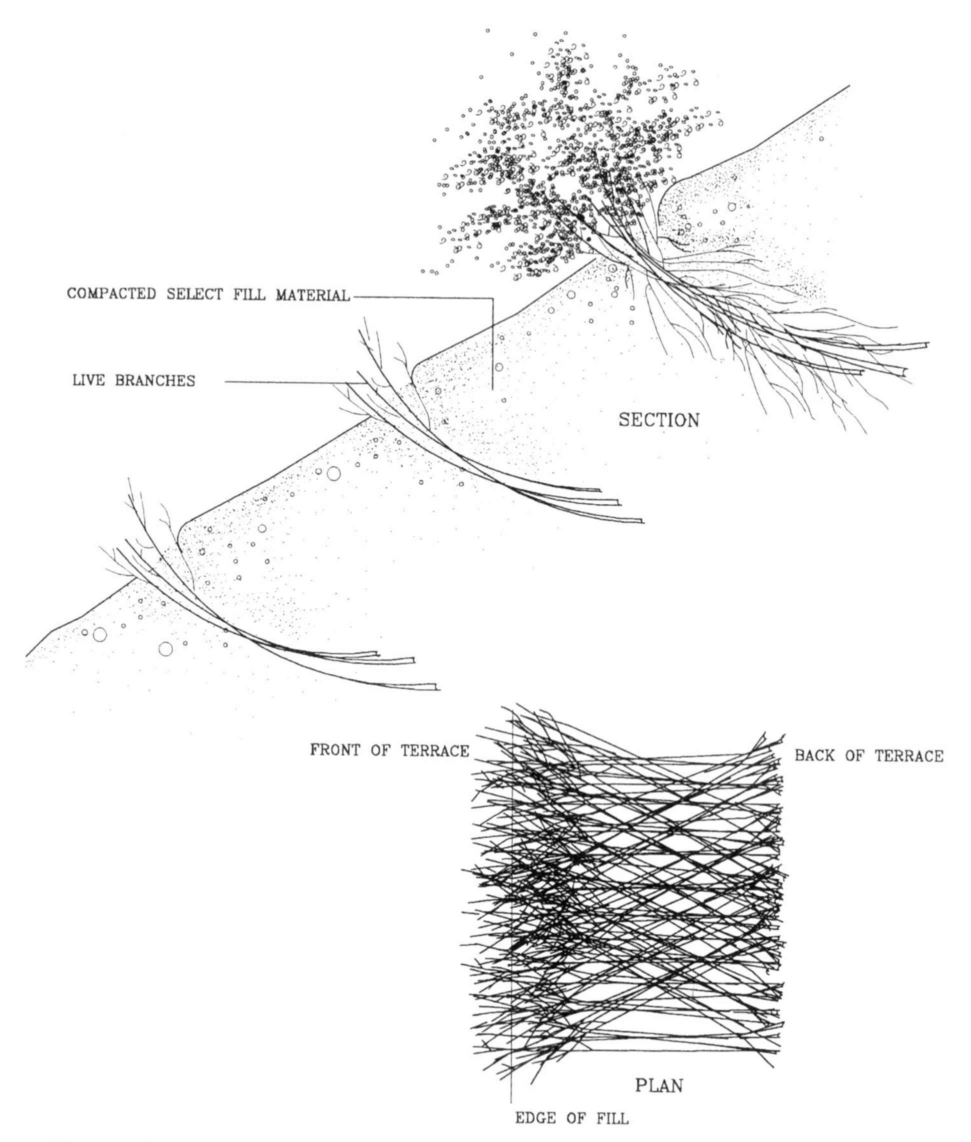

Figure 7-21. Schematic diagram of an established growing *fill slope* brushlayer installation showing alternating layers of live cut brush inserted between lifts of soil.

pacted. Live brushlayers can also be combined with natural geofabrics (e.g., coir fabric) or synthetic, polymeric geogrids to provide additional reinforcement at the outset (see Section 7.6 on Vegetated Geogrids)

Cut Slopes: The brushlayers are placed on narrow benches excavated into the slope, as shown in Figure 7-22. Unlike fill slopes, the length of the stems are normally shorter on cut slopes because the benches are excavated only 2 to

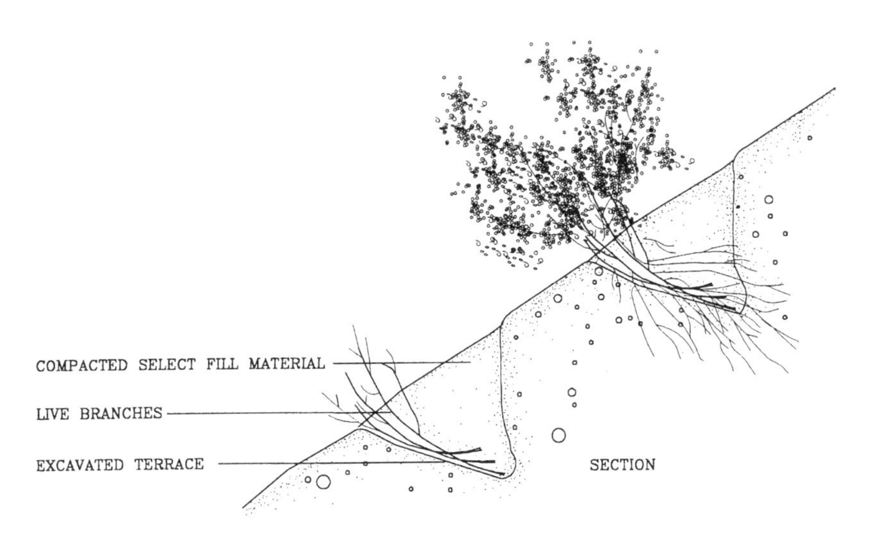

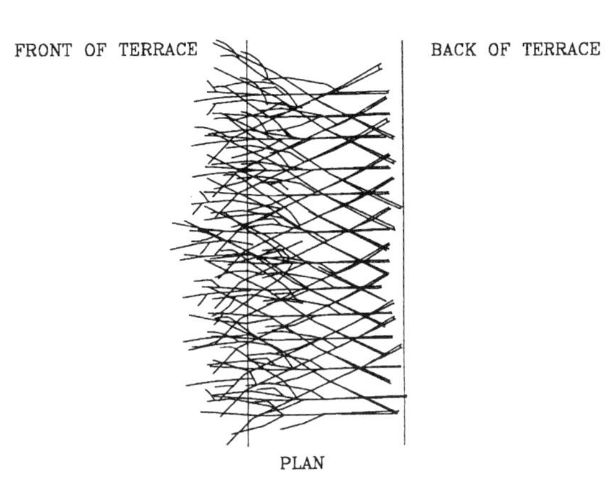

Figure 7-22. Schematic diagram of an established, growing *cut slope* brushlayer installation showing alternating layers of live cut brush placed on narrow benches or terraces excavated in the slope.

3 feet into the slope. Cut slope brush layers are recommended on slopes no steeper than 2 : 1 $(H : V)$.

7.5.2 Uses

Brushlayering can be used to stabilize a slope against shallow sliding or mass wasting in addition to providing erosion protection. Examples of brushlayer installations are shown in Figure 7-23. The orientation of the stems is more

(*a*)

(*b*)

Figure 7-23. Brushlayer installations. (*a*) Fill or embankment slope. (*b*) Cut slope.

effective than live fascines from the point of view of earth reinforcement and mass stability. Brushlayering works better on fill as opposed to cut slopes because much longer stems can be used in the former method. Brushlayers can be used to stabilize and reinforce the outside edge or face of drained earthen buttresses that are placed against cut slopes or embankment fills.

7.5.3 Preparation

Long branches are cut from willow, alder, and dogwood. The length of the branches will vary with the type of brushlayer (cut or fill slope) and the desired depth of reinforcement. Branches up to 12 feet in length can be used on fill slope brushlayer installations.

7.5.4 Installation

Brushlayer installations begin at the base of the slope and proceed upward. The installation procedure differs for fill as opposed to cut (or natural) slopes. The following guidelines and procedures are quite general and generic to both methods:

- The surface or terrace on which the brushlayers are placed should slope back into the slope slightly (approximately 10 to 20 degrees off horizontal).
- Place the live branch cuttings either on an excavated bench (*cut or natural slope*) or on a lift of soil (*fill or embankment slope*) in a crisscross or overlapping configuration, as shown in Figure 7-24.
- Branch growing tips should be aligned toward the outside face of the slope.
- Place backfill on top of the branches and compact to eliminate air spaces. The brush tips should extend slightly beyond the face of the slope to slow runoff and filter out sediment.
- The brushlayer rows should vary approximately from 3 to 10 feet apart along the slope, depending on the slope angle, site and soil conditions, and position on the slope. Nominal spacing guidelines are presented in Table 7-2. Spacings between rows should be reduced at the bottoms of steep, high slopes for added safety. Exact spacings can be calculated using force-equilibrium methods based on a desired mass stability safety factor, tensile properties of the brush, and other soil/site variables.
- Natural geofabrics or geogrids such as coir netting can also be wrapped around the soil layers to provide additional restraint and reinforcement. Alternatively, polymeric geogrids or synthetic geotextiles can be used if greater strength and durability are required. Guidelines for these modified brushlayer intallations are presented in Section 7-6.
- Long straw or similar mulching material should be placed between rows on 3 : 1 ($H : V$) or flatter slopes. Jute or coir fabric or hold-down netting should be used on steeper slopes. The fabric can be anchored in place by wrapping it around the front face of soil lifts in a fill slope application.

(*a*)

(*b*)

Figure 7-24. Placement of live cut branches in overlaping or crisscross manner during brush-layer construction. (*a*) Fill slope. (*b*) Cut slope.

TABLE 7-2. Recommended Brushlayer Spacings on Slopes

Slope Steepness ($H : V$)	Approximate Slope Distance Between Brushlayer Rows	
	On Angle Wet Slopes (ft)	On Contour Dry Slopes (ft)
1.5 to 2 : 1	3–4	4–5
2 : 1 to 2.5 : 1	3–4	5–6
2.5 : 1 to 3 : 1	4–5	6–8
3 : 1 to 4 : 1	5–6	7–10

7.6 VEGETATED GEOGRIDS

7.6.1 Description

A vegetated geogrid installation consists of live cut branches (brushlayers) interspersed between layers of soil and wrapped in natural or synthetic geotextile materials, as shown in Figure 7-25. The brush is placed in a crisscross or overlapping pattern so that the tips of the branches protrude just beyond the face of the fill, where they retard runoff velocity and filter sediment out of the slope runoff. The stems extend back into the slope in much the same manner as conventional, inert reinforcements, for example, geotextiles and geogrids. Unlike conventional reinforcements, however, the brushlayers are living and root along their lengths and also act as horizontal slope drains.

Vegetated geogrid structures are constructed in much the same fashion as a conventional mechanically stabilized earth (MSE) structural fill. Brush is imbedded between successive wrapped lifts or layers of fill at the outside edge. Each layer of brush or live branches is covered with soil and lightly compacted. The geogrids provide additional reinforcement at the outset. Once the live cuttings become established, their root systems become entangled with the grids and bind the entire system together in a unitary, coherent mass.

7.6.2 Uses

Vegetated geogrid structures can be used to stabilize very steep slopes in addition to providing surface erosion protection. Vegetated geogrids can be used to stabilize and reinforce drained, earthen buttress fills. They provide an alternative to vertical retaining structures for grade separation purposes and in situations that require avoiding right-of-way encroachment at the base or top of slopes. Vegetated geogrids can also be used to protect slopes that are subject to periodic scour or tractive stresses, such as drainage channels or

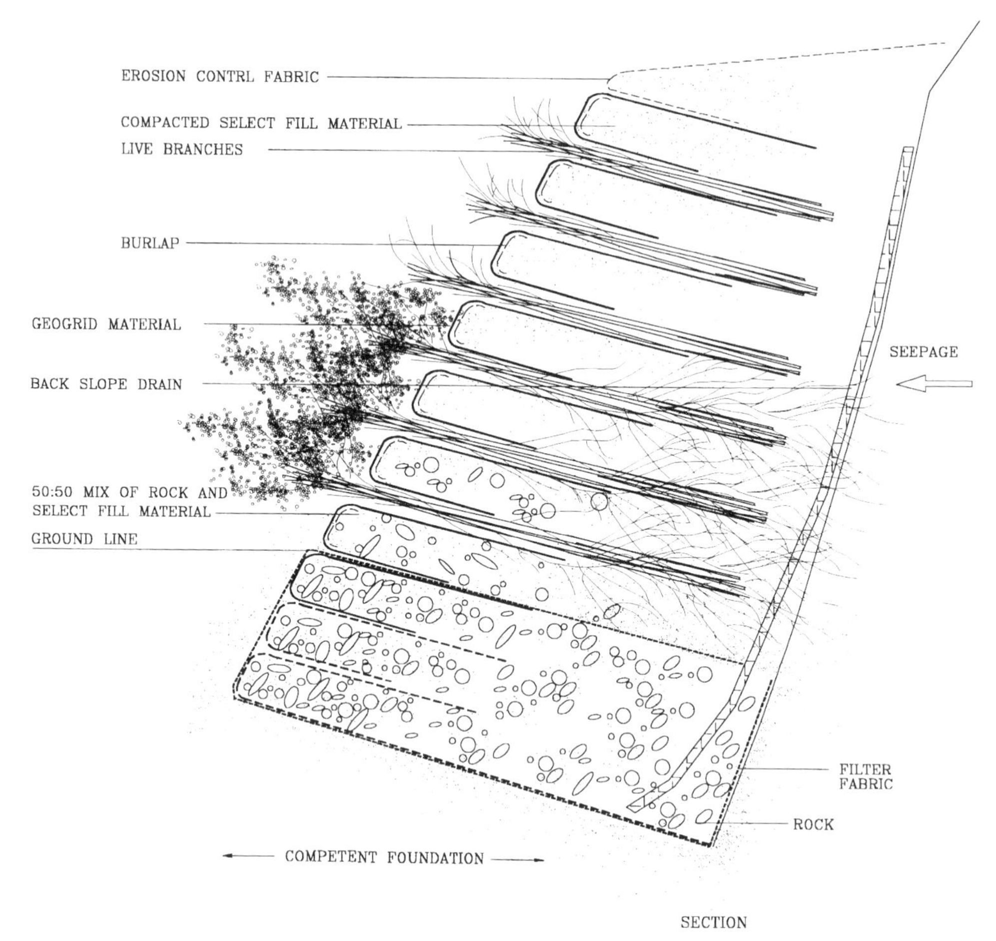

Figure 7-25. Schematic diagram of an established, growing fill brushlayer installation showing alternating layers of live cut brush inserted between lifts of soil, wrapped with either natural or synthetic (polymeric) geogrids.

upper portions of streambanks, as shown in Figures 7-26 and 7-27, respectively.

7.6.3 Materials and Preparation

Live materials consist of long branches, cut from willow, alder, and dogwood, that are $\frac{1}{2}$ to 2 inches in diameter. The length of the branches will vary with the type of application (embankment or buttress fill) and desired depth of reinforcement; ideally they should be long enough to reach the back of a buttress fill. The inert construction material consists of synthetic, polymeric geogrids. The geogrids can be selected according to their allowable unit tensile strength,

(*a*)

(*b*)

Figure 7-26. Drainage channel side slopes protected with vegetated geogrids. Coir fabric was used to wrap/reinforce successive lifts of soil. (*a*) After construction. (*b*) One year later.

Figure 7-27. Side slopes of streambank protected with vegetated geogrids. Synthetic polymeric geogrids were used to wrap successive lifts of soil. Live cut branches were also inserted between lifts.

sized in length (width), and spaced vertically to provide the main or primary reinforcement.

7.6.4 Installation

A vegetated geogrid installation begins at the base of the slope and proceeds upward. A vegetated geogrid structure should be supported on a rock toe or base and be battered or inclined at an angle of at least 10 to 20 degrees to minimize lateral earth forces (see Chapter 5). The following guidelines and procedures apply to the use of vegetated geogrids for constructing a buttress fill structure:

- A trench should be excavated to a competent horizon as well as below the likely depth of scour. This trench is backfilled with rock to provide a base for the vegetated geogrid structure. The top surface of the rock should be inclined with the horizontal to establish the desired minimum batter angle for the overlying, geogrid structure. Typically two courses or lifts of geogrid wrapped rock, 2 to 4 inches in size, are incorporated into the top of the rock toe, as illustrated in Figure 7-25.

- An earthen structure reinforced with synthetic geogrids and live brush is constructed on top of the rock base. Geogrids with a minimum roll width of 13 feet and aperture sizes or openings and unit tensile strengths equivalent to Tensar BX 1200 are generally suitable for this purpose. A geogrid strip is gathered near (or draped down) the front edge of the fill and staked down over the underlying lift with a minimum overlap of 3 feet. Wood construction stakes spaced every 3 feet along the length of the overlap are used for this purpose.
- Select fill material is placed on the geogrid and compacted in 3-inch lifts to a nominal thickness ranging from 12 to 30 inches. Thinner lifts are used at the base of the structure, where shear stresses are higher. Temporary batter boards are required at the front face to confine the select fill during the installation process and to form an even face. Burlap strips at least 4 feet are inserted between the fill and the geogrids at the front face and staked in place (see Figure 7-25) to contain the fines.
- The exposed sections of geogrids are pulled up and over the faces of the fill layers (see Figure 7-27) and staked in place. The geogrids should be pulled as uniformly as possible before staking, using a machine to develop initial tension in the geogrid. A tractor or winch pulling on a long bar with hooks or nails along its length works well for this purpose. The tensioned geogrid should be secured in place using wood construction stakes every 3 feet.
- One to two inches of select fill material is placed on top of each wrapped geogrid layer. Three layers of live cut branches are then placed with two to four feet of select fill material between each layer.
- The process is repeated with succeeding layers of fill, live brush, and geogrids until the specified height or elevation is reached.
- The recommended fill lift thickness between geogrid layers depends on soil and site variables, properties of the reinforcements, and desired safety factor. Spacings between lifts of fill should be reduced at the bottoms of steep, high slopes for added reinforcement and safety. The maximum vertical spacing and imbedded length of successive geogrid or reinforcement layers are determined from the specified safety factor, slope angle, soil shear strength, allowable unit tensile strength, and interface friction properties of the reinforcement layer. Conventional design guidelines for spacing and imbedment length of geogrid reinforcements in a slope can be adopted for this purpose (Thielen and Collin, 1993).

7.7 BRANCHPACKING

7.7.1 Description

Branchpacking consists of alternating layers of live branch cuttings and compacted backfill to repair holes and depressions in slopes. Long wooden stakes are driven into undisturbed ground beneath the void as well. A branchpacking system is shown schematically in Figure 7-28.

7.7.2 Applications

Branchpacking is used for the repair and filling of small slumps, slipouts, and headcuts in natural slopes, cuts, and embankments. It reinforces the backfill

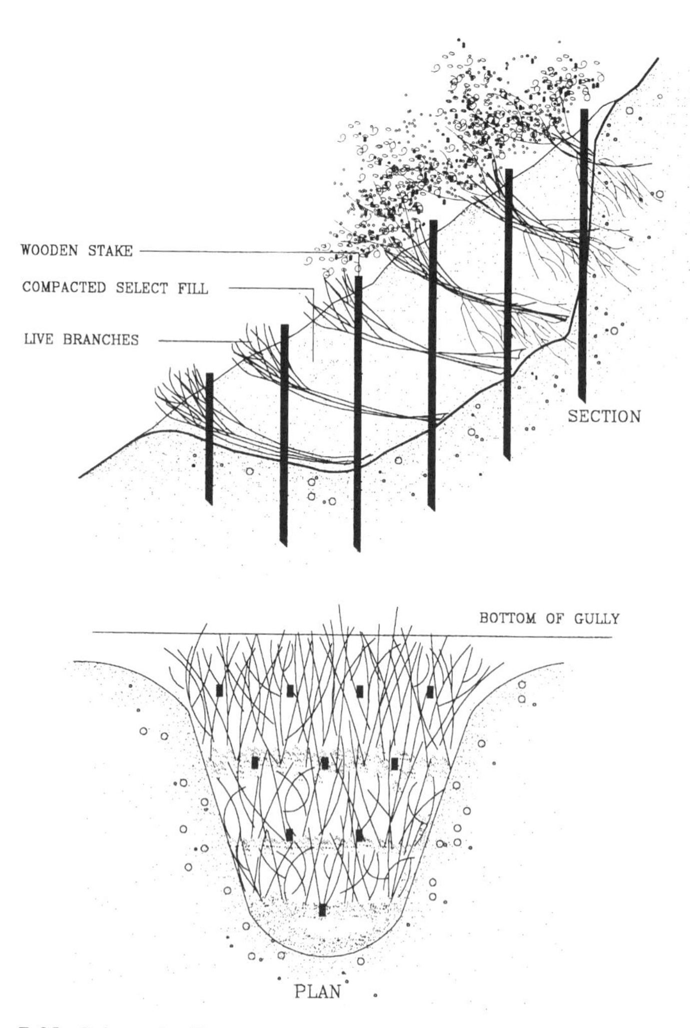

Figure 7-28. Schematic diagram of an established, growing branchpacking installation.

used to repair the defect and protects it against washout and scour. Branchpacking repairs in slump areas should be restricted to depressions no greater than 4 feet deep and 5 feet wide. Subsurface interceptor drains, for example, chimney or blanket drains, may be required if significant amounts of subsurface seepage enters the hole or void. The cause of the condition should also be addressed prior to or in conjunction with the branchpacking installation.

7.7.3 Preparation

Live branch cuttings can range from $\frac{1}{2}$ to 2 inches in diameter. They should be long enough to touch the undisturbed soil at the bottom (or back) of the void and protrude slightly beyond the rebuilt slope face. The wooden stakes should be 5 to 8 feet long and made from 3- to 4-inch diameter poles or 2×4 lumber, depending upon the depth of the hole or void.

7.7.4 Installation

Branchpacking installations begin at the lowest point in the void and proceed upward. The following guidelines and procedures should be followed when installing a branchpacking system:

- Starting at the lowest point, drive the wooden stakes vertically 3 feet into the ground, as shown in Figure 7-29. Set the stakes 1 to $1\frac{1}{2}$ feet apart.

Figure 7-29. Rows of vertical wooden stakes set in a ground cavity in preparation for branchpacking.

- A layer of living branches 4 to 6 inches thick is placed in the bottom of the hole, between the vertical stakes, and perpendicular to the back slope. Some of the basal ends of the branches should touch the back of the hole.
- Subsequent layers of live branches are placed with their basal ends lower than the growing tips.
- Each layer of branches must be followed by a layer of compacted soil to ensure intimate contact with the branch cuttings.
- The final installation should match the existing slope profile. Branches should protrude only slightly from the filled face. The fill soil should be moist (or slightly moistened) to insure that the live branches do not dry out. The appearance of newly constructed branchpacking is illustrated in Figure 7-30.

7.8 LIVE GULLY REPAIR FILL

7.8.1 Description

A live gully repair fill consists of alternating layers of live branch cuttings and compacted soil. This reinforced fill can be used to repair rills and small gullies. The method is similar to branchpacking but is more suitable for filling and repairing elongated voids in a slope such as gullies. The main details of a live gully repair system are shown schematically in Figure 7-31.

7.8.2 Applications

Repair and filling of rills and small gullies in natural slopes. The imbedded branches and secondary roots reinforce the backfill used to repair the gully and protect it against future washout and scour. Live gully fill repairs should be restricted to gullies that are a maximum of 2 feet wide, 2 feet deep, and 15 feet long. Subsurface interceptor drains, for example, chimney or blanket drains, may be required if significant amounts of subsurface seepage enters the gully at its head and sides. Water from the head or top should be intercepted and redirected away from the area.

7.8.3 Preparation

Live branch cuttings can range from $\frac{1}{2}$ to 2 inches in diameter. They should be long enough to touch the undisturbed soil at the bottom (or back) of the gully and protrude slightly beyond the rebuilt slope face.

7.8.4 Installation

Live gully repair installations begin at the lowest point in the gully and proceed upward. The live cut branches are inserted between successive lifts of

(*a*)

(*b*)

Figure 7-30. Branchpacking installation during construction. (*a*) Top view. (*b*) Frontal view.

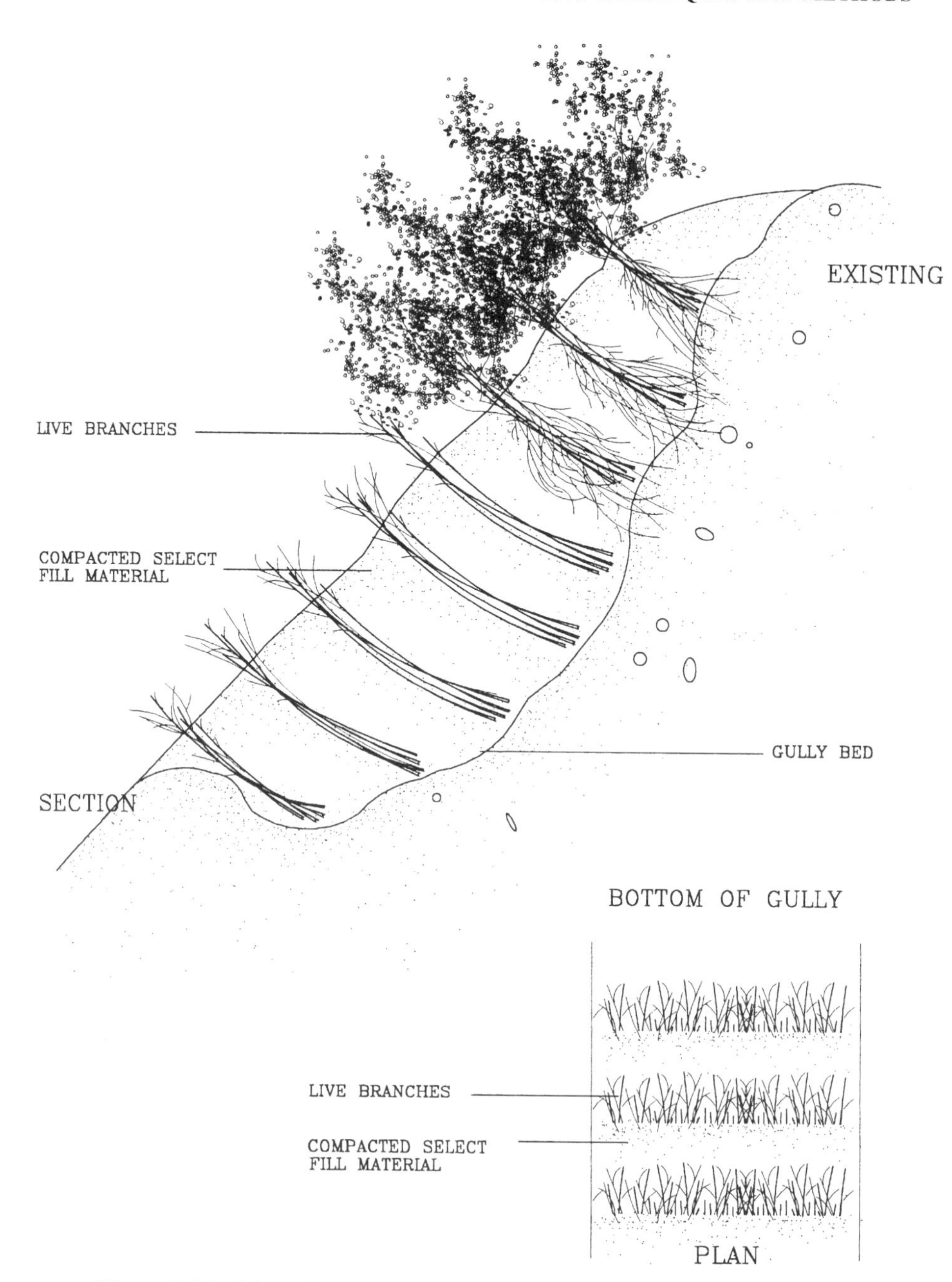

Figure 7-31. Schematic diagram of an established, growing live gully repair fill.

compacted soil. The following guidelines and procedures should be followed when installing a live gully repair system:

- Starting at the lowest point of the slope, place a 3- to 4-inch thick layer of branches at the lowest end of the gully and approximately perpendicular to the gully bottom (see Figure 7-31).
- Cover with a 6- to 8-inch thick layer of fill soil and compact.
- Place the live branches in a crisscross fashion. Orient the growing tips toward the slope face with the basal ends lower than the growing tips.
- Follow each layer of branches with a layer of soil; work and compact the soil to ensure intimate contact with the branches and to eliminate large voids in the fill.

7.9 VEGETATED (LIVE) CRIB WALLS

7.9.1 Description

A vegetated crib wall consists of a hollow, box-like interlocking arrangement of structural beams. In *live* crib walls the structural members are usually untreated log or timber members. The structure is filled with a suitable backfill material (or cribfill) and layers of live branch cuttings, which root inside crib as shown in Figure 7-32.

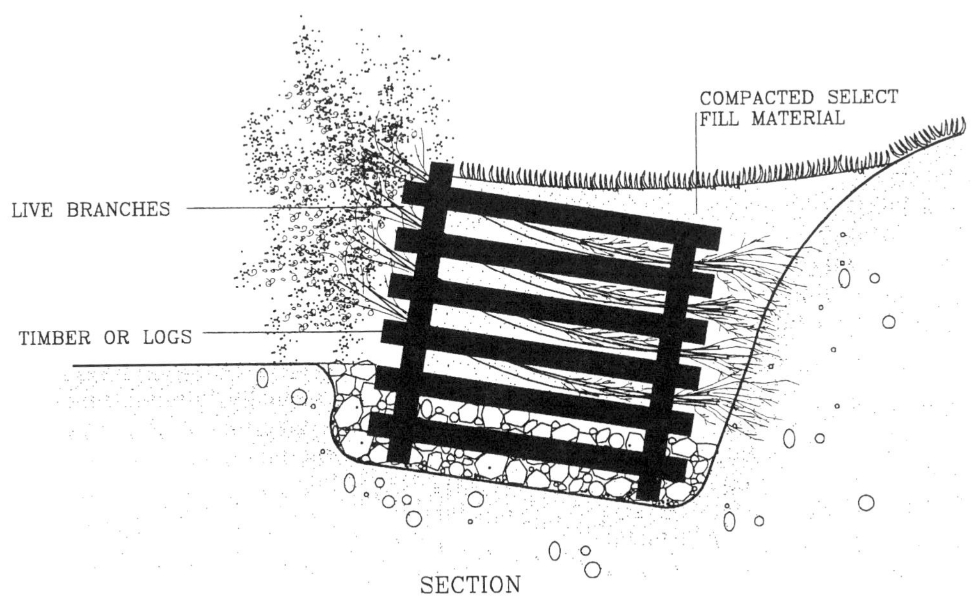

Figure 7-32. Schematic illustration of an established, growing vegetated (live) crib wall.

7.9.2 Objective

The live cut branches placed inside a live crib wall should extend beyond the crib into the backfill and native soil behind the structure, as shown in Figure 7-32. These cuttings will root inside the cribfill and backfill behind. Once the live cuttings root and become established, the subsequent vegetation gradually takes over the structural functions of the wood members. The roots and plant stems help to bind the cribfill and backfill into a coherent, unitary mass.

7.9.3 Effectiveness and Applications

A vegetated crib wall system provides the following potential uses and advantages:

- It is helpful at the base of slopes where a low toe wall can be used to reduce the steepness of a slope and stabilize the toe against scour and undermining.
- Such a system has a more natural appearance and is less visually intrusive than a structural treatment alone.
- It avoids encroachment by use of a more vertical protective structure.
- *Live* crib walls using *untreated* timbers are not intended to resist large, lateral earth stresses. They should be constructed to a maximum height of 6 feet, including the foundation.

7.9.4 Materials

Live materials consist of live branches cuttings that are $\frac{1}{2}$ to 2 inches in diameter and long enough to reach the back of the wooden crib structures, as shown in Figure 7-32. The *inert* construction materials consist of logs or timbers ranging from 4 to 6 inches in diameter.

7.9.5 Installation

The following guidelines and procedures should be followed when constructing a *live* crib wall system. The sequence of construction steps is shown schematically in Figure 7-33.

- Starting at the lowest point of the slope, excavate loose material 2 to 3 feet below the ground surface until a stable foundation is reached. The footing base should be inclined into the slope so that the structure will have a batter (inclination off vertical) of at least $1:6$ $(H:V)$. This battering provides additional stability to the surface.
- Place the first course of logs or timbers at the front and back of the exca-

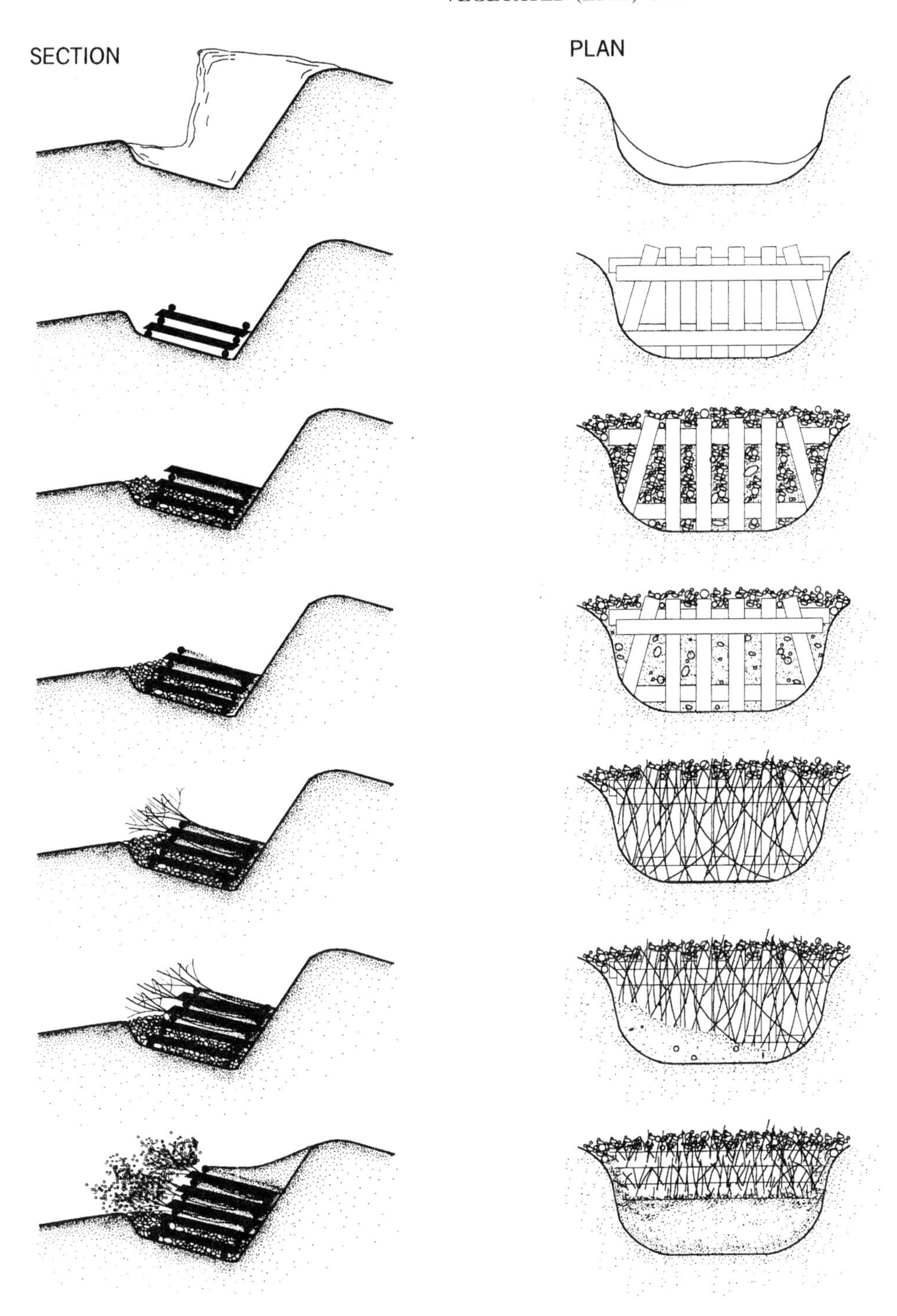

Figure 7-33. Sequence for live crib wall construction.

vated foundation, approximately 5 to 6 feet apart and parallel to the slope contour.

- Place the next course of logs or timbers at right angles (perpendicular to the slope) on top of the previous course to overhang the front and back of the previous course by 3 to 6 inches.
- Each course of the live crib wall is placed in the same manner and secured to the preceding course with nails or reinforcing bars. This construction procedure will result in an interlocking, box-like structure, as shown in Figure 7-34.
- When the live crib wall structure reaches the existing ground surface (at the base), place live branch cuttings on the cribfill perpendicular to the slope; then cover the cuttings with more cribfill and compact.
- The live branch cuttings or branches should be placed at each course to the top of the live crib wall structure with the growing tips oriented toward the front face (see Figure 7-34). Follow each layer of branches with a layer of compacted soil to ensure soil contact with the branch cuttings.
- Some of the basal ends of the cuttings should reach to undisturbed soil at the back of the live crib wall with growing tips protruding slightly beyond the front of the live crib wall. The front view a live crib wall during construction is shown in Figure 7-35.

Figure 7-34. Live crib wall under construction, showing box-like construction and placement of live branch cuttings.

Figure 7-35. Appearance of live crib wall during construction.

7.10 LIVE SLOPE GRATING

7.10.1 Description

A live slope grating consists of a lattice-like array (see Figure 7-36) of vertical and horizontal timbers that are fastened or anchored to a steep slope. The structural members are typically untreated log or timber members. The grating is constructed in such a manner so as to support itself from the bottom. The openings in the structure are filled with a suitable backfill material and layers of live branch cuttings, which are placed in a similar manner to brushlayering. A profile view of a slope grating system is shown in Figure 7-37.

7.10.2 Objective

A slope grating is a system used to establish vegetation on very steep slopes (steeper than $1\frac{1}{2}:1$) that normally cannot be revegetated or that can be revegetated only with great difficulty in the absence of measures used to anchor or hold the vegetation in place. The grating structure itself does little to armor or buttress the slope (i.e., it is not a true revetment); instead its main purpose is to facilitate the establishment of vegetation on steep, barren slopes and to protect them against weathering and slaking.

7.10.3 Effectiveness and Applications

A slope grating offers several advantages over other methods:

- It requires little excavation and clearance at the foot of the slope.

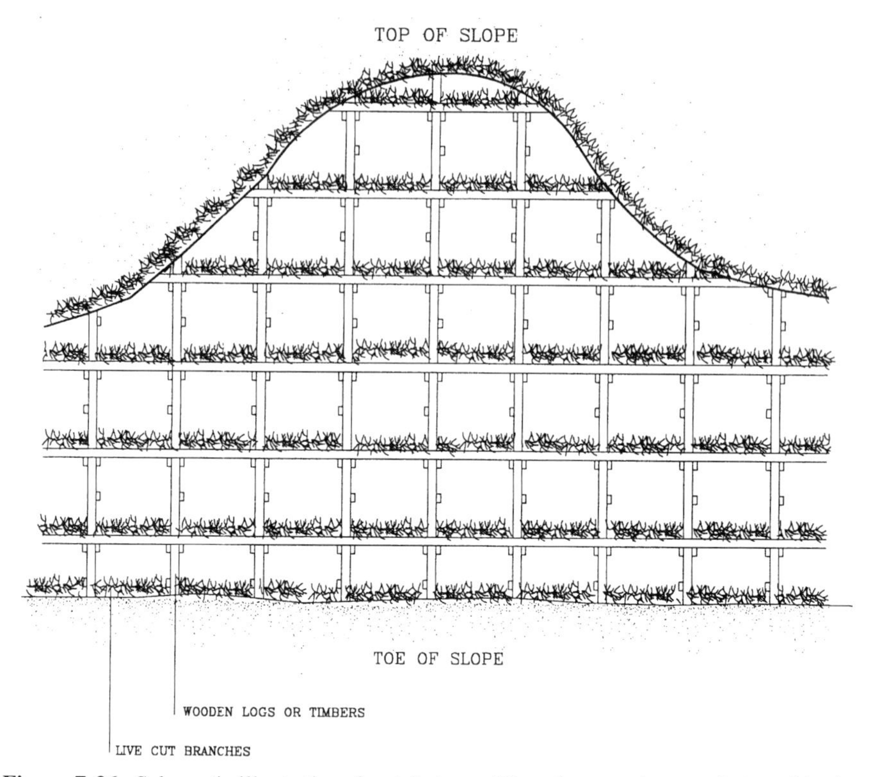

Figure 7-36. Schematic illustration, frontal view, of live slope grating consisting of lattice-like array or grid of horizontal and vertical timbers that are anchored to a steep slope.

- It permits establishment of vegetation on very steep slopes (up to 1 : 1) without the need for slope flattening.
- It requires very little importation of select backfill and cribfill.
- When filled with earth it covers and protects underlying exposed bedrock (especially certain shales) from slaking and disintegration.

7.10.4 Materials

Live materials consist of branch cuttings that are $\frac{1}{2}$ to 2 inches in diameter and long enough to reach the back of the slope and to extend just beyond the wooded slope grating face, as shown in Figure 7-37. The wooden framework consists of untreated wooden poles or logs 4 to 6 inches in diameter.

7.10.5 Installation

The following are general guidelines and procedures for constructing a *live* slope grating system:

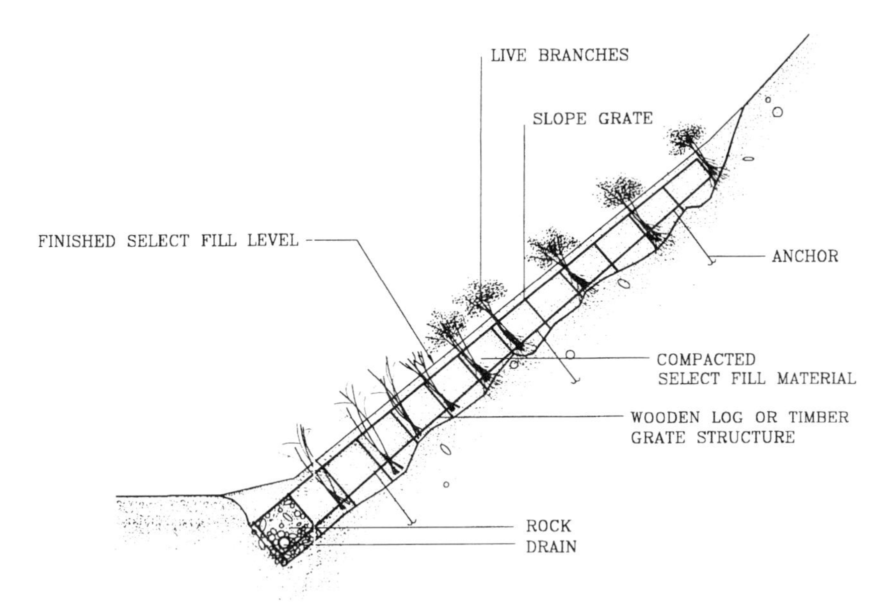

Figure 7-37. Profile view of an established growth slope grating system showing placement of live cuttings or branches in box-like compartments in the grating.

Grating Construction

- A trench is excavated at the bottom by hand or machine. The trench should be approximately $2\frac{1}{2}$ to 3 feet deep and the bottom sloped in so that the wooden frame or grating can be placed as flush as possible against the slope face (see Figure 7-37).
- The slope grating framework has a prism shape with a thickness of 24 to 36 inches at the bottom tapering to approximately 18 to 24 inches at the top to accommodate backfill and live cuttings in the compartments formed by the lattice construction.
- The grating framework can be constructed in several ways. A 6-inch diameter log, or alternatively, a 6×6 inch timber, is first placed horizontally in the bottom of the trench against the slope face. Long "vertical" poles or posts are placed in the trench, butted against the horizontal timber at the bottom, and laid back against the face of the slope on approximately 3 to 5 foot center to center spacings. These vertical poles should extend the entire length of the finished slope grating unit. The poles should be 6 inches in diameter at the bottom and taper to no less than 4 inches at the top. Alternatively, 6×6 inch dimensioned timbers may be used instead. Once in place, the vertical members are securely fastened with spikes to the horizontal timber in the trench bottom.
- Short separators or support posts are attached in pairs to both sides of the vertical poles. These typically consist of 2-foot long, 2×4 inch sawn

timber nailed at right angles to the vertical poles or posts every 4 feet up their length, as shown in profile view in Figure 7-37. These separator posts are used to support the top part of the grating system and establish the needed depth (perpendicular to the slope) for the grating compartments.

- A top layer of "vertical" poles or timbers are fastened between the support posts. The support posts are nailed to the topmost "vertical" poles or timbers in similar fashion as the bottom ones. The top layer is nailed in such a way to provide a compartment "depth" of approximately 24 inches at the base of the grating and 24 to 18 inches at the top end. The bottom end of these "vertical poles" should butt firmly against the side of the trench or natural ground at the base of the slope.
- The horizontal members that form the compartments in the grating (shown in plan view in Figure 7-36) consist of 4-inch diameter logs or alternatively 4×4 inch dimensioned timbers. These are placed on top of the "vertical" poles or timbers above the connections to the support posts (see Figure 7-37) and securely fastened by means of spikes to the underlying "vertical" members. The support posts can also be toe-nailed into the horizontal members. A photograph showing the constructed framework or skeleton of a slope grating is shown in Figure 7-38.
- A rock drain wrapped in a filter fabric is placed in the bottommost compartment at the base of the grating along its entire width. Depending upon conditions, the grating can be secured to the slope at selected points by means of suitable anchors such as screw plate or "duckbill" type anchors.

Figure 7-38. Constructed skeleton or framework of a live slope grating.

Fill and Branch Placement

- The compartments of the grating are backfilled with a soil capable of supporting vegetation and live cut branches. The branches are placed in an overlapping or crisscross fashion similar to brushlayering. The branches are placed at either the bottom and/or midheight of the compartments (see Figure 7-36) and are oriented approximately perpendicular to the face of the grating (see Figure 7-37).
- The branches are placed in two layers, separated by 2 to 4 inches of select fill material or soil. Soil or fill placed above the branch layers should be placed and compacted in 4-inch lifts. The basal end of the branches should reach all the way back to the parent slope material and the growing tips should just extend beyond the fill.
- The bottom 20 feet above the base of the wooden grating should be live brushed every 2 feet, that is, bottom and midheight of each compartment. Above this line or elevation the live brush can be placed in a row every 4 feet at the center or midheight of each compartment. Views of a live grating system at various stages of fill and branch placement are shown in Figures 7-39 and 7-40. The appearance of a completed slope grating during the first growing season is shown in Figure 7-41.

Figure 7-39. Placement of live cut branches and fill in a live slope grating.

Figure 7-40. Live slope grating system near end of construction. Note method for hoisting materials on right of photo.

Figure 7-41. Live slope grating system early in the first growing season.

7.11 SELECTION CRITERIA

7.11.1 Selection by General Slope Type and Location

Soil bioengineering methods can be selected according to their utility or suitability for two main slope types, namely upland slopes/hillsides versus streambanks/coastal bluffs. The main difference in these two cases is the requirement for some type of toe protection or shoreline defense in the latter to guard against scour and undermining of the slope.

Suitable Methods for Hillside or Upland Slopes: The following methods are possible candidates for consideration. Their relative degree of construction complexity is noted as well.

- Live staking — Very low
- Live fascines — Moderate
- Brushlayering — Moderate to high
- Branchpacking — Moderate to high
- Live crib walls — Moderate to high
- Vegetated geogrid — High
- Live slope grating — High

7.11.2 Selection by Soil and Site Conditions

Suitable soil bioengineering measures, or combinations of measures, can also be selected based on soil and site conditions. These include the following considerations:

- Slope height and inclination
- Soil depth
- Soil type (relative erodibility and shear strength)
- Cut versus fill slope
- Failure depth
- Failure mechanism (surficial erosion versus mass wasting)
- Need for toe protection
- Lateral earth forces
- Site hydrology

Selection guidelines based on these criteria are presented in Table 7-3. These criteria should be augmented by good judgement and experience. The use of pilot test sites in the field to compare and test the effectiveness of particular systems may also aid final selection.

Selection for Environmental and Recreational Goals: Soil bioengineering methods can also be selected according to their environmental and recreational

TABLE 7-3. Suitability of Different Soil Bioengineering Methods Based on Soil and Site Conditions

Factor or Failure Process	Intensity or Type of Condition	Soil Bioengineering Methods						
		Live Staking	Live Fascine	Brush-Layering	Branch-Packing	Live Crib Wall	Live Slope Grating	Vegetated Geogrid
Slope gradient	*Steep*		X	X	N/A	X	X	X
	Moderate		X	X	N/A	X	X	X
	Gentle	X	X		N/A	X		
Slope height	*High*	X	X	X	N/A		X	X
	Low	X	X	X	N/A	X	X	X
Soil depth	*Deep*	X	X	X	X	N/A	N/A	X
	Shallow	X	X			N/A	N/A	
Soil erodibility	*High*		X			N/A	X	X
	Moderate		X	X		N/A	X	X
	Low		X	X	X	N/A	X	X
Soil strength	*Moderate*	X	X	X	N/A	N/A	N/A	N/A
	Low	N/A	X	X	N/A	N/A	N/A	N/A
Slope type	*Cut*	X	X	X	X		X	
	Fill	X	X	X	X	X		X
Surficial erosion		X	X		X		X	
Mass movement	*Shallow*	X	X	X	X	X		
	Moderate			X				X

[a] N/A ≅ Not applicable.

goals. Some methods lend themselves better to desired visual effects, access, and wildlife habitat benefits, as described in Table 7-4.

7.12 RELATIVE COSTS OF SOIL BIOENGINEERING MEASURES

The range of unit costs for different soil bioengineering measures are listed in Table 7-5. The costs shown are the "installed" costs (1994 dollars) of var-

TABLE 7-4. Suitability of Different Soil Bioengineering Methods Based on Environmental and Recreational Goals

Goals and Benefits	Soil Bioengineering Methods						
	Live Staking	Live Fascine	Brush-Layering	Branch-Packing	Live Crib Wall	Live Slope Grating	Vegetated Geogrid
Recreation	Fair to good	Fair to good	Good	Negligible	Negligible	Fair	Fair
Wildlife habitat	Negligible	Good to very good	Good to excellent	Fair	Fair to good	Good to very good	Good to very good
Aesthetic value	Good to very good	Good to very good	Excellent	Fair to good	Good to excellent	Good to excellent	Good to very good

TABLE 7-5. Unit Costs for Soil Bioengineering Measures (in 1994 Dollars)

Method	Installed Unit Cost[a]
Live staking	$1.50 → 3.50 per stake
Joint planting	$2.00 → 9.00 per stake
Live fascine	$5.00 → 9.00 per lineal foot
Live crib wall	$10.00 → 25.00 per square foot of front face
Brushlayer—cut	$8.00 → 13.00 per lineal foot
Brushlayer—fill	$12.00 → 25.00 per lineal foot
Vegetated geogrid	$12.00 → 30.00 per lineal foot
Live slope grating	$25.00 → 50.00 per square foot of front face

[a] Installation includes: (1) harvesting, (2) transportation, (3) storage, and (4) placement.

ious soil bioengineering measures that were employed on actual projects (Sotir, 1995). These unit costs can be used to obtain relative comparisons between different methods or a suitable inflation rate can be applied to obtain approximate current costs. Costs tend to increase in relation to the relative difficulty or complexity of construction. Thus live staking, the simplest and least complex method, is also the least expensive to install.

7.13 SELECTED CASE STUDIES (APPLICATIONS)

Application Number 1 Cut Slope Stabilization

Location: Greenfield Road, Colrain, Massachusetts.

Treatment Objectives: To repair and stabilize a failing cut slope along a highway right-of-way in a visually pleasing manner. To integrate soil bioengineering treatments with conventional engineering methods.

Nature of Problem: Roadway widening and encroachment during upgrading of two-lane highway resulted in high, over-steepened slopes with marginal stability. Shallow surface failures (sloughing of soil along underlying bedrock contact) in addition to deep seated slumps (in sections with little or no bedrock exposure).

Site Conditions

- Cut slope constructed at 1.5 : 1 ($H:V$) grade, approximately 1200 feet long with 20- to 60-foot high slopes
- Silty, sandy residual soil with variable thickness, overlying quartz-mica schist bedrock
- Unfavorable bedrock attitude with bedding planes and soil/bedrock contact dipping into cut
- Water table in overlying soil and active groundwater seepage from slope face and fractures in underlying bedrock.

Treatment Considerations: The fractured nature of the bedrock, as well as its uncertain depth and unfavorable attitude, ruled out the use of structural retaining walls resting on the bedrock interface. A drained rock buttress at the toe of the cut would have satisfied mass stability requirements but would have left the top part of the cut exposed and vulnerable to seepage erosion. This limitation could have been overcome by running the rock all the way up the slope in the form of a rock blanket. This solution, however, was visually stark and environmentally incompatible.

Visual and aesthetic concerns—strongly expressed by local residents—favored treatments that were visually nonintrusive and that blended in with natural surroundings. A drained, earthen brushlayer fill or facing was proposed initially to satisfy this requirement. Project engineers were concerned about the adequacy of shear resistance and drainage at the base of an earthen fill. A compromise solution was eventually devised that consisted of placing a 10-foot high rock buttress at the toe of the cut, which in turn supported a drained, brushlayer fill above. Shear strength tests on the fill material in combination with slope stability analyses were employed to design this hybrid system. The stability analyses showed that the rock buttress at the bottom intercepted the critical failure surface, which passed through the toe of the slope.

Soil Bioengineering Treatments

- A drained, earthen *brushlayer fill* (or facing) was supported on a 10-foot high, crushed rock buttress at the toe of the cut slope. A drainage course was placed behind the brushlayer fill to intercept and divert seepage down into the rock toe.
- *Live fascines* were placed along the crest of the cut above the brushlayer fill.
- A *coir ground net* with long straw mulch, grass seeding, and live stakes was used on gentler slopes underlain by bedrock close to the surface.

Evaluation: The slope is now stable, well vegetated, and blends in with the natural surroundings. A few minor problems occurred initially that were repaired without difficulty. Rilling through the brushlayers was observed in places, due to surface runoff from the crest of the slope. This problem was eliminated by constructing a brow ditch along the crest that intercepted runoff from above. Some cracking, of no consequence, was observed at the top and back of the brushlayer fill; it was caused by settlement, as opposed to mass movement. This settlement was probably exacerbated by the use of frozen earthen fill during construction.

The rooting and establishment of live cuttings in the brushlayer fill and live fascines were quite satisfactory. An exception occurred near the base of the brushlayer fill, where alder was used that was not selected and handled with the same degree of care given to the willow cuttings used in the rest of the

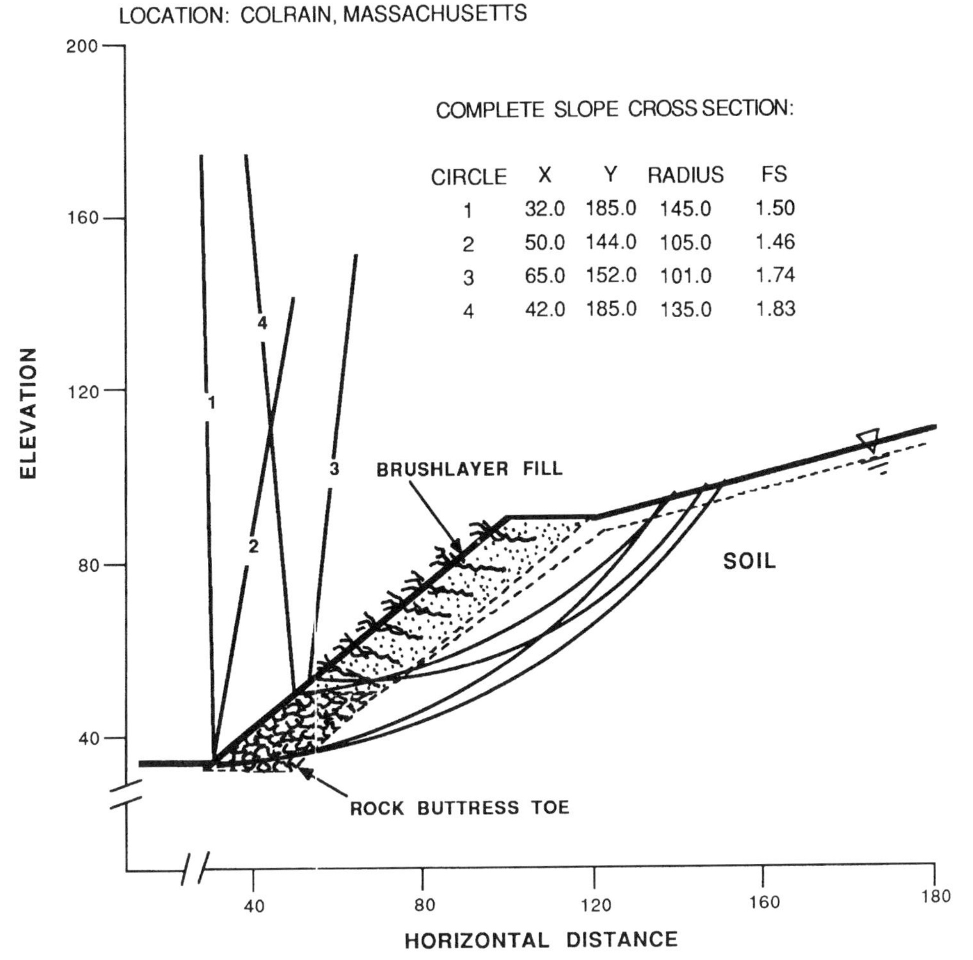

Figure 7-42. Factor of safety of cut slope stabilized by composite drained rock and earthen brushlayer fill. Greenfield Road, Colrain, Massachusetts.

fill. Even so, the brushlayers have served their intended function and purpose. The brushlayers have provided an opportunity for native vegetation to invade and establish on the slope. As a result, the process of plant succession is well underway and, after three years, the project site had already assumed a natural and pleasing appearance.

Figure 7-43. Brushlayer fill under construction during winter season. First course of live cut brush is being laid on top of rock toe buttress.

References

Gray, D. H., and R. Sotir (1992). Biotechnical stabilization of a highway cut. *Journal of Geotechnical Engineering* (ASCE) **118**(GT3): 335–353.

Gray, D. H., and R. Sotir (1995). Biotechnical stabilization of steepened slopes. *Transportation Research Board Record No. 1474*, National Academy Press, National Research Council, pp. 28–38.

Figure 7-44. Brushlayer fill installation immediately after construction. The earthen fill and brushlayers are supported on a rock toe buttress.

Figure 7-45. Brushlayer fill during first growing season. Brushlayers have rooted and leafed out.

Figure 7-46. Brushlayer fill 5 years after construction. The slope has remained stable and is well vegetated. Native, woody shrubs and trees have begun to invade and establish on the slope.

Application Number 2 River Bank Stabilization

Location: Buffalo Bayou, Houston, Texas.

Treatment Objectives: To repair and stabilize a high river bank slope adjacent to a bayou that had failed as a result of frequent flood discharges. To demonstrate the viability of a soil bioengineering protection system capable of withstanding severe erosion and mass wasting conditions in lieu of massive structural retaining or hard armor bank protection.

Nature of Problem: The combination of natural flooding and controlled releases resulted in the abrupt rise and fall of water level in the bayou in addition to prolonged periods of high water. These hydrologic conditions, combined with sandy and silty soils with little cohesion, resulted in widespread erosion and streambank failure.

Site Conditions

- Actively receding streambank approximately 280 feet (84 m) long and 25 to 35 feet (7.5 to 10.5 m) high on outside bend
- Loss of approximately 15 feet of land at top of bank and creation of near vertical scarps as a result of mass slope failures and streambank erosion
- Instability of over-steepened slope aggravated by the presence of fine sands and seepage of water from the bank face between 200 to 2000 gallons (0.75 to 7.5 m^3) per day

Treatment Considerations: Previous treatments, consisting of bank armoring with stacked cement filled bags tied together with steel reinforcing bars, were not only unsightly but also ineffective in preventing slope failures at this site. Successful treatment required redirecting the stream away from the foot of the bank, rebuilding and securing a toe berm, and stabilization of the slope face. Bank return flows during rapid drawdown after bankfull floods caused significant seepage erosion and slumping of the bank face. Accordingly, the interception and diversion of seepage water from the bank was also essential to successful stabilization. In addition to these functional considerations, visual and aesthetic concerns dictated a treatment that blended with the natural riparian surroundings.

A multi-pronged approach was adopted to deal with the various site conditions and requirements. A fill slope with a grade of 0.5 : 1 ($H:V$) was reconstructed upon a foundation of wrapped concrete rubble installed in a 7-foot (2-m) deep toe trench. The fill was constructed in 2-foot (61-cm) lifts wrapped with a synthetic geogrid and burlap. Live brush layers approximately 6 inches thick and long enough to extend from the undisturbed soil at the back and just beyond the reconstructed fill face were placed on each wrapped soil layer.

Because continued seepage and saturation of the fill would have substantially reduced the slope factor of safety, it was necessary to install appropriate drainage. This was accomplished by using vertical strip or chimney drains at the rear of the fill installed on 5-foot (1.5-m) centers that intercepted and collected the seepage water and conducted it into a drain beneath the fill. Water collected in the trench drain in turn was discharged to the bayou via pipes located at the upstream and downstream ends of the project.

A short live boom (groin) was installed at the upstream end to protect a sewer outfall and to keep the low-flow thalweg (the deepest part of the stream) away from the bank.

Soil Bioengineering Treatments

- A drained, earthen buttress fill with a grade of 0.5 : 1 ($H:V$) was reinforced with *vegetated geogrids*. The buttress fill was constructed upon a foundation of wrapped concrete rubble placed in a 7-foot (2-m) deep toe trench.
- *Live fascines* were placed on a midslope bench of reinforced fill.
- *Live staking* was installed on edges, sides, and selected areas of the project site.
- A *live boom* (a rock groin capped with a prism of earth reinforced with crisscrossing live fascines) was used to redirect the stream away from the bank.

Evaluation: The site has remained stable since construction, and the soil bioengineering installation has developed into a dense riparian buffer as native and naturalized species of herbaceous and woody vegetation invade the site. The site survived a major bankfull flood in the first and second year after construction. The supple stems of woody vegetation on the bank bent over during the flood

Figure 7-47. Appearance of the site before stabilization. Slope failures and streambank erosion resulted in near vertical scarps in high bank and substantial recession of the crest.

Figure 7-48. Slope protection system under construction. Vegetated geogrid reinforced fill was placed on top of rubble-filled trench at base. Note the vertical chimney drains at back, which discharge into the trench.

Figure 7-49. Project site immediately after construction. Benched fill reinforced with vegetated geogrids rests atop toe trench filled with wrapped concrete rubble.

Figure 7-50. Stabilized site and soil bioengineering installation a year after construction. Bank is stable, well vegetated, and has survived a major bankfull flood.

and protected the bank against scour erosion. Minor maintenance and pruning may be required periodically to keep the bank from becoming too overgrown and to remove large trees from the top or midslope portions of the bank.

References

Nunnally, N. R. and R. Sotir (1994). Soil bioengineering for streambank protection. *Erosion Control* **1**(4): 39–44.

Sotir, R. (1995). Soil bioengineering experiences in North America: in *Vegetation and Slopes*, edited by D. H. Barker. London: Thomas Telford, pp. 190–201.

Application Number 3 Gully Washout Repair

Location: Olf Silverhill Airfield, Alabama.

Treatment Objectives: To repair and stabilize a large gully that was expanding rapidly and encroaching on adjacent properties. To integrate the effectiveness of soil bioengineering treatments with conventional structural measures.

Nature of Problem: The gully was eroding by a process of headcutting, lateral widening, and channel lowering. Secondary or tributary gullies along the sides were encroaching onto adjacent private lands. Active erosion threatened to undermine a previously constructed drop structure at the head of the gully and was generating substantial offsite sediment.

Site Conditions

- Large, heavily eroded gully approximately 1500 feet long, 40 feet deep, with bed widths ranging from 20 to 100 feet
- Loamy, fine sand low in natural fertility and organic content; moderate to severe erosion hazard rating
- High rainfall in region coupled with high infiltration rates caused water to move rapidly into gully during storms, saturating the banks and causing severe seepage erosion at gully head and sides
- Large failing, existing concrete drop structure at head of gully to control headcutting

Treatment Considerations: The site was chosen by the Department of the Navy as a demonstration project to better understand test and integrate the effectiveness of biotechnical systems for repair and rehabilitation of badly eroded or gullied sites. The soil bioengineering system design for the site was required to be integrated into a conventional, structural protection system, which consisted of the initial drop structure at the gully head plus two additional grade stabilization structures designed by the USDA Natural Resources Conservation Service. The latter were constructed on the channel bottom and were designed to prevent any further downcutting and lowering of the channel bed.

Figure 7-51. Slope conditions before construction in the vicinity of the drop structure at the gully head, illustrating severe head cutting and side slope erosion.

Figure 7-52. Gully head area in the final stages of reconstruction, showing brushlayers on left bank.

Figure 7-53. View of left bank of gully head area three months of after soil bioengineering installation, showing brushlayers and branchpacking tied into wingwall of existing drop structure.

Figure 7-54. Live crib wall under construction, adjacent to gabion toe-wall and downstream of installed grade stabilization structure.

Figure 7-55. Live crib wall at toe slope and brushlayers during construction.

Figure 7-56. View of side slope in vicinity of the second grade stabilization structure showing brushlayers on downstream bank 3 months after installation.

The main function of the soil bioengineering system was to stabilize the gully sides and to provide some additional protection and roughness to the channel bottom between the three drop structures. Woody plant material for the soil bioengineering work was gathered locally and consisted primarily of black willow (*Salix nigra*) and coastal plain willow (*Salix caroliniana*). A grass and legume mix for secondary erosion control was also developed by the USDA Soil Conservation Service for seeding and interplanting after installation of the woody soil bioengineering systems.

Soil Bioengineering Treatments

- *Live crib walls* were constructed below an installed grade stabilization structure adjacent to each gabion toe-wall. The purpose of each was to protect the toe of the slope in a critical reach of the channel.
- *Brushlayers* were placed on small benches or terraces excavated on the gully sides in cut sections or placed simultaneously with lifts of earth in fill sections. The live cut brush extended 2 feet into the slope in the former and up to 8 feet in the latter.
- *Live fascines* were constructed on contour between brushlayers to provide extra protection against rill erosion on the gully sides.
- *Branchpacking* was used to reinforce and stabilize a fill section in a bank cavity at the gully head and to tie the grade stabilization structure at this location to the brushlayer installation on the gully side.
- *Vegetated geogrids* were used in critical sections with steeper, higher banks and/or where higher velocity flows were expected.
- *Live siltation construction* or trenches backfilled with live brush angled downstream were constructed in rows across the channel bottom.

Evaluation: For the most part the gully has been successfully repaired and stabilized. The soil bioengineering system was integrated into the conventional structural system without any difficulty. Lack of adequate rainfall, coupled with sandy and droughty soils at the site, severely tested plant survival. Despite drought conditions after installation of the soil bioengineering measures, plant survival was high enough and the inherent mechanical stabilization sufficient to stabilize the site. The brushlayers that extended deeper into the slope in the fill sections had a higher rate of survival and establishment than did the shallower cuttings placed in cut sections or in the live fascine trenches. Lack of adherance to plant material handling and installation specifications in some areas also affected survival and establishment.

References

Soil Bioengineering Corp. (1991). Soil bioengineering gully washout repair, Silverhill airfield, Baldwin County, Alabama. *NCEL Contract Report CR91.012*, Naval Civil Engineering Laboratory, Port Hueneme, CA, 39 pp.

Application Number 4 Backslope Stabilization

Location: Kananaskis Highway, Alberta, Canada.

Treatment Objectives: To repair and stabilize a large, open cut backslope adjacent to a highway that was experiencing both surficial erosion and mass movement. To evaluate the effectiveness of a variety of soil bioengineering treatments under harsh climatic and soil conditions.

Nature of Problem: The site had remained in a bare, denuded condition in spite of several attempts at hydroseeding with grass mixtures. In addition to surface erosion problems, several shallow slumps had occurred in the mid to lower portion of the cut 7 to 10 m up the face of the slope.

Site Conditions

- Open cut with slope face area of approximately 1000 m^2 and slope gradient of 1 : 1 becoming steeper at the top. The cut was 50 meters along the base and 30 meters up slope to the highest point.
- Sterile, sandy, gravely glacial outwash subsoil with low organic content, poor nutrient status, and high pH. Hydroseeded grasses unable to establish on the slope.
- North, northwest exposure with very cold winter and strong drying wind conditions the year round, which exacerbated plant establishment and erosion problems.

Figure 7-57. Kananaskis Highway 68 backslope prior to construction (1986).

Figure 7-58. Appearance of Kananaskis Highway backslope during the first growing season after construction.

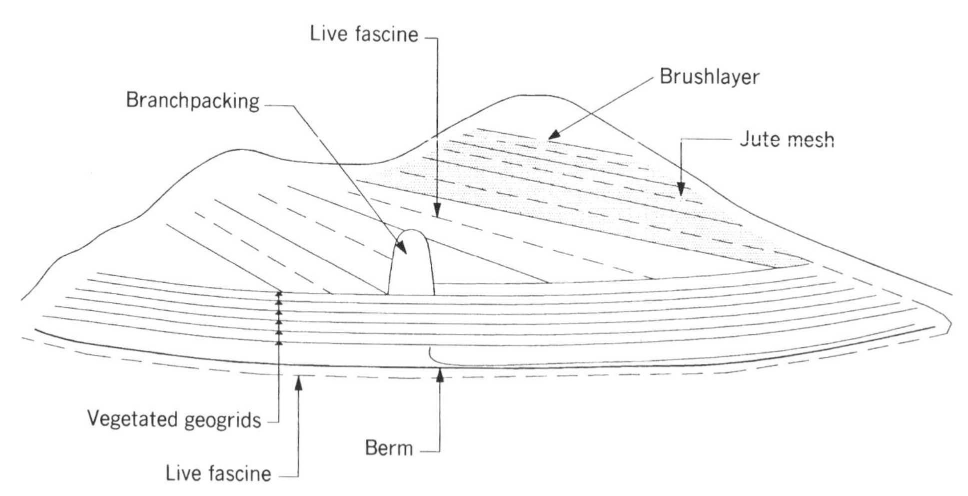

Figure 7-59. Schematic illustration of the Kananaskis backslope and the various soil bioengineering methods used. Live stakes were installed over the entire site except in the vegetated geogrid area and the branchpacking area. Rooted plants were installed over the upper slope area, above the vegetated geogrids, except in the branchpacking area.

Figure 7-60. Close-up view showing construction of vegetated geogrid, buttress fill at the toe of the backslope.

Figure 7-61. Close-up view showing placement of brushlayers on terraces excavated into slope in upper portion of cut.

Figure 7-62. Close-up view of live fascine and jute net installation on the Kananaskis Highway backslope.

Treatment Considerations: The site was chosen by Alberta Transportation and Utilities to evaluate soil bioengineering stabilization of slopes in a far northern climatic region. The site was judged to be representative of typical erosion problems experienced along the province's highway system in that area. A variety of treatments were tried on the slope to ascertain the performance of each method. The sterile nature of the soils exposed in the backslope, lack of available calcium, and a pH of 9 made vegetative establishment very difficult. In an attempt to improve conditions, loose material excavated from the site during initial grading operations was hauled to a nearby borrow site and mixed with a slightly better borrow soil there. The resulting mix had a slightly lower pH of 7.8, which was still above the optimum level for live cuttings and plants used in the soil bioengineering treatments.

Vegetation used for the soil bioengineering treatment was harvested offsite at locations ranging from 3 to 80 km away. The harvested plant material was transported to the site in flatbed trucks and stored in beaver ponds prior to use. Plant types used included many willow (*Salix*), wolf willow (*Elaeagnus*), balsam poplar, and some alder. Rooted plantings consisting of both nursery stock and local shrubs were employed, as well as broadcast seeding of grass and native flowers.

Soil Bioengineering Treatments

- *Vegetated geogrids* were used to stabilize the toe. Successive lifts of amended, excavated soil were wrapped with synthetic geogrids to create a buttress fill at the base of the backslope. Spruce bows were placed between the lefts.

- *Brushlayers* were placed on excavated benches (or terraces), and used to stabilize the upper portion of the backslope.
- *Live fascines* were installed between the brushlayers in the upper and middle section of the slope. Jute netting also was placed on the ground between live fascines, and anchored into the live fascine trenches as well.
- *Branchpacking* was used to stabilized a small slumped area in the middle part of the slope.
- *Live staking* was employed throughout the site to complement the other treatments and tie them all together.

Evaluation: The site was inspected and monitored over a 6-year period. The backslope remained stable and free of any significant erosion problems during this time. Plant establishment and survival in the soil bioengineering installations was minimal because of the poor soil and harsh and unfavorable site conditions. Results from this project underscore the importance of properly assessing soil pH and nutrient status and amending or ameliorating the soil as necessary. The residual mechanical reinforcing effect provided by these systems nevertheless provided significant benefits. Rooted transplants did relatively well. Installed construction costs on this site were relatively high (\$64(Can)/m^2), mainly because of time spent training work crews and transportation of the work force and plant materials to the site.

Reference

Tetteh-Wayoe, H. (1994). Soil bioengineering demonstration project in Alberta. Alberta Transportation and Utilites Report No. ABTR/RD/-94/02, Research and Development Branch, Edmonton, Alberta, 28 pp.

7.14 REFERENCES AND ADDITIONAL READING

Gray, D. H., and A. T. Leiser (1982). *Biotechnical Slope Protection and Erosion Control.* New York: Van Nostrand Reinhold.

Gray, D. H., and R. Sotir (1992a). Biotechnical stabilization of a highway cut. *Journal of Geotechnical Engineering* (ASCE) **118** (GT10) 335–353.

Gray, D. H., and R. Sotir (1992b). Biotechnical stabilization of cut and fill slopes. *Proceedings*, ASCE-GT Specialty Conference on Slopes and Embankments, Berkeley, CA, June, Vol. 2, pp. 1395–1410.

Gray, D. H., and R. Sotir (1995). Biotechnical stabilization of steepened slopes. Transportation Research Board Record No. 1474, National Academy Press, National Research Council, pp. 23–38.

Kraebel, C. J. (1936). Erosion control on mountain roads. USDA Circular No. 380, 43 pp.

Schiechtl, H. M. (1980). *Bioengineering for Land Reclamation and Conservation*. Edmonton, Canada: University of Alberta Press, 404 pp.

Sotir, R. B. (1991). A review of recent soil bioengineering projects in North America. *Proceedings*, NSF Workshop on Biotechnical Stabilization, Ann Arbor, MI, August.

Sotir, R. B. (1995). Unit cost data for soil bioengineering measures obtained from projects supervised by Robbin B. Sotir & Associates, Personal communication.

Sotir, R., and D. H. Gray (1989). Fill slope repair using soil bioengineering systems. *Proceedings*, XX International Erosion Control Association Conference, Vancouver, pp. 473–485.

Thielen, D. L., and J. G. Collin (1993). Geogrid reinforcement for surficial stability of slopes. *Proceedings*, Geosynthetics '93 Conference, Vancouver, pp. 229–244.

USDA Soil Conservation Service (1992). Chapter 18: Soil bioengineering for upland slope protection and erosion reduction. Part 650, 210-EFH, *Engineering Field Handbook*, 53 pp.

8 Biotechnical Stabilization: Guidelines for Vegetative Treatment of Revetments and Retaining Structures

8.0 INTRODUCTION

Biotechnical slope protection and erosion control entail the integrated or conjunctive use of plants and structures. Plants can be introduced and established in and around structural systems, as summarized previously in Table 4-1 under the category of "mixed construction" methods. Plants can be introduced, for example, on the benches of tiered retaining wall systems or, alternatively, they can be inserted and established in the interstices or frontal openings of porous revetments, cellular grids, and retaining structures. They can also be introduced in and around check dams.

Guidelines and procedures are presented in this chapter for the incorporation of live vegetation in a variety of structural systems, both retaining structures and revetments. Live cuttings are used primarily for this purpose, but rooted plants (or transplants) and seeding can be used as well. The conjunctive use of vegetation and structural/mechanical augmentation in ground cover systems, that is, *biotechnical ground covers or reinforced grass*, are discussed in Chapter 9.

8.1 ROLE AND FUNCTION OF STRUCTURAL COMPONENTS

Structural components of biotechnical earth support and slope protection systems must be capable of resisting external forces causing sliding, overturning, and bearing capacity failure. In addition these structures must resist internal forces that cause shear, compression, and bending stresses. In the case of revetments or slope armor systems, the armor units must have sufficient weight and interlocking to resist displacement under wave action or tractive stresses exerted by flowing water. Design guidelines to insure both the external and the internal stability of the structural components are discussed in greater detail in Chapter 5. Only a brief review is presented in this section of the basic role and function of the structural components.

8.1.1 Retaining Structures

Retaining structures are designed to resist fairly large, lateral earth forces without excessive displacement or rotation. Gravity structures resist these external forces primarily as a result of their weight. Crib, gabion, cantilever, and reinforced earth walls are examples of gravity structures. Structures with porous faces or frontal openings lend themselves to vegetative treatment. Vegetation can also be introduced on the benches of tiered, or stepped-back retaining structures.

8.1.2 Revetments

Revetments are designed primarily to armor a slope against scour and erosion from wave action and streamflow. They are not designed to resist large lateral earth forces and normally are placed on slopes no steeper than 1.5 : 1 ($H : V$). Revetments resist displacement by the weight and interlocking characteristics of the armor units. A revetment must be placed on a suitable filter course or blanket to avoid washout of fines or soil behind the armor units.

8.1.3 Articulated Block Walls

Articulated block walls and rock breast walls are not designed to resist large lateral earth forces. They provide some lateral earth support and protection to the toe of slopes. Articulated block walls resist external forces primarily by their weight. The resultant earth force acting on the wall is reduced considerably by battering or inclining the wall against a slope. Local shear displacement between the blocks (or rocks) is resisted by friction and interlocking (or articulation) between the structural units.

8.1.4 Slope Gratings (Three-Dimensional Cellular Grids)

A slope grating or cellular grid provides some restraint or lateral earth support; however, its main purpose is to facilitate and permit the establishment of vegetation on steep slopes, that is, slopes steeper than 1.5 : 1 ($H : V$). A slope grating system covers bare slopes that are vulnerable to rapid weathering and slaking disintegration when exposed, for example, certain shales

8.2 SELECT SOIL BACKFILL FOR PLANT MATERIALS

8.2.1 Engineering versus Agronomic Requirements

Select soil backfill refers to the fill material that will be used around live cut plant material or other vegetation that is introduced into and around structures. It should be free of any material or substance that could be harmful to plant growth and should be nutrient tested prior to use. There may be some conflict between engineering specifications for a backfill as opposed to agronomic specifications. Coarse, well-graded sands and gravels are the most suitable, and clays the least suitable from an engineering point of view. Gravel or coarse sand, however, are not suitable material for use as a fill around live plant materials, nor are heavy plastic clays. Select soil backfill need not be organic topsoil, but it must be able to support plant growth.

8.2.2 Techniques for Achieving Compatibility

There are several ways around this seeming conflict between engineering and agronomic requirements. In many cases the requirements are congruent and compatible, for example, the need for well-drained backfill and the exclusion of heavy, plastic clays from consideration. The main requirement from an agronomic point of view—which conflicts somewhat with engineering requirements—is the need for sufficient fines in the backfill to provide some moisture and nutrient retention and availability for live plants. Possible means of resolving these conflicting requirements include modifying the specifications for the backfill, amending selected zones, and placing cuttings carefully.

Modification of Backfill Specifications: It may be possible to relax engineering backfill specifications to allow the presence of nonplastic fines in the backfill or cribfill without compromising engineering performance. Some commercially available crib wall retaining systems already allow this modification.

Amended Zones: Fines and other soil amendments can be added to selected pockets or zones where the live cuttings will be placed or inserted without affecting the overall engineering integrity or performance of the rest of the backfill.

Placement of Cuttings: In many instances it may be possible to insert or place the cuttings through the structure so that basal ends are imbedded in native, undisturbed earth behind the structure. This procedure is possible, for example, with relatively thin retaining walls, such as articulated block walls. It can also be used with rock breast walls and revetments such as riprap and gabion mattresses.

8.2.3 Soil Testing and Preparation

Agronomic soil samples of the proposed backfill should be taken prior to live woody plant installation. Nutrient testing by an approved laboratory should include analyses for a full range of nutrient and metal contents: nitrogen, phosphorus, potassium, pH, as well as any toxins present that would be harmful to plant growth. The laboratory reports should also include any recommended fertilizer and lime amendment requirements for woody plant material.

8.3 VEGETATED RIPRAP (JOINT PLANTING)

8.3.1 Description

Joint planting refers to the insertion of live cuttings (stakes) in the openings or joints between the rock in a riprap revetment, as shown in Figures 8-1 and 8-2. Alternatively, the cuttings can be tamped into the ground at the same time the rock is being placed on the slope face. The latter approach facilitates installation of the cuttings but also complicates rock placement and increases the likelihood of damage to the cuttings if the rock is tailgated or dumped in place.

8.3.2 Objective

Live cuttings placed in this manner should extend into the soil beneath the stone armor, as illustrated in Figure 8-1. The objective is to have these live cuttings root in the soil beneath the riprap, thus reinforcing the bank, anchoring the riprap, and improving drainage by extracting soil moisture.

8.3.3 Effectiveness

A vegetated riprap revetment (joint planting) provides the following advantages:

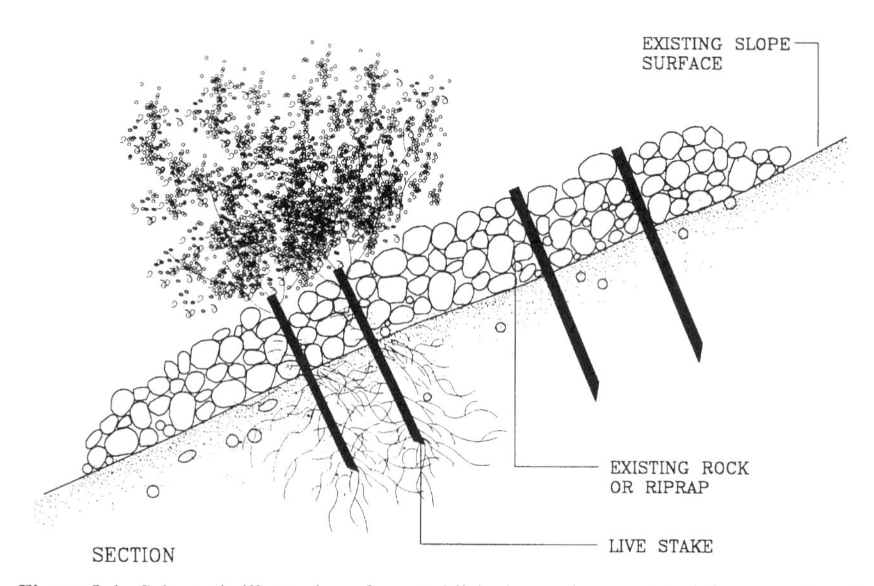

Figure 8-1. Schematic illustration of an established, growing vegetated riprap revetment.

Figure 8-2. Photo of vegetated riprap revetment showing cuttings that have rooted and sprouted between the armor rocks.

- It improves the performance of the armor layer by preventing the washout of fines and by reinforcing the underlying native soil.
- It has a more natural appearance and is less visually intrusive than a structural treatment alone.
- It provides some riparian cover and wildlife habitat.
- It helps to slow water velocities near the bank and trap sediment.

8.3.4 Materials

Live materials consist of cut stakes that are 1 to $1\frac{1}{2}$ inches in diameter and long enough to reach beyond the base of the riprap, as shown in Figure 8-1. Willow cuttings work best for this purpose. The cuttings must be fresh and must be kept moist after they have been prepared into appropriate lengths. They should be installed the same day that they are prepared. The *inert* construction materials consist of rocks, which should be sized to resist dislodgment by waves or currents and a filter course, which should be designed to prevent washout of fines in the native soil beneath the revetment.

8.3.5 Installation

The following general guidelines and procedures can be followed for constructing a vegetated rock revetment on a slope:

- Grade the bank back to a slope no greater than $1\frac{1}{2}:1$ $(H:V)$ and secure the filter fabric on the slope. Place the rock armor layer on top of the filter course, being careful not to damage or puncture the filter layer.
- Tamp the live cuttings into the openings (joints) between the rock. It may be necessary to use an iron bar or rod to create a pilot hole. The latter may also be necessary for penetration through the filter fabric.
- The live stakes should be oriented as perpendicular to the slope as possible with the growing tips protruding slightly above the finished face of the rock, as shown in Figure 8-1. The basal end of the stake must fit snugly in the hole beneath the revetment.
- Tamping the cuttings into the ground is best accomplished with a dead blow hammer (i.e., a hammer with the head filled with lead shot). Avoid stripping the bark during tamping as this will stress and kill the stake.
- Place the stakes in a random configuration in the revetment at a density ranging from two to four cuttings per square yard. The exact placement locations will depend on the positions of openings between the rocks.

8.4 VEGETATED GRADE STABILIZATION STRUCTURES

8.4.1 Description

Vegetated grade stabilization structures refers to the ancillary planting of live cuttings (stakes) around gully grade stabilization or control structures. The latter include check dams, drop structures, and gully plugs.

8.4.2 Objective

The cuttings are inserted in various arrays to complement and enhance the performance of the structures. The live stakes also facilitate the growth and establishment of native vegetation in a gully, which is an essential component of gully healing and stabilization.

8.4.3 Materials

Live materials consist of cut stakes that are 1 to $1\frac{1}{2}$ inches in diameter, similar to stakes used in regular live staking. Willow cuttings usually work best for this purpose. The cuttings must be fresh and must be kept moist, after they have been prepared into appropriate lengths. They should be installed the same day that they are prepared.

8.4.4 Installation

The following general guidelines and procedures can be followed for staking a grade stabilization structure with live willow cuttings.

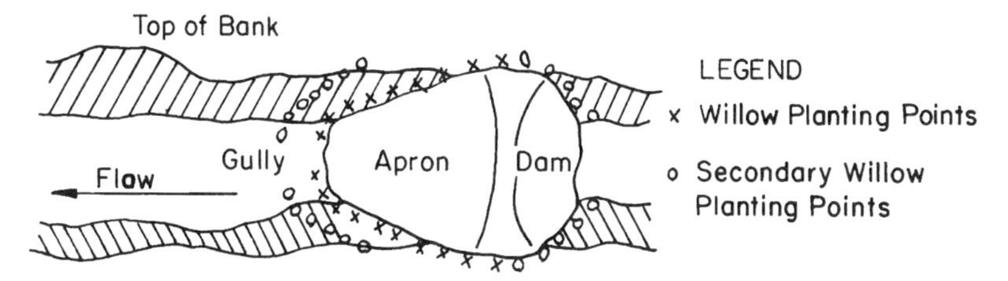

Figure 8-3. Plan of gully check dam showing points for installing live stakes.

Staked Check Dams: For use in conjunction with check dams, the live cuttings should be planted around the dam, as shown in Figure 8-3. Normally, the stakes should be spaced 12 to 18 inches apart. On large dams the cuttings should also be planted at additional points marked "secondary" in Figure 8-3. This secondary planting consists in extending the willows part way around the upstream face of the check dams, and in planting another row about 2 feet downstream from the apron. An example of ancillary use of live cuttings around a series of board check dams is shown in Figure 8-4.

Staked Gully Plugs: The head of a gully is a particularly vulnerable location. Drop structures and rock plugs can be placed at these locations to stop headward erosion of the gully. A plug is more effective if vegetation can become established in and around it soon as possible. In order to strengthen gully head

Figure 8-4. Example of conjunctive use of live willow stakes and board check dams to control gully erosion.

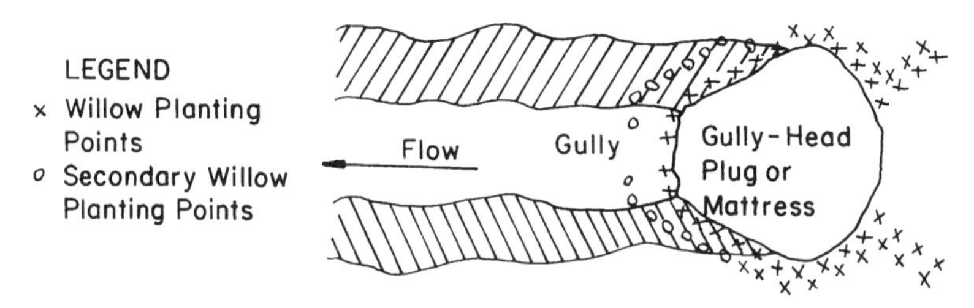

Figure 8-5. Plan of gully head plug or structure showing points for installing willow cuttings.

plugs, willow stakes can be planted in the pattern shown in Figure 8-5. The stakes should be spaced 12 to 18 inches apart.

8.5 VEGETATED GABION MATTRESSES

8.5.1 Description

Gabion mattresses, sometimes referred to as "Reno" mattresses, are shallow, rectangular containers fabricated from a triple twisted, hexagonal mesh of heavily galvanized steel wire. These containers are laid on a slope or streambank, tied together, and then filled with rock to form a revetment or armor layer. Live stakes are then inserted in the open spaces between rocks that have been previously placed in the gabion baskets, as shown in Figures 8-6 through 8-8.

8.5.2 Objective

The live stakes placed in this manner should extend into soil backfill behind the gabion baskets, as illustrated in Figure 8-6. The objective is to have these live cuttings root in the soil beneath the gabions, thus helping to anchor the gabion revetment to the ground and to bind the revetment together in a coherent unitary mass.

8.5.3 Effectiveness

A vegetated rock gabion mattress provides the following advantages:

- It improves the performance of the armor layer by preventing the washout of fines and by reinforcing the underlying native soil.
- It has a more natural appearance and is less visually intrusive than a structural treatment alone.
- It provides some riparian cover and wildlife habitat.
- It helps to slow water velocities near the bank and trap sediment.

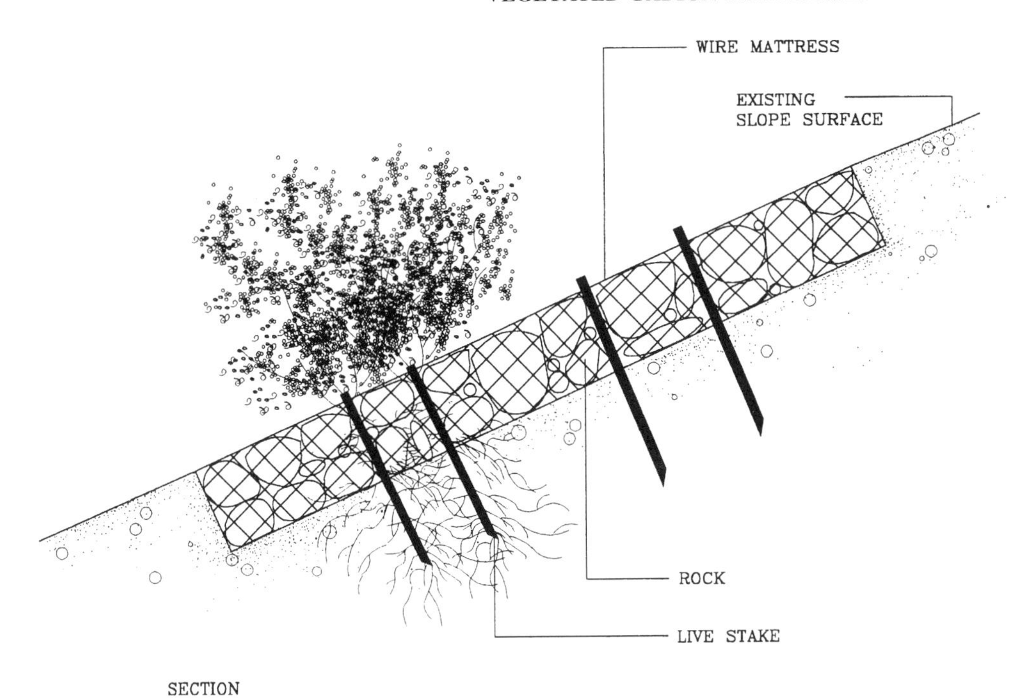

Figure 8-6. Schematic illustration of an established, growing vegetated gabion mattress.

Figure 8-7. Gabion revetment protecting streambank along Carmel River, Monterey, California. Photo taken during construction. (Photo courtesy of Terra Aqua Gabions)

Figure 8-8. Same gabion revetment (Figure 8-7) after insertion of willow cuttings. Photo taken 3 months after construction. (Photo courtesy of Terra Aqua Gabions)

8.5.4 Materials

Live materials consist of cut stakes that are 1 to $1\frac{1}{2}$ inches in diameter and long enough to reach beyond the bottom of the rock baskets, as shown in Figure 8-6. Willow cuttings work best for this purpose. The cuttings must be fresh and must be kept moist, after they have been prepared into appropriate lengths. They should be installed the same day that they are prepared. The *inert* construction materials consist of the gabion wire baskets, binding wire, and gabion rock fill. Gabion mattresses are normally no thicker than 1 foot.

8.5.5 Installation

The following general guidelines and procedures can be followed for constructing a vegetated gabion mattress system on a streambank:

- Grade the bank back to a slope no greater than $1\frac{1}{2}:1$ $(H:V)$ and place the wire baskets on the bank following manufacturer's installation guidelines for rock gabion baskets.
- After filling the baskets with rock, insert cuttings (stakes) through openings in the rock. It may be necessary to use an iron bar or rod to create a pilot hole. The basal end of the stake must fit snugly in the hole beneath the revetment; loose stakes will dry out and fail to root.

- Tamping the cuttings into the ground is best accomplished with a dead blow hammer (i.e., a hammer with the head filled with shot or sand). Avoid stripping the bark during driving as this will stress and kill the stake. Using an iron bar to create a pilot hole will minimize this problem.
- Place the stakes in a random configuration in the revetment at a density ranging from two to four cuttings per square yard. The exact placement locations will depend on the positions of openings between the rocks.
- Soil may be drifted into the gabion revetment after installation of the cuttings. A sandy soil with good drainage characteristics that fills the spaces between the rocks works best. This soil helps the establishment of the cuttings, accelerates the colonization of the revetment by native plants, and lends a more natural appearance.

8.6 VEGETATED GABION WALLS

8.6.1 Description

Gabions are rectangular containers fabricated from a triple twisted, hexagonal mesh of heavily galvanized steel wire. Gabions are normally supplied folded flat and bundled together for easy handling. The empty gabion baskets are placed in position and wired to adjoining baskets in various stacking arrays or configurations, as shown in Figures 8-9 and 8-10. They are then filled with rocks. Live cut branches can be placed in horizontal layers *in between* the rock-filled gabion baskets, as shown schematically in Figure 8-9. Some cuttings can also be inserted *through the baskets* (see Figure 8-11) themselves while they are being backfilled with rock, but this procedure is more difficult to implement.

8.6.2 Objective

The branches placed in this manner should extend into the soil backfill behind the gabion baskets, as noted in Figure 8-9. The objective is to have these live cuttings and branches root into the soil behind the gabions, and eventually within the interior of the gabions themselves, thus helping to bind the structure and backfill into a coherent, unitary mass.

8.6.3 Effectiveness

A vegetated rock gabion wall system provides the following advantages:

- It is helpful at the base of slopes where a low toe-wall can be used to reduce the steepness of a slope and protect the toe against scour and undermining.
- It has a more natural appearance and is less visually intrusive than a structural treatment alone.
- It avoids encroachment by use of a more vertical protective structure.

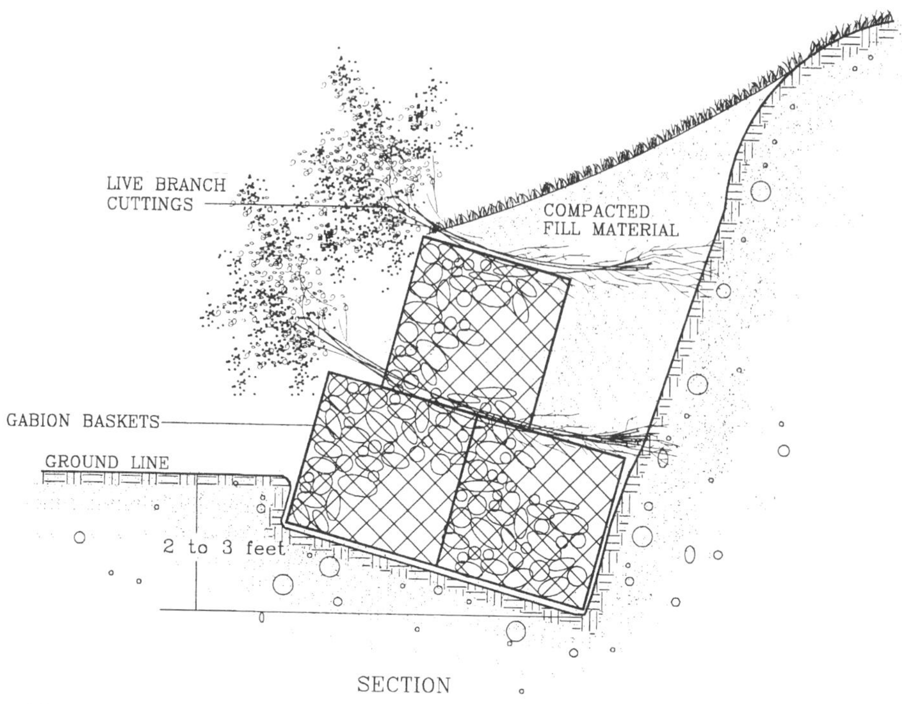

Figure 8-9. Schematic illustration of an established, growing vegetated, gabion wall system showing insertion of live cuttings between rock-filled wire baskets.

Figure 8-10. Photo of gabion wall constructed of rock-filled wire baskets that are assembled in various stacking arrangements.

Figure 8-11. Close-up view of vegetated gabion wall. Live willow cuttings that were inserted through gabions into backfill have rooted and leafed out.

8.6.4 Materials

Live materials consist of branches or cuttings that are $\frac{1}{2}$ to 1 inch in diameter and long enough to reach beyond the back of the rock baskets, as shown in Figure 8-9. The *inert* construction materials consist of the gabion wire baskets, binding wire, gabion rock fill, and select backfill. The latter must be capable of supporting plant growth.

8.6.5 Installation

The following general guidelines and procedures can be followed for constructing a vegetated gabion wall system:

- Starting at the lowest point, excavate a 2- to 3-foot deep footing area below natural grade. The footing base should be inclined into the slope so that the structure will have a batter (inclination off vertical) of at least $1:6$ $(H:V)$. This battering provides additional stability to the structure.
- Place the first course of fabricated wire baskets on the prepared footing base and fill them with rock. Two or possibly more baskets placed side by side may be required in the initial course, depending on the height of the wall, as shown in Figure 8-9.
- *As a rule of thumb, gabion walls require a minimum width to height ratio of 0.5 for external stability. Battering and the use of counterforts (units that extend perpendicularly from the wall into the backfill) tend to reduce*

the minimum width to height ratio requirement. The structural wall design should be approved by a qualified geotechnical engineer.

- Spread a thin layer of earthen backfill atop each successive tier or course of rock-filled closed wire baskets. Select backfill should also be placed behind the gabion baskets to an elevation that is level with the top of the gabions.
- Place the live branches or cuttings on the soil/rock-filled wire baskets. They should be placed at right angles to the wall with the growing tips to the front and butt ends in the backfill behind the wall. The tips of the cuttings should extend a few inches beyond the front and the butt ends, well into the backfill behind the gabions.
- The live cuttings or branches should then be covered with another thin layer of soil. Finally, the soil or backfill should be compacted and or tamped firmly to ensure good contact with the cuttings.
- The remainder of the vegetated rock gabion erection procedure is repeated, as each course is added, until the required height is reached.

8.7 VEGETATED ROCK WALLS

8.7.1 Description

A vegetated rock wall is essentially a rock breast wall in which live cuttings are inserted between the rocks and also placed in the soil bench formed above the wall, as shown schematically in Figure 8-12.

8.7.2 Objective

Unlike conventional retaining structures, rock breast walls are placed against relatively undisturbed earth and little if any backfill is placed behind the wall. Rock breast walls are not designed to withstand large lateral earth stresses. Their main purpose is to finish off and protect the toe of a slope against undermining. They also help to decrease the steepness of a slope at its base and facilitate the establishment of vegetation.

8.7.3 Effectiveness

A vegetated rock breast wall provides the following advantages:

- It is helpful at the base of slopes, where a low toe-wall can be used to reduce the steepness of a slope and protect the toe against scour and undermining.
- It has a more natural appearance and is less visually intrusive than a structural treatment alone.
- It facilitates establishment of vegetation at the toe of the slope.
- Plant roots provide articulation and help to bind the rock together into a coherent, unitary mass and reduce the danger of local shear failure.

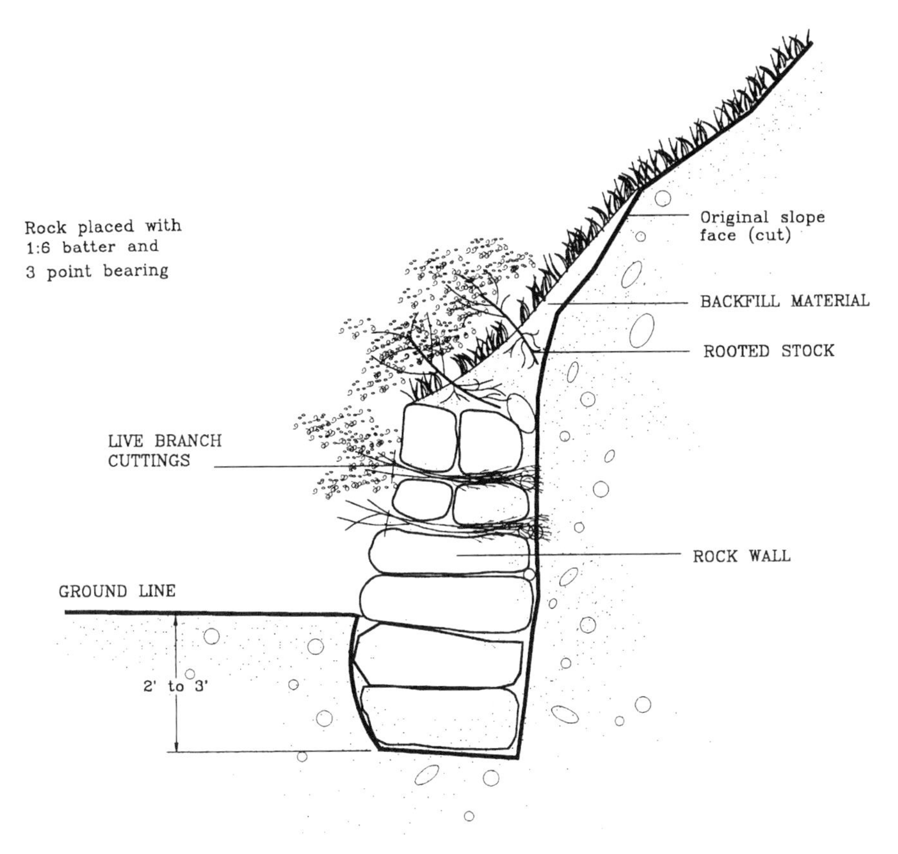

Figure 8-12. Schematic illustration of an established growing vegetated rock wall.

8.7.4 Materials

Live materials consist of branches or cuttings that are $\frac{1}{2}$ to 1 inch in diameter and long enough to reach beyond the rock structure into the fill or undisturbed soil behind, as shown in Figure 8-12. The *inert* construction materials consist of rocks and fill material for the wall construction. The rocks used should normally range from 8 to 24 inches in diameter. Large boulders should be used for the base.

8.7.5 Installation

The following general guidelines and procedures can be followed for constructing a vegetated rock wall system:

- Starting at the lowest point of the slope, remove loose soil until a stable base is reached. This will normally entail excavating a 2 to 3 foot deep footing area or trench below natural grade. The footing base should be sloped down slightly so that the rock will have a batter (inclination off vertical) of at least $1:6$ $(H:V)$. This battering provides additional stability to the structure.

- Excavate a minimum amount from the toe of the existing slope to provide a suitable recess for the wall. The back face of the excavation should also have a slight batter or inclination, as shown in Figure 8-12.
- Provide a well-drained, coarse aggregate base in locations subject to deep frost penetration.
- Place rocks with at least a three-point bearing on the foundation material or underlying rock courses. Place large rocks or boulders at the bottom and smaller rocks at the top. Also place the rocks so that their center of gravity is as low as possible, with their long axis slanting inward toward the slope if possible.
- When a rock wall is constructed adjacent to an impervious surface, place a drainage system at the back to provide an appropriate drainage outlet.
- The overall height of the wall should not exceed 5 feet. *Increasing the wall batter angle improves its stability. Higher walls require the use of very large rocks or boulders or a very thick base and their design should be approved by a qualified geotechnical engineer.*
- Live branch cuttings can be placed between the rocks during construction or tamped/inserted between the rocks after construction. The basal ends of the cuttings should extend as far back as possible into the native soil or fill behind the wall.
- Live cuttings and/or transplants can also be placed in a sloping bench of suitable backfill above the wall, as shown in Figure 8-13.

Figure 8-13. Rock breast wall protecting toe of road cut. Live fascines and grass seed were placed on the bench above the wall.

8.8 VEGETATED CRIB WALLS

8.8.1 Description

A vegetated crib wall consists of a hollow, box-like interlocking arrangement of structural beams. In *conventional* cribwalls the structural members are fabricated from concrete, wood logs, and dimensional timbers (usually treated wood). In *live* cribwalls (see Chapter 7) the structural members are usually untreated log or timber members. The structure is filled with a suitable backfill material (or cribfill) and layers of live branch cuttings that root inside the crib. The structure (or crib) eventually rots away, but its function is replaced by the cribfill, which becomes indurated with roots and which behaves as a coherent gravity structure itself.

8.8.2 Objective

The frontal spaces between the stretchers in conventional crib walls provide openings through which vegetative cuttings or rooted plant material can be inserted and established in the cribfill. The vegetation provides an attractive screen or landscaping touch on the face of the crib wall, as shown in Figures 8-14 and 8-15. The vegetation also helps to prevent the cribfill from washing out through the frontal openings.

Figure 8-14. An open-front crib wall supporting a roadway. Volunteer vegetation has become established in the frontal bays or openings of the wall.

Figure 8-15. Close-up view of open-front, concrete crib wall utilizing "dog-bone" header design. Note establishment of vegetation in cribbing.

8.8.3 Effectiveness and Applications

A vegetated crib wall system provides the following potential uses and advantages:

- It is helpful at the base of slopes where a low toe-wall can be used to reduce the steepness of a slope and stabilize the toe against scour and undermining.
- It has a more natural appearance and is less visually intrusive than a structural treatment alone.
- It avoids encroachment on limited rights-of-way by use of a more vertical protective structure.
- It minimizes problems with graffiti and defacement of retaining walls.

8.8.4 Design and Installation Guidelines

Conventional, open-front crib walls are typically constructed out of concrete or dimensioned, treated timbers. The crib wall structure is formed by joining a number of cells together in various stacking arrangements and filling them with suitable fill. In crib structures, the members are essentially assembled "log cabin" fashion. The frontal, horizontal members are termed stretchers; the lateral members, headers.

The crib wall must be designed to withstand expected lateral earth forces and must satisfy external stability requirements (see Chapter 5). The crib structure must also satisfy internal stability, that is the structural members must be capable of sustaining expected shear, bending, and compressive forces. A number of commercial crib wall systems and standard designs are available that satisfy these requirements. *The design/construction of higher crib walls should be approved by a qualified geotechnical engineer.*

8.8.5 Examples of Vegetated, Open-Front Crib Wall Systems

A number of commercial cribwall systems (Hilfiker, 1972; Schuster et al., 1973; AWPI, 1981, Jaecklin, 1983; Randall and Wallace, 1987) are available that are constructed from either concrete or treated wood. Each has particular advantages with regard to cost, availability, constructability, and ease of vegetative treatment. Crib wall systems that lend themselves to the introduction of vegetation in frontal openings include the following:

- Hilfiker Concrib Walls—Conventional concrete crib walls
- Evergreen Wall System—Interlocking, stacked, modular units
- AWPI Crib Wall—Treated timber cribbing with vertical support posts
- Perma-Crib System—Conventional log cribbing

Of these crib wall systems, only the Evergreen wall was designed specifically to accommodate the introduction of plants and vegetation into openings in the front face. Shelves created by the stretches can be planted with vegetation that eventually covers the entire wall. Less than 50 percent of the exposed face is concrete. Examples of the Evergreen wall systems are shown in Figures 8-16

Figure 8-16. Open-front, Evergreen retaining wall constructed to support vertical cut along roadway. Use of vertical retaining wall avoids encroachment of road right-of-way on hillside. (Photo courtesy of Evergreen Systems, AG.)

Figure 8-17. Vegetation has been introduced into the frontal bays or openings of Evergreen retaining wall. Vegetation mutes harsh or stark appearance of the wall. (Photo courtesy of Evergreen Systems, AG.)

and 8-17. Engineering requirements on the cribfill are less stringent in the case of the Evergreen wall.

8.9 VEGETATED CELLULAR GRIDS

8.9.1 General Description

A cellular grid is essentially a lattice-like array of structural members that is fastened or anchored to a slope, as shown in Figure 8-18. The structural members may be either concrete, timber, or a three-dimensional expandable, polymeric web. The spaces within the lattice or honeycomb array are planted with suitable vegetation. The grid structure itself does little to armor or buttress the slope (i.e., it is not a true revetment); instead its main purpose is to facilitate the establishment of vegetation on steep, barren slopes. Examples of various types of cellular grid structures are shown in Figures 8-18 through 8-20.

8.9.2 Applications and Effectiveness

A cellular grid offers several advantages over other methods:

- It requires little excavation and clearance at the foot of the slope.

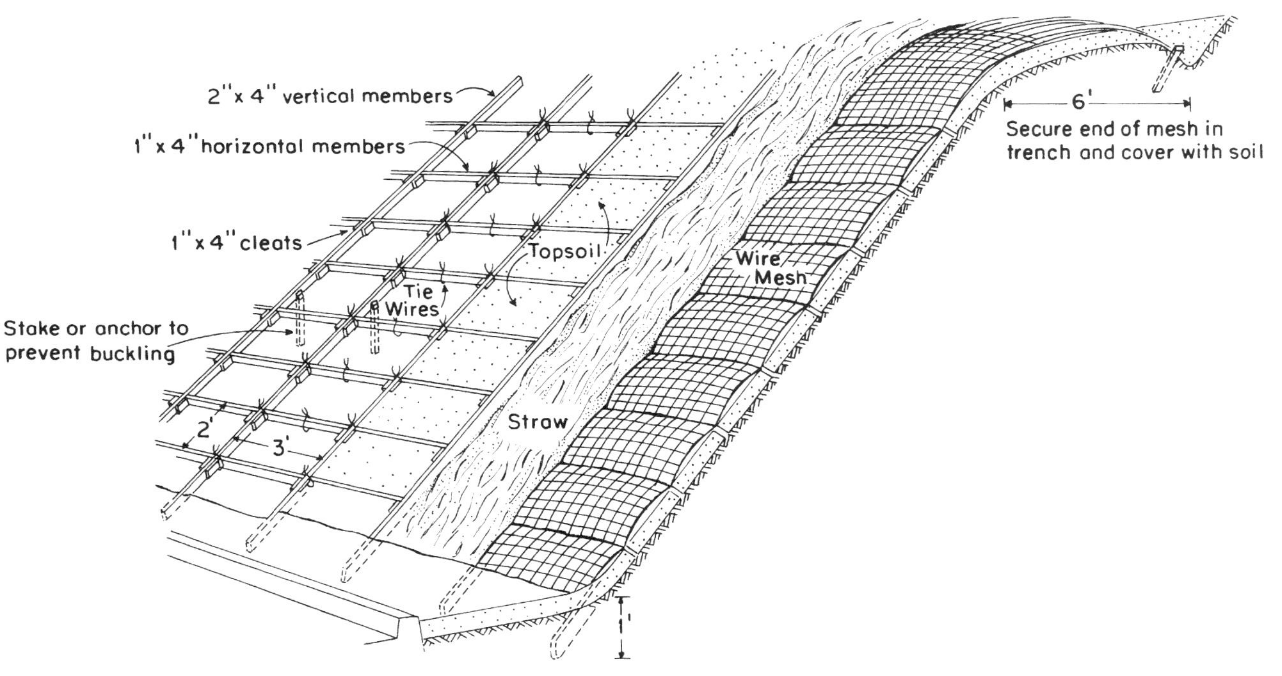

Figure 8-18. Anchored timber grid used to hold topsoil and slope plantings.

Figure 8-19. Vegetated, concrete cellular grid system.

Figure 8-20. Three-dimensional, expandable, "honeycomb" cellular grid. (Used with permission of Presto Products, Inc.)

- It permits establishment of vegetation on very steep slopes (up to 1 : 1) without the need for slope flattening.
- It does not require the importation of select backfill and cribfill.
- When filled with earth it covers and protects underlying exposed bedrock (especially certain shales) from slaking and disintegration.

Cellular grid structures cannot be placed on rough, broken, or severely rilled slopes. The slope must be relatively smooth beforehand, or it must be scaled and graded to this condition.

8.9.3 Installation Guidelines

The simplest cellular grid consists of a ladder-like array constructed from 2 × 4 timbers as shown in Figure 8-18. The openings in the grid are 2 × 3 feet. The grid structure is fastened to stakes driven into the slope. Topsoil and seeds can be introduced into the spaces so as to completely cover the framework. The surface is then covered with straw, which is held down by a jute or coir netting. The latter is secured tightly to the grid with tie wires, as shown in Figure 8-18.

8.10 TOE-WALL WITH SLOPE FACE PLANTINGS

8.10.1 General Description

Plantings cannot easily be established on slopes steeper than $1\frac{1}{2}:1$ $(H:V)$. A low retaining structure at the foot of a slope often makes it possible to flatten the slope slightly, establish vegetation, and minimize the amount of cut slope reworking, as shown schematically in Figure 8-21. The wall also protects the toe of the slope against scour and undermining at the base. Plants on the face of the slope protect it against surface erosion.

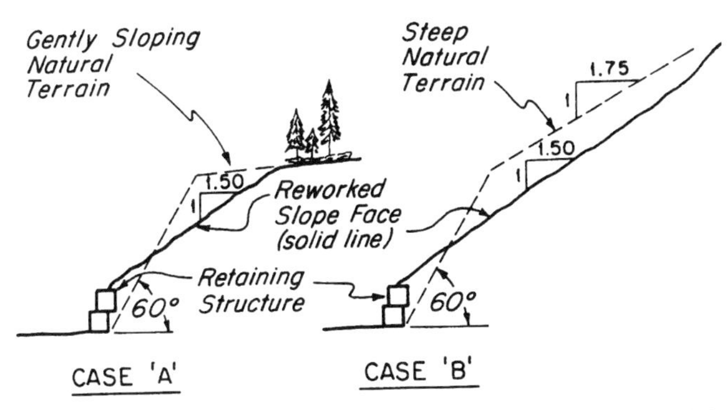

Figure 8-21. Influence of toe structure on amount of cut slope reworking and slope gradient. (From White and Franks, 1978.)

8.10.2 Suitable Wall Types and Designs

Several basic types of retaining structures can be employed as low toe-walls. The simplest type is a gravity wall that resists lateral earth pressures by its weight or mass. The following types of retaining structures can be classified as gravity walls and used as toe structures:

- Masonry and concrete walls
- Crib and bin walls
- Gabion walls
- Cantilever and counterfort walls
- Reinforced earth and geogrid walls
- Articulated block walls

A slope above the wall creates higher lateral pressures against the wall than does a level surface. This lateral earth pressure can be reduced substantially by inclining or battering the wall against the slope, as discussed in Chapter 5. Battering is essential in the case of articulated block walls (see Figure 8-22), which have a relatively small width/height ratio and, therefore, cannot resist very high lateral earth forces. Articulated block walls also depend upon their articulation to resist local shear failure between the units. Vegetation can be introduced into interstices between blocks or in special cavities in the blocks themselves, as shown in Figure 8-22. Examples of other types of toe-wall structures are shown in Figure 8-23.

Figure 8-22. Articulated block wall used as a toe-wall. Battering or inclination of the wall reduces the lateral earth force against the structure.

(*a*)

(*b*)

Figure 8-23. Examples of vertical toe-walls used in combination with vegetated, sloping backfills. (*a*) Treated timber wall. (*b*) Masonry wall.

A toe-wall must be sized and designed to satisfy both external and internal stability (see Chapter 5). It is important to provide adequate drainage behind the wall to prevent the buildup of water pressure against the back of the wall.

8.11 TIERED WALL WITH BENCH PLANTINGS

8.11.1 General Description

A tiered retaining wall system provides an alternative to a low toe-wall with face plantings. Shrubs and trees can be planted on the benches to screen the structure and lend a more natural appearance, as shown schematically in Figure 8-24. This alternative effectively allows vegetation to be planted on slopes that would otherwise be too steep. Roots that permeate the backfill or slope behind the structure will also tend to reinforce and indurate the soil. This root reinforcement should more than offset the surcharge from the weight of the vegetation.

8.11.2 Applications and Effectiveness

A tiered retaining structure provides the following benefits and advantages:

- It provides numerous landscaping opportunities on the horizontal benches between structures.

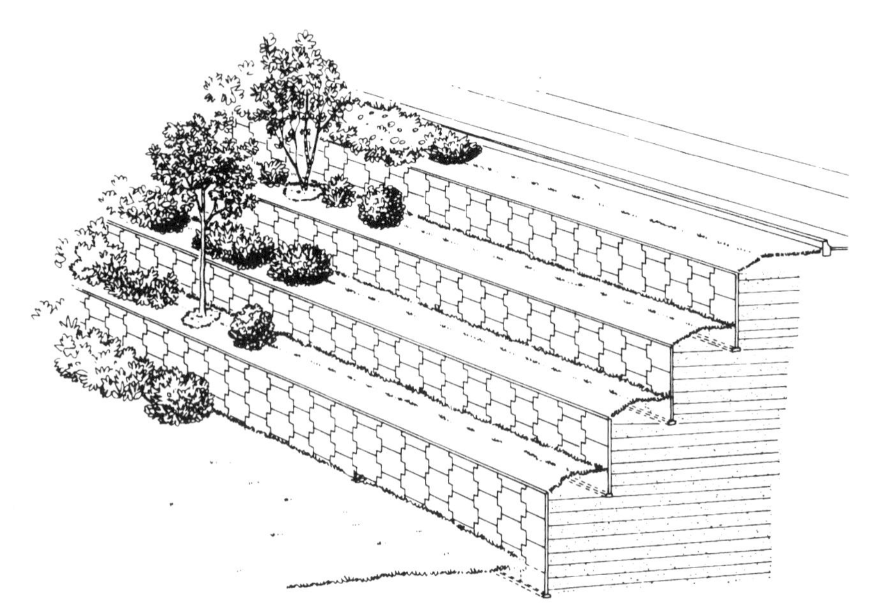

Figure 8-24. Schematic illustration of tiered or stepped-back retaining structure with landscaped benches, reinforced earth type design. (From Walkinshaw, 1975.)

- It permits establishment of vegetation on very steep slopes without the need for excessive slope flattening.
- Woody vegetation growing on the steps or benches screens or hides most of the tiered retaining structure system and lends a more natural appearance to the slope (see Figures 8-25 and 8-26).

8.11.3 Suitable Wall Types and Designs

Virtually any of the retaining structures cited in Section 8.10 can be used in a tiered wall system. Design is a little more complicated because the upper wall units exert a vertical surcharge stress that translates into increased lateral stress on the lower wall units. For this same reason it may be advisable to select shrubs as opposed to trees in order to minimize surcharge from the weight of

Figure 8-25. Tiered, reinforced-earth type retaining wall system with landscaped benches, Vail Pass, Colorado. (Used with permission of The Reinforced Earth Company.)

Figure 8-26. Highway embankment fill supported by tiered, concrete retaining wall system (crib wall above, cantilever wall below). Bench between walls has been planted and landscaped with native shrubs and trees.

the vegetation and possible windthrow problems. A tiered wall system provides numerous opportunities for adding vegetative and landscaping benefits on steep slopes and embankments, as shown in Figures 8-25 and 8-26.

8.12 REFERENCES CITED

AWPI (1981). *Standard Designs for Treated Timber Cribbing*. Washington, DC: American Wood Preservers Institute.

Hilfiker, W. K. (1972). Reinforced Concrete Cribbing. U.S. Patent #3,631,682 (January 4, 1972).

Jaecklin, F. P. (1983). Retaining beautifully. *Architect and Builder* (May).

Randall, F. A. and M. Wallace (1987). The many types of concrete retaining walls. *Concrete Construction* **32**: 589–597.

Schuster, R. L., W. V. Jones, R. L. Sack, and S. M. Smart (1973). A study and analysis of timber crib retaining walls. USDA Forest Service Final Report No. PB-221-447, 186 pp.

Walkinshaw, J. L. (1975). Reinforced earth construction. Federal Highway Administration Final Report FHWA-DP-18, U.S. Department of Transportation, 70 pp.

White, C. A., and A. L. Franks (1978). Demonstration of erosion and sediment control technology: Lake Tahoe region of California, California State Water Resources Control Board, Final Report, Sacramento, CA, 393 pp.

9 Biotechnical Ground Covers

9.0 INTRODUCTION

The type and extent of ground cover is one of the most important factors controlling the severity of surficial erosion. A dense herbaceous or grass cover comprises one of the best defenses against soil erosion. This is generally true provided the velocity of water flowing over the surface is not of sufficient duration and intensity to degrade the vegetative cover. In the latter case a more resistant cover must be employed or, alternatively, the vegetation can be reinforced or stabilized with two-dimensional, open-weave fabrics, meshes, and nets or with three-dimensional geosynthetic mattings. In addition to their reinforcing and anchoring function, some of these erosion control products also behave as mulches that enhance the establishment of vegetation. Three-dimensional geocellular containment systems can also be used in conjunction with vegetation to provide local confinement and stabilization.

Under conditions of high velocity and long duration flow, a hard armor system may be required. Open or porous structural armor and revetment systems can also be used in conjunction with vegetation. Gabion mattresses, articulated concrete blocks, and riprap fall into this category. Woody plants growing in a porous, structural revetment improve interlocking between structural units and increase resistance to lift-off forces and tractive shear stresses.

"Reinforced grass," a term coined by Hewlett et al. (1987), refers to a grass surface that has been artificially augmented with an open structural coverage (meshes, geosynthetic mattings, interlocking concrete blocks, etc.) to increase its resistance to erosion above that of grass alone.

9.1 CLASSIFICATION

9.1.1 Ground Cover Materials

Ground covers range from inert facings (cement or stone layers) to organic mulches (hay, wood chips, straw, etc.) and live materials (seedings and sod, cuttings, and transplants). Different types of ground cover materials for erosion control are characterized and classified in Figure 9-1. For many installations vegetation alone will provide adequate long-term erosion protection. Establishment of vegetation on steep or difficult sites, on the other hand, may require

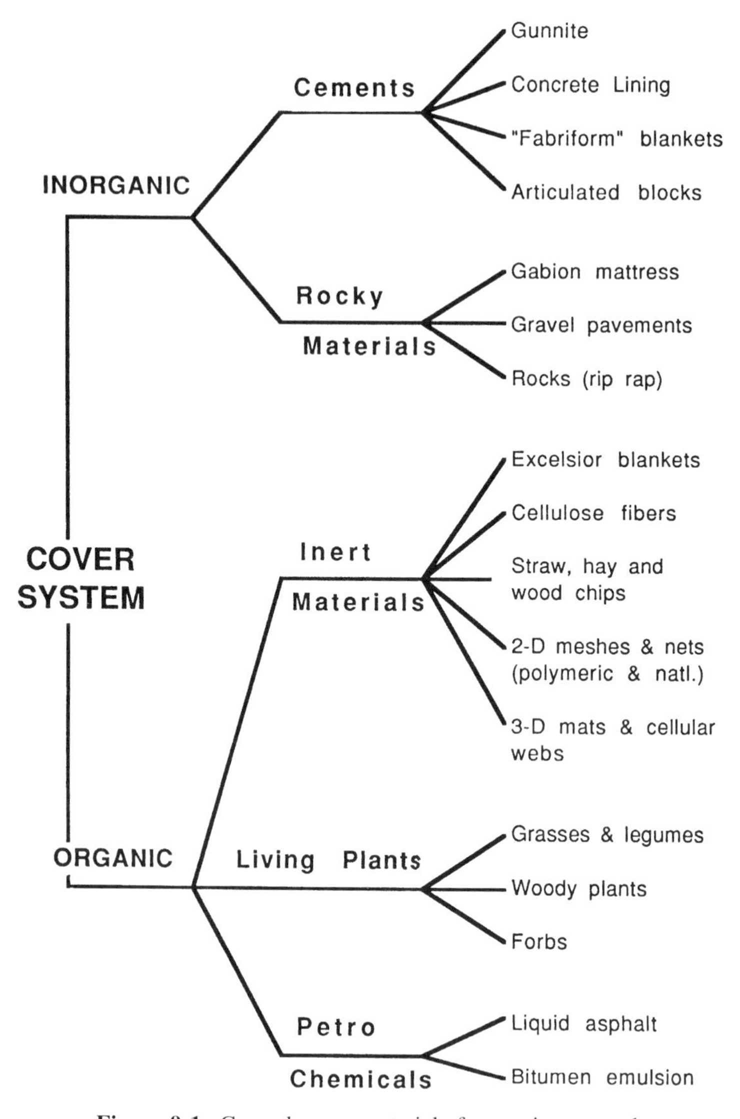

Figure 9-1. Ground cover materials for erosion control.

a variety of techniques, such as the integrated use of vegetation and structural reinforcing or armor units.

9.1.2 Structural Reinforcement Elements

The main types of structural elements used to reinforce and/or armor grass and other vegetated surfaces can be grouped as follows:

1. Geosynthetic:
   - Two-dimensional geotextiles and meshes
   - Three-dimensional and cellular confinement systems
2. Stone:
   - Gabion mattresses
   - Riprap
3. Concrete:
   - Articulated concrete blocks
   - Cable tied concrete blocks
   - Poured cellular grids

9.1.3 Classification by Product Type

Many new ground cover and erosion control products have come on the market in the past decade, and a multitude of terms have been used to describe these products. Some sort of systematic classification system based on attributes and function of these products is required. A hierarchical description and classification proposed by Lancaster and Austin (1994) has been adopted for this purpose.

Conventional Mulching: Mulches have been used for years to provide temporary erosion protection and to aid in vegetation establishment. The main benefits of mulches include:

- Moderation of soil temperature by shading soil and retention of soil moisture by capturing moisture and mitigating evaporative losses
- Protection against raindrop splash by interception of raindrops
- Contribution of organic matter to the soil from decomposition of mulch

Loose Mulches: Hay and straw are the most common dry mulch materials. They are either applied by hand or blown over the slope, as shown in Figure 9-2. Application rates on flat to gently sloping terrain typically range from 1.5 to 2.0 tons/acre (3370 to 4490 kg/ha). Long straw mulches with fiber lengths ranging from 4 to 8 inches (10 to 20 cm) are the most effective (Kay, 1983) because they provide the best interlocking and surface coverage (see Figure 9-3). Loose mulches are subject to blowing or washout and should be anchored or crimped into the soil. A variety of methods can be used for this purpose ranging from disc harrows to studded rollers. On steeper slopes crimping techniques may be supplanted by the use of viscous over-sprays or tackifiers. These are comprised of asphaltic emulsions, latex solutions, vegetable gums, and other compounds that attach the mulch fibers to themselves and to the ground.

(*a*)

(*b*)

Figure 9-2. Straw mulch application on a slope. (*a*) Using a mulch blower. (*b*) Spread by hand.

Figure 9-3. Effective use of a long straw mulch in combination with flexible down drains on a highway embankment slope.

Hydraulic Mulches: Hydraulic mulches are composed of wood cellulose, paper pulp, and recycled newsprint and/or cardboard fiber. They can be conveniently applied in a one-step application in which seed, fertilizer, soil amendments, and mulch may be placed in a single pass of a hydraulic mulcher. Hydraulic mulching can be used to reach steep, inaccessible slopes, and a tackifier or synthetic fibers can be added to the hydraulic slurry to improve the tenacity of the fibers and their adhesion to the ground surface. The main limitation of hydraulic mulches is their short fiber length, generally less than 0.2 inch (5 mm), which is required for their passage through the pumps of a hydraulic seeder/mulcher.

Rolled Erosion Control Products (RECPs): Rolled erosion control products are generally flexible nets or mats manufactured from both natural and synthetic materials that can be brought to a site, rolled out, and fastened down on a slope. They are manufactured in diverse forms and combinations of materials, have widely differing properties, and are designed for specific functions or site conditions. Materials used in their manufacture include wood excelsior, straw, coconut fiber, polyolefins, polyvinyl chloride (PVC), and nylon. Products manufactured from these materials include *erosion control nets, open-weave textiles, erosion control blankets* (ECBs), and *geosynthetic mattings*. The combination of materials and prescribed structure used in their manufacture enables designers to incorporate some of the best attributes of long-fiber mulches with the tensile strength and reinforcing effect of dimensionally stable nets, meshes, and geotextiles.

These products may be used for either temporary or permanent erosion control, for aiding the establishment of a vegetative cover, or for reinforcing and enhancing the effectiveness of mature grass cover under difficult or harsh site conditions (e.g., steep, highly erodible open slopes; channel bottoms and side slopes with moderately high velocities; and shoreline environments subjected to some wave action).

Erosion Control (Holdown) Nets: This class consists of relatively thin, two-dimensional woven natural fibers or geosynthetic biaxially oriented process (BOP) nettings that are used to anchor loose-fiber mulches such as straw or hay. These nets are rolled out over a selected area and stapled or staked in place. They do not offer the same degree of protection as erosion control blankets (ECBs) because of the superior structural integrity provided by stitching the netting to a layer of coconut, straw, or wood fiber in the latter. On the other hand, they normally exhibit better performance when used in conjunction with straw as compared to hydraulic mulches. Erosion control nettings are suitable for use under moderate site conditions, where geotextiles and ECBs are not warranted.

Open-Weave Textile Meshes: Erosion control meshes or geogrids are woven from either natural or polyolefin yarns. Polypropylene meshes are examples of manufactured polyolefin products. Coir (coconut fiber) and jute nets are examples of natural products. The latter consist of meshes woven with thicker fibers and smaller openings that enable them to provide erosion control with or without the use of an underlying, loose mulch layer. Meshes made from high-quality coir fibers also tend to display higher tensile strengths than other natural fiber nettings, which makes them useful on steep slopes or where a netting is required for supplementary reinforcement and slope facing in soil bioengineering applications, for example, vegetated geogrids or brushlayers. Coir fabrics, for example, may exhibit unit tensile strengths ranging from 100 to 150 lb/inch (18 to 26 kN/m), depending on the closeness of weave and fabric unit weight. Coir fiber contains about 45 percent by weight lignin, which gives it a woody texture and relatively long durability. Netting manufactured from coir fibers should have a useful life of anywhere from 5 to 10 years, depending upon moisture and soil conditions. Natural fabrics or meshes have good mulching properties because they can adsorb and retain large amounts of moisture. Jute netting, for example, has a water absorption capacity in excess of 450 percent of the dry fabric weight. Both jute and coir are natural fibers that ultimately biodegrade and add organic matter to the soil. Extensive tests by Kay (1978, 1983) on 2:1 and 5:1 slopes showed that straw (at 3000 lb/acre) under jute netting provided very cost-effective protection.

Synthetic fabrics or nettings, on the other hand, do not absorb appreciable amounts of water, but exhibit high tensile strengths per unit weight. For example, a 75 gm/m^2 open-weave synthetic fabric exhibits 17 to 51 lb/inch (3 to 9 kN/m) unit tensile strength. These products also photodegrade and ultimately become part of the soil.

Erosion Control Blankets (ECBs): Erosion control blankets are constructed from a variety of degradable organic and/or synthetic fibers that are woven, glued, or structurally bound with nettings or meshes. Erosion control blankets are manufactured typically of fibers such as straw, wood, excelsior, coconut, polypropylene, or a combination, stitched or glued to or between geosynthetic BOP netting or woven natural fiber netting, as shown in Figure 9-4.

The netting may be composed of either natural or UV degradable synthetic material in applications where temporary erosion control and mulch retention are required during vegetation establishment. Blanket effectiveness, durability, and longevity are strongly dependent on alterations in netting and fiber components; thus, ECBs span a very broad application range. Some blankets are available with a seed-impregnated, recycled cellulose medium incorporated into their structure at their base. ECBs must be placed in intimate contact with the ground surface for optimum effectiveness. Otherwise, erosion rills will form under the blanket, as shown in Figure 9-5. ECBs are normally fastened to the ground using staples and pins in a number and pattern recommended by the manufacturer according to slope and site conditions. Devices for attaching ero-

Figure 9-4. Ground surface protected by erosion control blanket (ECB) consisting of straw stitched between a synthetic BOP netting. Grass grows up through netting, which is UV degradable.

Figure 9-5. Failure of an ECB on a highway embankment slope. Concentrated discharge at the top in combination with poor ground surface contact caused erosion rills to form beneath the blanket.

sion control mats or blankets to the ground surface are shown in Figure 9-6. Additionally, the placement of ECBs and other rolled erosion control products into anchor trenches will enhance the stability of the installation.

ECBs are designed to assist in vegetation establishment and to provide temporary erosion protection. They are generally limited to areas where natural, unreinforced vegetation will eventually provide long-term stabilization and protection. They should be considered for use on sites requiring greater, more durable, and/or longer lasting protection than mulching alone or mulches with erosion control netting. Applications for newly installed ECBs range from gradual to steep slopes (up to 1 : 1 $H : V$), low-to moderate flow velocities (less than 3 m/sec for durations under 2 hours), and low-impact shore lines.

Geosynthetic Mattings: Geosynthetic mattings consist of various UV-stabilized synthetic fibers and filaments processed into permanent, high-strength, three-dimensional matrices. These mattings can be regarded as a type of permanent, "soft armor" alternative to rock riprap. They are designed for permanent and critical hydraulic applications such as drainage channels, where expected discharges result in velocities and tractive shear stresses that exceed the limits of mature, natural vegetation. The reinforcing mechanisms of these mats are described in greater detail in the next section. In brief, a three-dimensional mat functions as an open, stable matrix for the entanglement of plant roots,

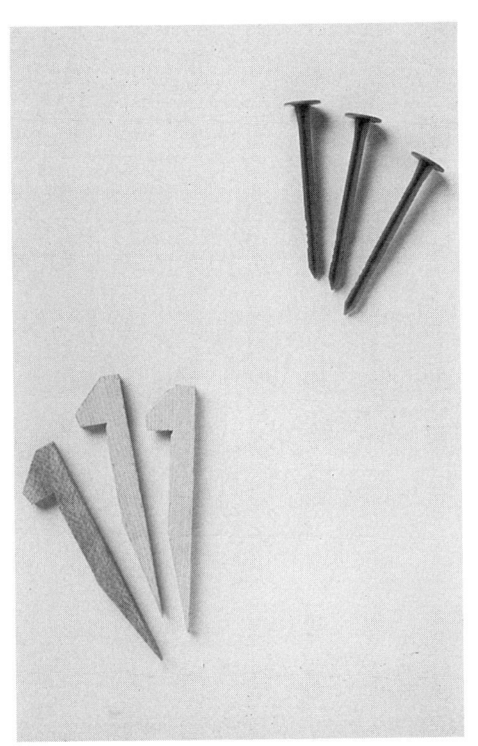

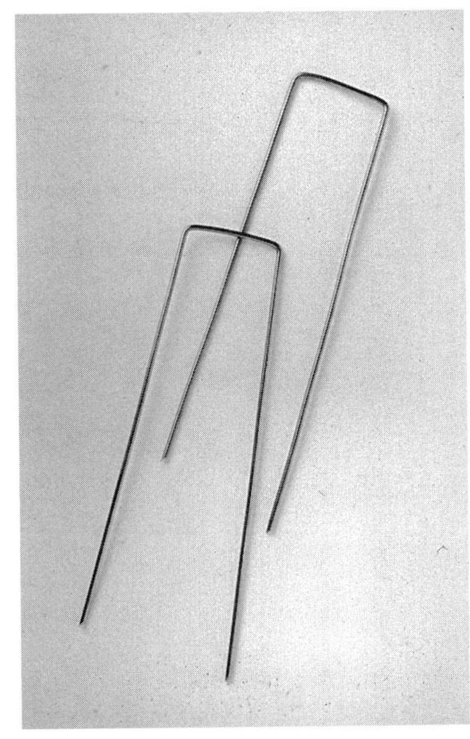

Figure 9-6. Devices for attaching rolled erosion control products (e.g., blankets and mats) to ground surface include wooden pegs, plastic pegs, and wire staples. (Used with permission of the North American Green Company.)

stems, and soil, which together form a coherent, living matrix. Geosynthetic mats can differ substantially in their construction, method of deployment, and intended function. Two distinct types of mats can be identified, namely, *turf reinforcement mats (TRMs)* and *erosion control revegetation mats (ECRMs)*, respectively.

1. Turf Reinforcement Mats (TRMs): Turf reinforcement mats consist of a three-dimensional web of mechanically or melt bonded polymer netting, monofilaments, or fibers that are entangled to form a strong and dimensionally stable mat. TRMs provide sufficient thickness and internal void space to permit soil filling/retention and the development of plant roots within the matrix, as shown schematically in Figure 9-7. TRMs are typically installed by rolling out and fastening the geosynthetic mat to the ground surface and then filling with a fine soil and a prescribed seed mix.
2. Erosion Control Revegetation Mats (ECRMs): Erosion Control Revegetation Mats are denser, lower profile mats designed to provide better short-

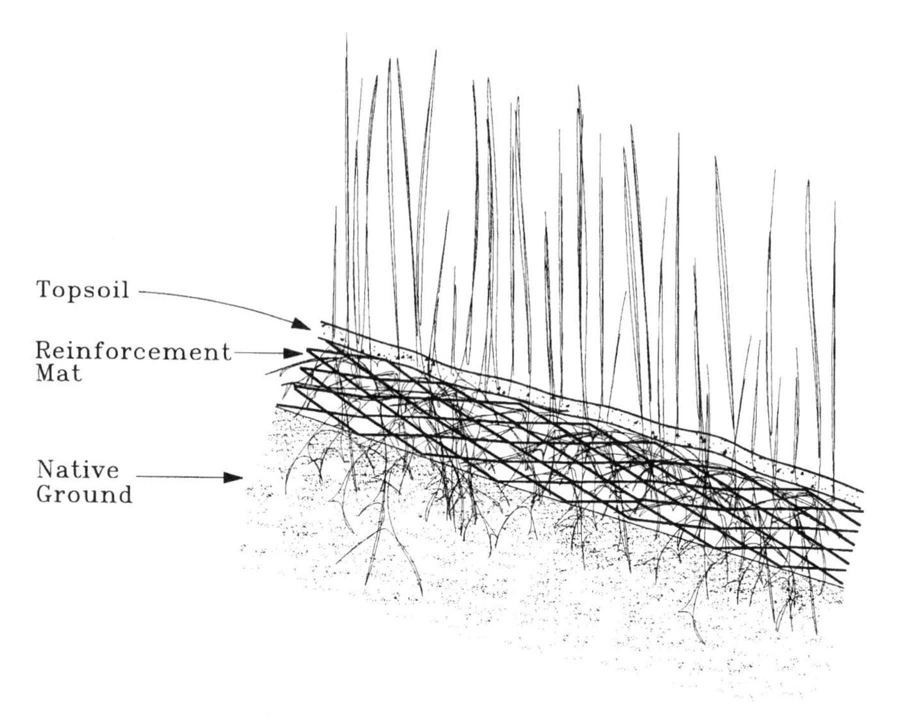

Figure 9-7. Schematic illustration of a turf reinforcement mat (TRM) showing entanglement of plant roots and soil with mat fibers to form a living, coherent mat that armors and protects the ground surface.

term ground cover and erosion protection. The ground surface is seeded prior to deployment and placement of an ECRM. Natural sedimentation processes fill in the mat over time and allow a more gradual development of a reinforced vegetated lining. ECRMs provide sufficient ground cover to reduce initial erosion and the necessary thickness and tensile strength for long-term reinforcement of vegetation stem and root systems. ECRMs probably provide superior temporary erosion protection than TRMs, but the latter can be expected to provide more root entanglement and better long-term protection.

Geocellular Containment Systems (GCSs): Strictly speaking, a GCS is not a rolled, geosynthetic erosion control product. It consists of individual three-dimensional cells up to 20 cm deep. The system is usually manufactured from polyethylene or polyester strips. When expanded into position, the cells have the appearance of a large honeycomb, as shown in Figure 9.8. The cells can be backfilled with soil, sand, or gravel, depending on the application. For conjunctive use with vegetation, the soil-backfilled cells are also seeded, fertilized, and often covered with a temporary mulch cover.

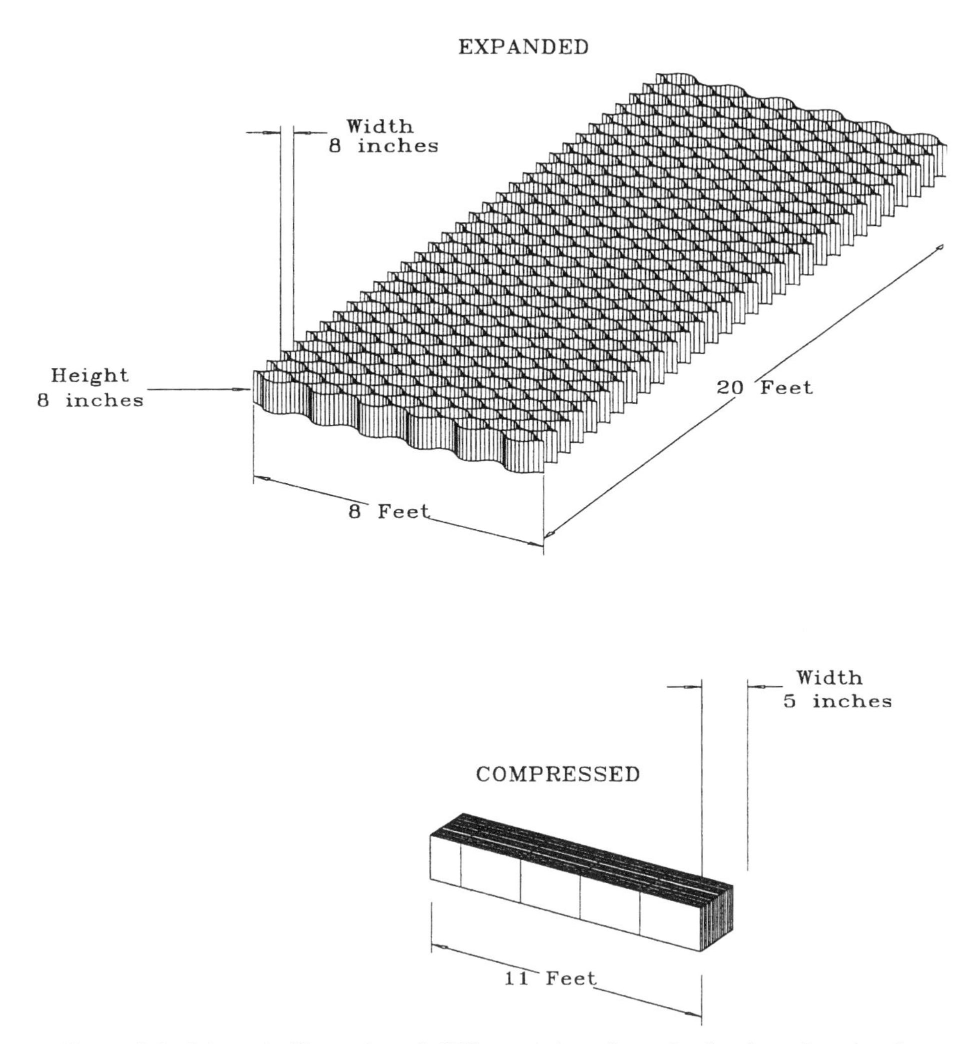

Figure 9-8. Schematic illustration of GCS consisting of a web of polymeric strips that expands into a large honeycomb-like array.

Results of large-scale triaxial texts conducted by Bathhurst and Karparapu (1993) on isolated geocells have demonstrated that cellular confinement imparts apparent cohesion to cohesionless, compacted granular material on the order 3500 to 4000 psf (169 to 190 kPa). The permeation of filled cells with plant roots further increases the coherence of the fill and underlying soil. GCSs allow the use of on-site sands and granular fills in place of more costly imported materials. The cellular geometry preserves natural drainage while mitigating the effect of hydraulic forces at work on cut and fill slopes or in drainage channels.

9.1.4 Classification by Application/Performance

Ongoing efforts to classify rolled erosion control products according to application and performance are described by Lancaster and Austin (1994). They note that the various products span applications based on permanent versus temporary control, and low- versus high-velocity flows. They devised a classification system that separates the various products into the following categories: (1) low-velocity degradable RECPs, (2) high-velocity degradable RECPs, and (3) long-term, nondegradable RECPs; the classification is shown in Table 9-1.

Revegetation techniques or establishment aids and reinforced grass systems can be classified into either temporary or permanent measures. Theisen (1991) has named these two categories as "TERMS" and "PERMS," respectively.

"TERMS"

Characteristics: Consist of degradable natural and/or synthetic components that provide *temporary* erosion control and facilitate vegetative establishment. These short-term materials eventually degrade, leaving only vegetation for long-

TABLE 9-1. Performance Classification of Rolled Erosion Control Products (RECPs)

Required Performance Characteristics	Suitable Type of RECPs	Properties and Typical Applications
Low-velocity degradable	Single net, organic fiber ECBs	One- to two-season longevity
	Biodegradable natural fiber and photodegradable geosynthetic nets	Limited capacity in resisting damage and resisting erosion under severe conditions
		Slopes of moderate grade, length, and runoff; channels where potential for damage during installation is minimal
High-velocity degradable	Double net ECBs or high strength nettings and/or increased quantities of organic fibers	Similar to low-velocity degradables in installation and function, but designed for more severe site conditions
	Dense, open-weave geotextiles or meshes (e.g., coir, jute, polypropylene)	Heightened durability and longevity (1–5 years)
		Steeper slopes and high-velocity channel linings where natural, unreinforced vegetation is expected to provide permanent stabilization
Long-term nondegradable	High-strength geosynthetic mattings and cellular containment e.g., ECTRMs, TRMs, GCSs	Provide immediate, high-performance erosion protection followed by permanent reinforcement of established vegetation
		Steep slopes with very erodible soils; channel linings subjected to relatively high, long-duration flows

Source: Adapted from Austin and Driver (1995).

term resistance. They are suitable for sites with relatively gentle slopes and sites with short duration, low to medium flow velocities.

Examples

Straw, hay, and hydraulic mulches

Tackifiers and soil stabilizers

Erosion control meshes and nets (ECMSs)

Erosion control blankets (ECBs)

"PERMS"

Characteristics: Consist of biotechnical composites on the one hand or porous, structural facings (revetments) on the other. They are better suited for steep sites, harsh sites, and/or sites subjected to long duration, medium to high flow velocities where vegetation requires *permanent* reinforcement.

Biotechnical Composites: "Soft armor" systems composed of nondegradable elements that furnish temporary erosion protection, enhance vegetative establishment, and ultimately become intimately entangled with living plant tissue (roots) to extend the performance limits of vegetation. This reinforced vegetation provides "permanent" protection against medium to high flows.

Examples

Erosion control revegetation mats (ECRMs)

Turf reinforcement mats (TRMs)

Vegetated geocellular containment systems (GCSs)

Hard Armor Systems: Composed of hard, inert structural units that provide resistance to long-duration, high flow velocities. These systems provide permanent erosion protection in areas subject to high waves and/or scour attack where site conditions exceed performance limits of biotechnical composite systems. Vegetation can be planted in porous structural armor or facing systems. The vegetation mutes the harsh, unnatural appearance of the armor layer. In addition, the plants improve the performance of the armor (Hewlett et al., 1987) by improving interlocking between structural units and increasing resistance to lift off and local shear forces.

Examples

Articulated, concrete block systems (ACBs)

Gabion mattresses

Riprap

Performance Evaluation: The performance of these ground cover systems has been evaluated in laboratory tests using flumes and rainfall simulators (Rustom and Weggel, 1993) and in field trials using controlled surface flows on the upstream face of a dam (Hewlett et al., 1987). Rustom and Weggel (1993) evaluated the performance of 12 different natural and geosynthetic rolled erosion control products by subjecting them to simulated rainfall on the upper reaches of a 2.5 : 1 ($H : V$) slope. They measured the surface runoff and sediment yield from both bare soil and protected slopes resulting from nominal rainfall intensities of 2, 5, and 8 inches/hour (50, 125, and 200 mm/hour). They recorded sediment yield during both the initial, rising limb of the hydrograph and also during the equilibrium phase. Their analysis focused on the behavior of each erosion control system and how it acted to control raindrop impact, limit soil detachment, reduce flow velocities, and minimize sediment yield.

All of the products tested demonstrated some ability to reduce erosion from the test plots. All reduced the sediment yield to at least below 60 percent of the sediment yield from an unprotected soil. Some products exhibited superior performance to others under the particular conditions of the test program. All products were evaluated on the basis of their performance in the absence of vegetation, unfilled with soil, and fastened to the upper reaches of a slope where raindrop impact is important and shear stresses imposed by overland flow somewhat less important. Admittedly, some products are designed for different conditions than those under which they were tested; nevertheless, these test results provide a comparison of product performance immediately after installation before they are filled with soil or before vegetation becomes established.

9.2 PRINCIPAL SYSTEM COMPONENTS

The principal components of a reinforced ground cover system used in a waterway are shown in Figure 9-9 for "soft armor" (e.g., biotechnical composite) and "hard armor" (e.g., concrete block) systems, respectively. These components and their properties include:

Armor Layer: Formed generally by the structural reinforcement in concert with the grass plants and soil with which they interact. The armor layer resists the erosive action of the water flow, and thus protects the more easily eroded subsoil.

Underlayer: Usually a geotextile or granular material used in conjunction with concrete reinforcement. The underlayer mainly provides erosion control and filtration.

Subsoil: Protected from direct attack by the armor layer, it must both support grass growth and be stable under seepage flow. Rooting in the subsoil strengthens it and also helps to anchor the armor layer to the subsoil.

Anchors and Shear Pins: May be used to restrain the armor layer from pulling away from or slipping down slope.

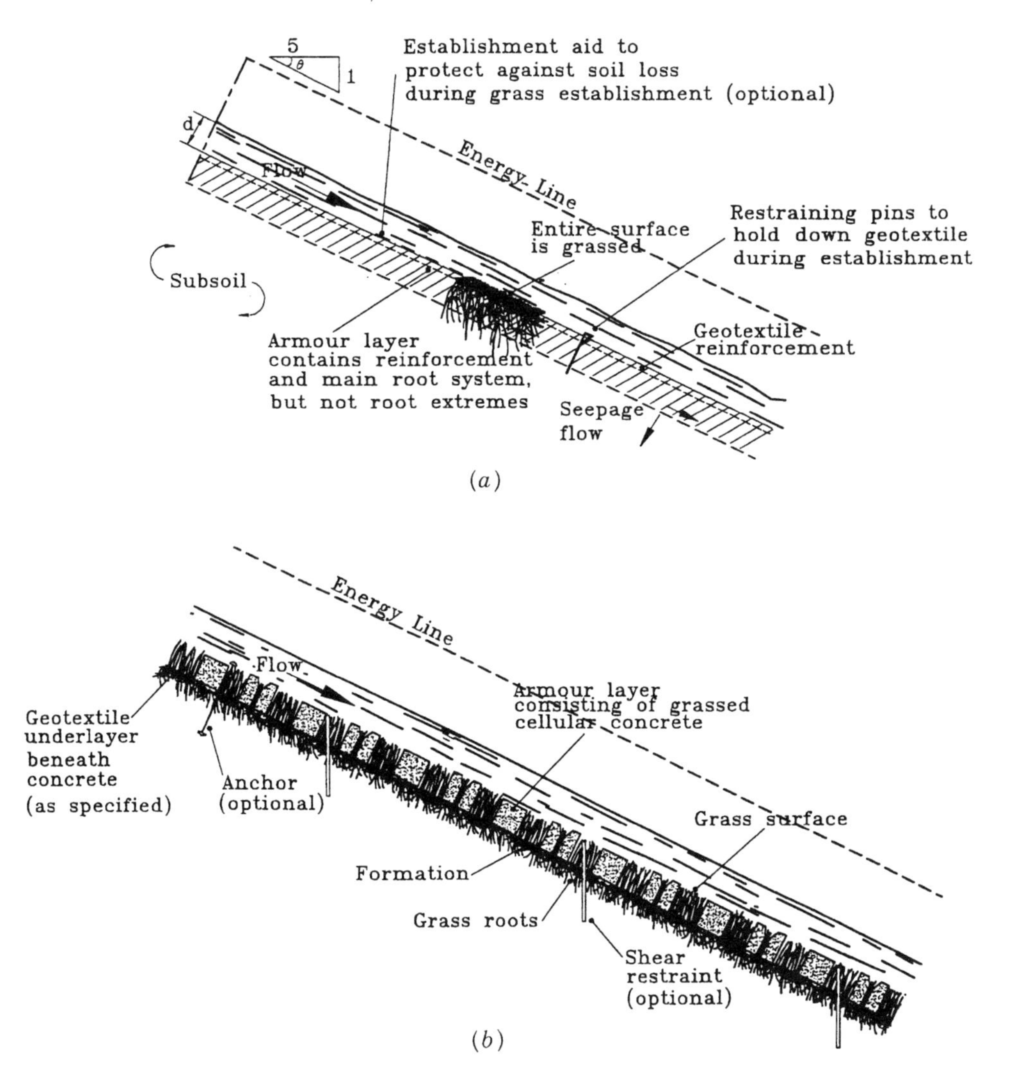

Figure 9-9. Sections through "reinforced grass" waterway protection systems. (*a*) Biotechnical composite. (*b*) Concrete blocks. (From Hewlett et al., 1987. Design of reinforced grass waterways. CIRIA Report 116. Used with permission of Construction Industry and Information Association.)

9.3 REINFORCING/STABILIZING ROLE OF NETS AND MESHES

With net or mesh reinforcement the entire surface is grassed, and the roots bind around the strands to form a strong, continuous biotechnical composite. Such a system allows the surface to provide greater resistance to raindrop impact and to sustain higher velocities than could be tolerated by grass alone. This feature may be important in the case of grassed waterways, upper portions of streambanks, and other areas that are subject to periodic inundation by high-velocity water flows. Some or all of the following advantages over plain grass are provided:

- Improved protection of soil surface by retarding development of weak spots in grass cover
- Assistance to root structure in restraining surface soil particles from erosion by flowing water
- Improvement in lateral continuity between grass plants and reduction in risk or local failure.
- Entanglement of plant roots, soil, and structural elements into a coherent, living mat

A partial listing of proprietary, two-dimensional meshes/nets and three-dimensional cellular grids/mats is given in Table 9-2. These meshes and nets are manufactured from either polymeric materials on the one hand (e.g., *LANDLOK® Erosion Mat, Enkamat®*, and *Tensar Mat®*) or natural fibers on the other (e.g., coir and geojute).

With any rolled erosion control product, it is essential to minimize seepage flow between the mesh and ground surface. The flexibility of the mesh or net and the method of installation must be adequate to avoid bridging between the mesh and soil.

TABLE 9-2. Examples of Rolled Erosion Control Products

Trademark or Registered Name	Manufacturer/ Supplier	Brief Description
	Fabrics and Meshes (Two-Dimensional)	
POLYJUTE®	Synthetic Industries	Woven polypropylene mesh
Terram 42A	ICI Fibres Geotex Gp.	Woven polyester mesh
Geojute	Belton Industries	Woven jute mesh
DeKoWe 700 & 900	Belton Industries	Woven coir mesh
S75, SC150, C125	North American Green Inc.	Polymeric netting sewn together with straw and coconut fibers
	Mats and Cellular Webs (Three-Dimensional)	
Enkamat®	Akzo Industrial Systems	Mat of heavy nylon monofilaments
Tensar Mat®	Netlon Ltd.	Multi-layered polyethylene mat
C350	North American Green Inc.	Coconut matting with permanent three-dimensional reinforcement net structure
LANDLOK®	Synthetic Industries	Three-dimensional mat of polyolefin fibers and nets
PYRAMAT™	Synthetic Industries	Three-dimensional woven geotextile
Armater®	Akzo Industrial Systems	Polyester expanding cellular grid
Geoweb®	Presto Products	Polymeric expanding cellular grid

9.4 FUNCTION OF VEGETATION

Structural facings and armor units enhance the performance of vegetation on steep slopes and sites subjected to long-duration, high-velocity flows by armoring the surface and reinforcing the vegetation. Conversely, vegetation can improve the performance of structural facings or revetments by increasing interlocking and increasing the resistance to local shear and lift-off forces. Field tests conducted by CIRIA (see Hewlett et al., 1987) have demonstrated the contributions of plant roots to improved performance of vegetated, structural channel linings in grassed waterways. Vegetated stone (riprap) channel revetments have performed as well or better (Shields et al., 1990) than bare, nonvegetated stone revetments channel revetments alone.

9.4.1 Rooting Effects

Concrete systems provide immediate protection to the subsoil against effects of rainfall and surface runoff. Effects of grass are secondary, but over time provide additional restraint to movement of concrete units. Reinforcement/restraint provided by the *roots* include:

- Anchorage and resistance to uplift by grass roots growing below cells or joints
- Interface friction (between armor layer and underlayer) as a result of penetration by stems/roots, which act as mini shear pins
- Increased shear strength or reinforced soil/root plugs below cells or joints
- Increased friction and interlocking between adjacent blocks when joints are wedged by root infill

9.4.2 Field Trial Results

Contributions of grass rooting and anchorage were determined during the course of field trials sponsored by CIRIA. A total of nine reinforced grass channels and a plain grass control channel were constructed in the upstream face of a 10-m high dam at the Jackhouse Reservoir in England in 1984.

Root Anchorage and Block Restraint: Lift-off tests were carried out on individual blocks in one of the cable-tied concrete block arrays after removal of the cables.

Peak force to lift an isolated 0.15-kN grassed block (all adjacent blocks removed)	0.19 kN
Tensile lift-off restraint or root anchorage stress (based on vegetated cell area of 0.063 m^2)	3.0 kN/m^2

Peak force to lift a 0.15-kN grassed block (adjacent blocks *not* removed)	1.4 kN
Root wedging/interblock friction restraint	1.2 kN

In Situ Root Shear Strength: Shear strength tests on potential failure planes beneath vegetated (grassed) concrete systems were measured by direct in situ shear tests with the following results:

Contribution to shear strength of grass roots associated with concrete block systems	3–5 kN/m^2

9.4.3 Laboratory Test Results

Most major manufacturers of rolled erosion control products have conducted extensive laboratory flume studies. Data are available that describe a particular product's maximum shear stress and flow velocity resistance under both vegetated and unvegetated conditions. Most testing has been conducted with flow durations lasting from 0.5 to 50 hours.

9.5 EFFECTIVENESS OF SLOPE EROSION CONTROL SYSTEMS

One way of gauging the effectiveness of a ground cover system for reducing surficial erosion is to compute soil losses *with* and *without* the system in place. Soil losses can be estimated by means of the universal soil loss equation (USLE). The annual soil loss from a site is predicted according to the following relationship:

$$A = R \cdot K \cdot LS \cdot C \cdot P \tag{9-1}$$

where: A = computed soil loss (tons/acre or kg/hectare) for a given storm period or time interval
R = rainfall factor
K = soil erodibility value
L = slope length factor
S = steepness factor factor
C = vegetation or cover factor
P = erosion control practice factor

The USLE provides a simple, straightforward method of estimating soil losses, and it provides an idea of the range of variability of each of the parameters, their relative importance in affecting erosion, and the extent to which each can be changed or managed to limit soil losses. The climatic factor (R), topographic factor (LS), and erodibility factor (K) only vary within one order of magnitude. The vegetation or cover factor (C), on the other hand, can vary over several orders of magnitude, as shown in Table 9-3. Moreover, unlike the

TABLE 9-3. Cover Index Factor (*C*) for Different Ground Cover Conditions

Type of Cover		Factor *C*	Percent Effectiveness[a]
None (fallow ground)		1.0	0.0
Temporary seedings (90% stand)			
Ryegrass (perennial type)		0.05	95
Ryegrass (annuals)		0.10	90
Small grain		0.05	95
Millet or sudan grass		0.05	95
Field bromegrass		0.03	97
Permanent seedings (90% stand)		0.01	99
Sod (laid immediately)		0.01	99
Mulch			
Hay, rate of application, tons/ac:			
	0.5	0.25	75
	1.0	0.13	87
	2.0	0.02	98
Small grain straw	2.0	0.02	98
Wood chips	6.0	0.06	94
Wood cellulose	1.5	0.10	90
Fiberglas	1.5	0.05	95

Source: From USDA Soil Conservation Service (1978).

[a] Percent soil loss reduction as compared with fallow ground.

other factors, the cover factor (C) can be radically decreased by the selection, method of installation, and maintenance of a particular cover system. Factor C values tend to change with time following certain types of surface treatment such as mulching, seeding, and transplanting. For example, factor C values for grass may decrease from 1.0 (for fallow, bare ground) to about 0.001 between time of initial seeding and full establishment with a dense grass sod.

The system effectiveness (or percent effectiveness) tabulated in the last column of Table 9-3 can be computed in the following manner:

$$\%\text{ effectiveness} = \frac{\text{soil saved}}{\text{soil loss without cover}} \times 100 \tag{9-2a}$$

$$= \frac{A_1 - A_2}{A_1} \times 100 \tag{9-2b}$$

where: A_1 = soil loss without cover (bare ground; $C = 1.0$)
A_2 = soil loss with ground cover

If all the other factors in the USLE are constant, then Equation 9-2 reduces to:

$$\%\text{ effectiveness} = (1 - C) \times 100 \tag{9-3}$$

The percent effectiveness for different covers tabulated in the final column of Table 9-3 was calculated using Equation 9-3. Effectiveness values are presented for loose hay (straw) mulch, sod, and temporary vegetation (cereal grasses or grains). Manufacturers of different rolled erosion control products have computed factor C or percent effectiveness values that can be compared against the numbers in Table 9-3.

9.6 DESIGN OF FLEXIBLE CHANNEL LININGS

9.6.1 Peak Hydraulic Loading Estimates

Selection of an appropriate ground cover system or channel lining in a waterway requires an estimate of the peak hydraulic loading in the channel:

$$\text{Peak hydraulic loading} = f\left\{\begin{array}{l}\text{design discharge, channel}\\ \text{cross section, gradient, and}\\ \text{hydraulic roughness}\end{array}\right.$$

Peak hydraulic loading (maximum velocity and depth of flow), together with duration of flow, determines which, if any, type of structural reinforcement is appropriate for a particular application.

Only relatively simple, idealized models for flow in open channels are available for this purpose. These models assume that the channel is straight and rigidly bounded by bed and banks that maintain their shape and gradient. They further assume a constant discharge of clear water and uniform flow depth.

The velocity and depth of flow can be estimated using Manning's equation:

$$V = \frac{1.49R^{2/3}S^{1/2}}{n} \tag{9-4}$$

where: V = mean velocity of flows (fps)
R = hydraulic radius (feet), which equals cross-sectional area of flow divided by wetted perimeter
S = slope of water surface (slope of channel bed) during uniform flow conditions
n = Manning's roughness coefficient

Alternative forms of the equations are:

$$Q = \frac{1.49AR^{2/3}S^{1/2}}{n} \quad \text{for discharge} \tag{9-5}$$

and

$$q = \frac{1.49 d^{5/3} S^{1/2}}{n} \quad \text{for discharge intensity in a wide channel} \tag{9-6}$$

where: Q = discharge (cfs)
A = area of flow (feet^2)
q = discharge per unit width of channel (cfs/foot)
d = depth of flow

A channel may be considered "hydraulically" wide when the velocity in the center is not affected by friction at the sides. This is usually the case when the width of the channel is larger than five times the flow depth (Chen and Cotton, 1988).

9.6.2 Influence of Grass on Hydraulic Roughness

Influence of grassed surface on hydraulic roughness depends upon physical characteristics of the grass sward (height, stiffness, and density) and its interaction with the flow. This interaction can be divided into the basic regimes according to hydraulic loading, as shown in Figure 9-10.

Hydraulic Roughness—Slopes Flatter than 1 in 10: The so-called "*VR* method," developed after extensive tests at the U.S. Department of Agriculture Labs in Stillwater, Oklahoma, can be used to determine the hydraulic roughness for channels with slopes flatter than 1 in 10. The *VR* method relates Manning's roughness coefficient, n, to flow in the form of a hydraulic loading parameter, *VR*, and to the physical characteristics of the grass, which are defined by height and density. The flow parameter *VR* is equal to discharge intensity q in a hydraulically wide channel.

The relationships are empirical and are shown graphically in Figure 9-11. The roughness characteristics of various types, density, and length of grass can be classified into different "retardance" categories. These physical characteristics are divided into three "retardance" groups (C, D, and E) in Figure 9-11, based on the average length of grass. In this context, Manning's coefficient includes the effect of all factors tending to retard flow and is referred to as a "retardance coefficient." A more complete classification of retardance categories and nomographs to calculate corresponding roughness or retardance coefficients has been published by Chen and Cotton (1988).

Hydraulic Roughness—Slopes Steeper than 1 in 10: On steep slopes, the grass tends to be laid down by the flow throughout the normal range of design discharges. Under these conditions, Manning's n appears to be independent of hydraulic loading and grass length, and to vary with waterway slope only. Based on the CIRIA field trials, a Manning's n value of 0.030 is recommended for slopes of 1 in 10 and a value of 0.020 for slopes of 1 in 3 and steeper, with intermediate values for slopes between these as indicated in Figure 9-12.

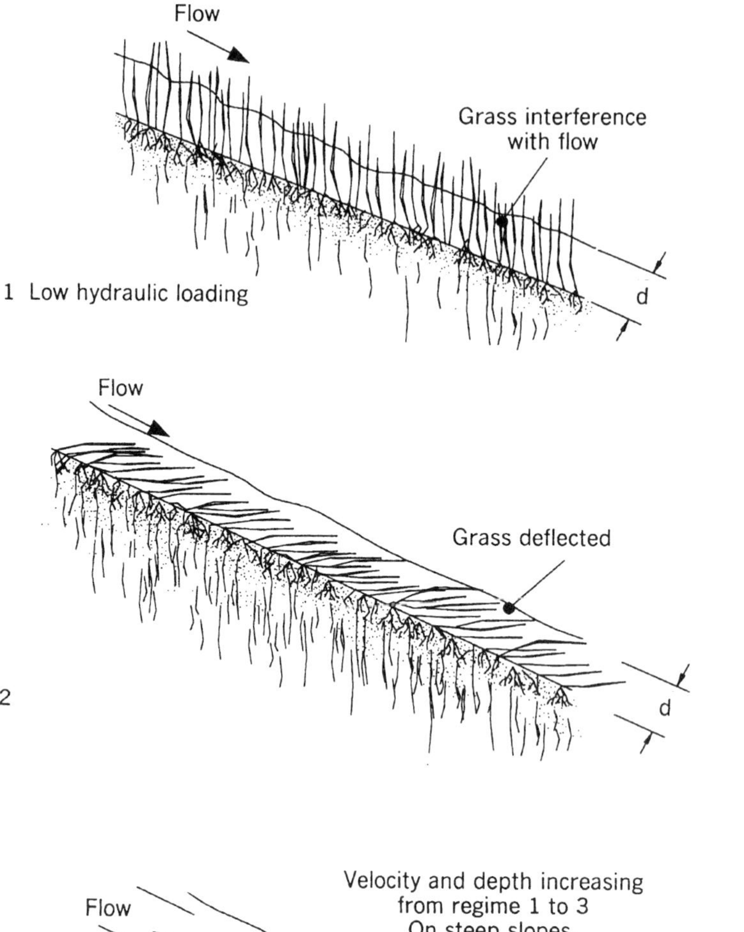

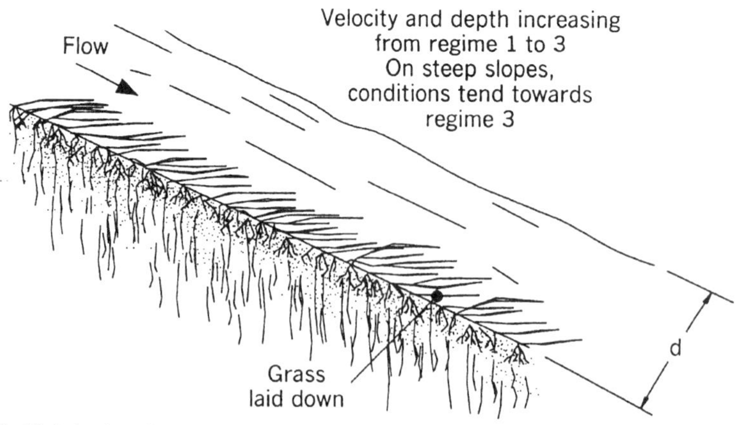

Figure 9-10. Effect of hydraulic loading on grassed surface. (From Hewlett et al. 1987. Design of reinforced grass waterways. CIRIA Report 116. Used with permission of Construction Industry and Information Association.)

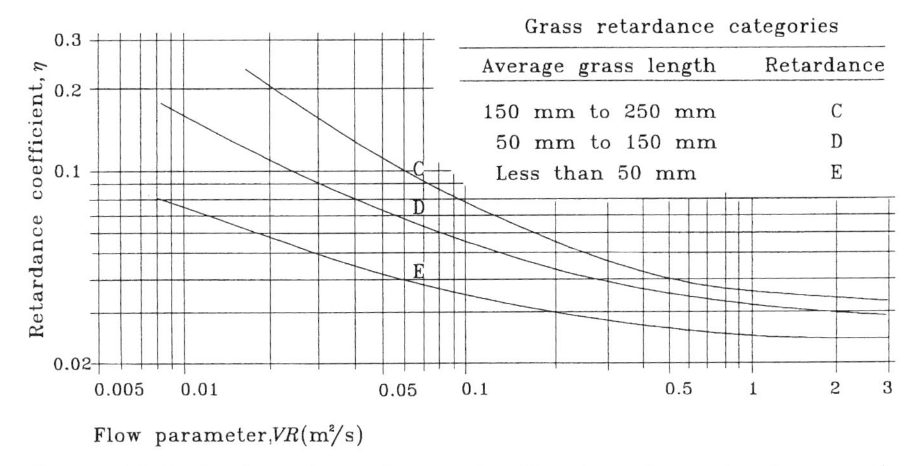

Figure 9-11. Hydraulic roughness of grasses in different retardance categories on gentle slopes. (From Hewlett et al., 1987. Design of reinforced grass waterways. CIRIA Report 116. Used with permission of Construction Industry and Information Association.)

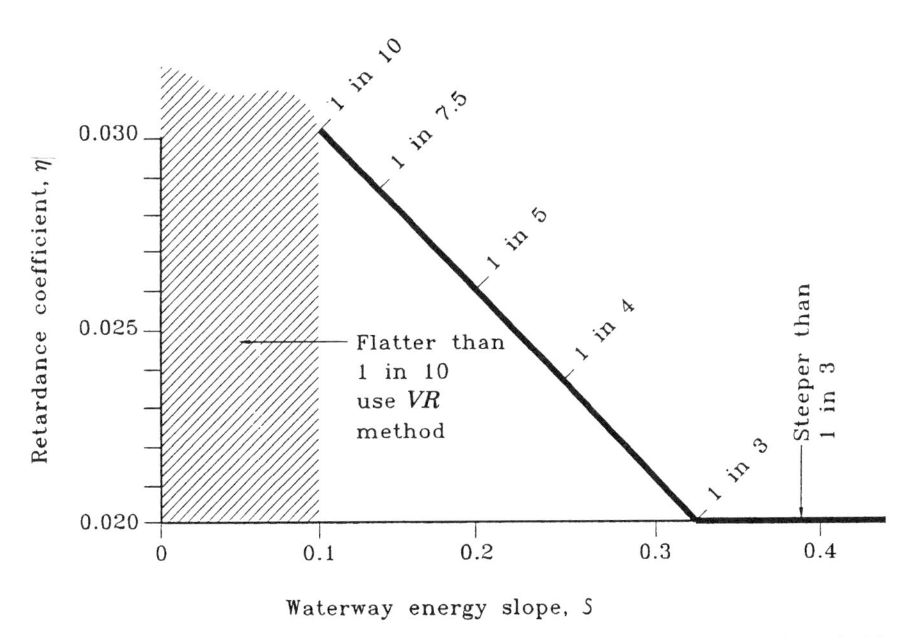

Figure 9-12. Recommended retardance coefficients for grassed slopes steeper than 1 in 10. (From Hewlett et al., 1987. Design of reinforced grass waterways. CIRIA Report 116. Used with permission of Construction Industry and Information Association.)

9.7 SELECTION OF GROUND COVER SYSTEM

9.7.1 Hydraulic Loading Considerations

The main hydraulic parameters are the hydraulic loading (viz., the velocity, depth, and duration of flow) and the erosion resistance of various ground cover systems. A variety of ground cover options (soft vs. hard armor) may be suitable at the site. The detailed hydraulics of each option should be investigated by the following procedure:

1. Select an initial soft or hard armor layer (see Table 9-2) and appropriate Manning's n value.
2. Solve Manning's equation by trial and error for design flow or discharge intensity using different depths of flow to determine velocity.
3. Compare this computed velocity with recommended velocity for a particular armor layer.

Detailed and systematic procedures for evaluating and selecting flexible channel linings can be found in a manual prepared by Chen and Cotton (1988) for the Federal Highway Administration. Flexible linings include riprap, gravel, and vegetation reinforced with geosynthetic mattings. Design procedures are given in the manual for rock riprap, wire-enclosed riprap (gabion blankets), gravel riprap, jute netting, fiber roving systems, curled wood shavings mats, ECBs (straw with netting), and geosynthetic mats. The design procedures are based on the concept of maximum permissible tractive force.

9.7.2 Allowable Flow Duration/Velocity Limits

The erosion resistance of reinforced grass systems (whether "soft" or "hard" armor type) can be considered in terms of hydraulic loading parameters of *velocity* and *duration* of flow. Velocity controls the tractive forces acting at the bed fluid interface. Duration of flow is an important variable because even moderate flow velocities can cause severe erosion damage over a long time period. A major precipitation event can produce significant flow velocities with durations lasting hours or days—rather than minutes.

Guidelines for different erosion control systems are shown in the velocity-duration diagram in Figure 9-13. To enhance the erosion resistance of plain grass, a reinforcement system must both assist in either protecting or stabilizing soil particles at the surface, and also improve lateral continuity between grass plants. Both three-dimensional mats and cellular webs and two-dimensional fabric meshes may fulfill these functions to varying degrees.

If the openings in a mesh or net are too large, they only provide the latter function, namely, lateral continuity between plants (Hewlett et al., 1987). Accordingly, large aperture nets sustain, rather than enhance, the protection provided by plain grass alone by providing lateral continuity and reducing the risk of localized erosion of grass plants. Enhancement of erosion resistance

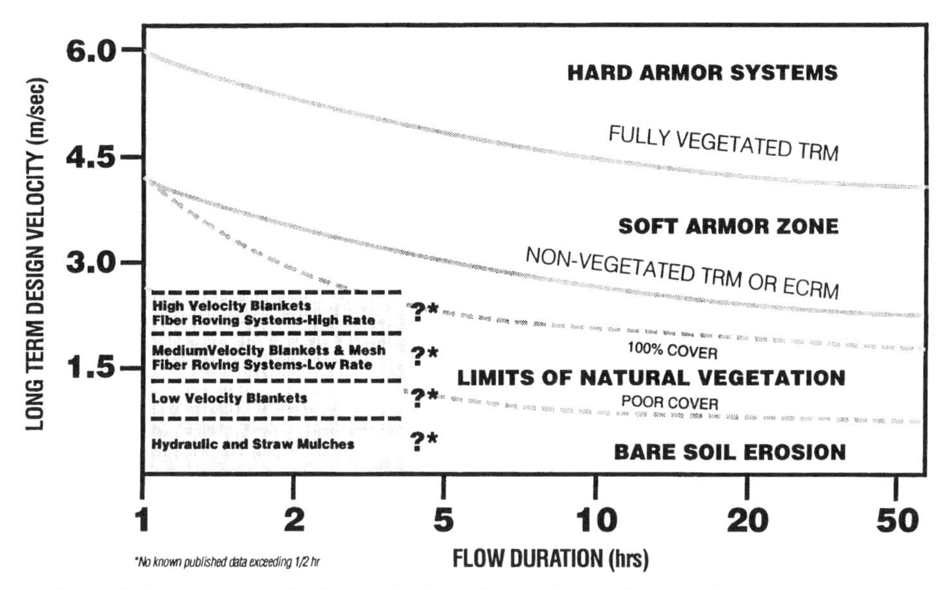

Figure 9-13. Suggested limiting velocities for erosion resistance of natural vegetation, "soft" and "hard" armor systems. (Used with permission of Synthetic Industries, Inc.)

over plain grass can only be achieved with good vegetative cover in the first place.

The "soft" armor zone begins just above the limits of natural vegetation. Performance data for reinforcing mats range from unvegetated TRMs and ECRMs (which exceed performance of natural vegetation) to the upper curve, which delineates maximum recommended velocities obtained from field and laboratory evaluation of vegetated TRMs (Hewlett et al., 1987; Chen and Cotton, 1988). Photos of a grassed waterway protected with a turf reinforcement mat (TRM) are shown in Figures 9-14 and 9-15.

"Hard" armor systems occupy the upper end of the graph in Figure 9-13. The presence of vegetation in porous facings tends to improve performance, as noted previously, because of improved interlocking and increased resistance to lift off and local shear. Articulated block systems have been used to protect the downstream faces of embankment dams (Koutsourais, 1994) against overtopping, as shown schematically in Figure 9-16. The downstream faces of earth dams are very vulnerable to severe scour and erosion because of steep-gradient, high-velocity flows that can occur during overtopping. Armoring combined with revegetation helps prevent erosion of the downstream face of a dam during overtopping. Older, outdated dams can often be rehabilitated to meet higher regulatory standards and to accommodate revised estimates of a probable maximum flood in this manner. A photo of a vegetated, articulated block armor layer protecting a drainage channel is shown in Figure 9-17.

Figure 9-14. "Reinforced grass" waterway. A turf reinforcement mat (TRM) has been used as a liner. Photo taken prior to filling with soil and seeding. (Used with permission of Synthetic Industries, Inc.)

Figure 9-15. "Reinforced grass" waterway. Turf reinforcement mat (TRM) after filling, seeding, and establishment of mature grass stand in the liner. (Used with permission of Synthetic Industries, Inc.)

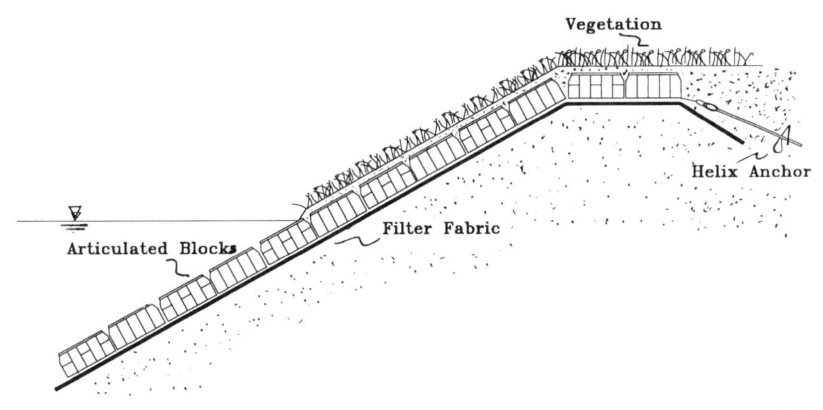

Figure 9-16. Cross section of an embankment slope protected along the crest and downstream slope with an articulated concrete block (ACB) system.

9.7.3 Allowable Tractive Stress Criteria

An alternative selection approach is to compute the tractive stress exerted by flowing water and to compare this stress with the permissible tractive stress for a particular lining system. The average tractive force on a channel, or shear stress, is equal to:

Figure 9-17. Close-up view of vegetated, articulated concrete block (ACB) system protecting sides of drainage channel.

$$\tau = \gamma RS \tag{9-7}$$

where:
τ = average shear or tractive stress (psf)
γ_w = unit weight of water (62.4 pcf)
R = hydraulic radius (feet)
S = average bed slope

The maximum shear stress, τ_d, for a straight channel occurs on the channel bottom bed and is given by the following equation:

$$\tau_d = \gamma_w dS \tag{9-8}$$

where:
τ_d = maximum shear or tractive stress (psf)
d = maximum depth of flow (feet)

The shear or tractive stress in channels is not uniformly distributed along the wetted perimeter. A typical distribution of shear stress in a trapezoidal channel tends toward zero at the corners with a maximum on the center line of the bed, and the maximum for the side slopes occurring about the lower third of the side, as shown in Figure 9-18. Flow around a bend also creates secondary currents, which impose higher shear stresses on the channel sides and bottom in the bend compared to a straight reach. Chen and Cotton (1988) provide a nomograph for calculating a correction factor for bend curvature that depends on the ratio of the bend radius to the channel bottom width. This correction only becomes significant when the ratio drops below 10.

The relationship between permissible shear stress and permissible velocity for a lining can be found by substituting Equation 9-8 into Equation 9-4, which yields:

$$V_p = \frac{0.189R^{1/6}}{n} \tau_p^{1/2} \tag{9-9}$$

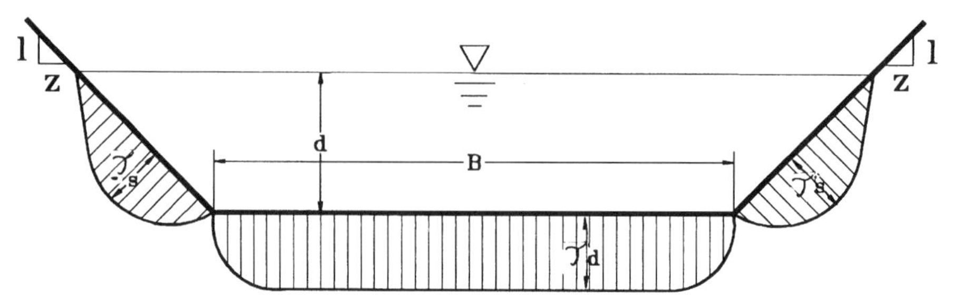

Figure 9-18. Typical distribution of tractive forces or shear stresses in a straight channel. (After Chen and Cotton, 1988.)

where: τ_p = permissible shear stress (psf)
V_p = permissible velocity (fps)

The tractive force method offers several advantages over the permissible velocity method. The calculations are simpler and the failure criteria for a particular lining is represented by a single critical shear stress value. This critical shear stress value is applicable over a wide range of channel slopes and shapes. Permissible velocities, on the other hand, are a function of lining roughness, channel slope, and channel shape. Accordingly, velocities are only approximately constant over a range of these parameters.

Permissible shear stresses for selected lining materials are listed in Table 9-4. These values are based on research conducted at laboratory facilities and in the field. They are presented herein only as guidelines. Manufacturers of rolled erosion control products have developed permissible shear stresses for their products, which should be consulted as well. Some products such as TRMs have variable permissible stresses, depending upon the stage of vegetative establishment in the mat.

TABLE 9-4. Permissible Shear or Tractive Stresses for Selected Lining Materials

		Permissible Unit Shear Stress	
Lining Category	Lining Type	(psf)	(N/m^2)
Temporary degradable RECPs	Jute net	0.45	22
	Straw with net:		
	Single net	1.55	74
	Double net	1.65	79
	Coconut fiber with net	2.25	108
	Fiber glass roving	2.00	96
Long-term nondegradable RECPs	Synthetic mats:		
	Unvegetated	3.00	144
	Partially established	4.0 → 6.0	192 → 288
	Fully vegetated	8.00	384
	Three-dimensional woven geotextiles	10.00	480
Vegetative	Class A	3.70	178
	Class B	2.10	101
	Class C	1.00	48
	Class D	0.60	29
	Class E	0.35	17
Gravel riprap	1-inch	0.33	16
	2-inch	0.67	32
Rock riprap	6-inch	2.00	96
	12-inch	4.00	192

Source: Adapted from Chen and Cotton (1988).

9.8 REFERENCES CITED

Austin, D. N., and T. Driver (1995). Classifying rolled erosion control products. *Erosion Control* **2** (1): 48–53.

Bathurst, R. J., and R. Karparapu (1993). Large-scale triaxial compression testing of geocell-reinforced granular soils. *Geotechnical Testing Journal* (GTJODI) **16** (3): 296–303.

Chen, Y. H., and G. K. Cotton (1988). Design of roadside channels with flexible linings. Federal Highway Administration Report HEC- 15/FHWA-1P-87-7.

Hewlett, H. W. M., L. A. Boorman, and M. E. Bramley (1987). Design of reinforced grass waterways. CIRIA Report No. 116, Construction Industry Research and Information Association, London, 118 pp.

Kay, B. L. (1978). Mulches for erosion control and plant establishment on disturbed sites. Agronomy Progress Report No. 87, University of California Davis Agricultural Experiment Station, Davis, CA.

Kay, B. L. (1983). Straw as an erosion control mulch. Agronomy Progress Report No. 140, University of California Davis Agricultural Experiment Station, Davis, CA.

Koutsourais, M. (1994). A study of articulated concrete blocks designed to protect embankment dams. *Geotechnical Fabrics Report* (8): 20–25.

Lancaster, T. and D. N. Austin (1994). Classifying rolled erosion-control products. *Geotechnical Fabrics Report* **12** (6): 16–22.

Rustom, R. N. and J. R. Weggel (1993). A study of erosion control systems: Experimental results. *Proceedings*, International Erosion Control Association Conference XXIV, Indianapolis, IN, Feb. 23–26, pp. 255–273.

Shields, F. D., L. T. Ethridge and T. N. Waller (1990). A study of vegetation on revetments, Sacramento River Bank Protection Project, Phase 1: Literature and pilot study. Technical Report No. HL-90-19, U.S. Army Engineering Waterways Experiment Station.

Theisen, M. (1991). The role of geosynthetics in erosion and sedimentation control. *Proceedings*, Fifth GRI Seminar on Geosynthetics in Filtration, Drainage & Erosion Control, Geosynthetics Research Institute, Drexel University, Philadelphia, December 12, pp. 188–203.

USDA Soil Conservation Service (1978). Predicting Rainfall Erosion Losses: a Guide to Conservation Planning. USDA Agricultural Handbook #537, Washington, DC.

10 New Developments and Future Directions in Biotechnical Stabilization

10.0 INTRODUCTION

Biotechnical stabilization is a dynamic and evolving field. New advances and developments continue to reshape the range of options and possibilities. Biotechnical ground cover systems described in Chapter 9, for example, were developed largely during the past decade. New ways of combining inert construction materials with living vegetation for slope protection and erosion control loom on the horizon (Barker, 1994; Gray et al., 1996). We examine a few of these nascent technologies in this chapter.

At the same time old uncertainties persist about the performance of living, constructed systems. The properties of plant materials and the characteristics of plant communities vary in time and space. It is not easy to ascertain, much less predict, the distribution and behavior of belowground root systems over time. This makes it difficult to analyze and quantify the contribution of vegetation to the stability of slopes. A special workshop was sponsored by the National Science Foundation (1991) to assess the state-of-the-art of biotechnical/soil bioengineering stabilization and to identify needed areas of research to clarify some of these problems and uncertainties. Salient findings from this workshop are reviewed herein.

There is no reason why biotechnical stabilization should concern itself solely with the conjunctive use of plant materials. Biological microorganisms, namely, fungal and bacterial colonies, are potential candidates for soil stabilization as well (Muir Wood et al., 1995). A coral reef is a quintessential biotechnical structure—the marine equivalent of a terrestrial hedgerow wall. A reef is built up with remains and excretions of microscopic marine organisms. The cementitious excretions of biological organisms can be thought of as a type of "biological glue" that plays a role similar to that of plant roots. Developments in this field are likewise reviewed briefly in this concluding chapter.

10.1 STATE-OF-THE-ART AND NEEDED AREAS OF RESEARCH

A Workshop on Biotechnical Stabilization was convened by the National Science Foundation in the early 1990s. The workshop focused on the role and function of natural vegetation and fibrous inclusions for reinforcing soils and stabilizing slopes. The main objective of the workshop was to provide a forum for the exchange of views and information, to assess the state-of-the-art, and to identify research issues and needs. Critical information and findings from the workshop were published in a Proceedings volume (National Science Foundation, 1991). The Proceedings also contain invited position papers and short summaries by leading researchers working in the area of fiber/root reinforcement and by practicing professionals in the area of biotechnical stabilization and soil bioengineering. The findings and recommendations of various workshop discussion or focus groups were also included. These focus groups were charged with identifying unresolved issues or problems and needed areas for future study. Conclusions and recommendations for future research were prepared by the workshop participants on such questions as the conjunctive use of vegetation with synthetic fabrics or geogrids, compatibility problems between structural/mechanical components and vegetation, performance evaluation criteria, predictive models, and ways of accounting for the reinforcing or stabilizing role of roots and other natural fibrous inclusions in the ground. The highlights of these recommendations and findings are presented below.

TOPIC 1: TEMPORAL AND SPATIAL ROOT DEVELOPMENT IN SOIL BIOENGINEERING SYSTEMS COMPARED TO NATURAL SLOPE VEGETATION

Information about Root Architecture and Engineering Properties

The root architecture/strength characteristic of a given plant species should be documented in different localities or environments and some forum established for the exchange of information. In situ root studies of a particular species should be conducted in different climates, soil types, and under different methods of establishment or installation.

Systematic Plant Selection Criteria

The question of using native versus nonnative species needs to be addressed. Is it possible to establish native plants, for example, on denuded, cut slopes that have little or no topsoil left and that may bear little resemblance to surrounding soils? What are the advantages and limitations of using exotic plants that are genetically engineered to have specific rooting properties and characteristics relative to endemic, locally grown plants?

Role and Function of Site Management

Can a site be managed in such a way as to encourage the growth of plants with favorable rooting habits and engineering characteristics? What is the influence of such practices as irrigation, drainage, selective cutting and thinning, coppicing, herbicide applications, fertilization, burning, interplanting, and so on?

Positive and Negative Aspects of Plant Competition

Monocultures are generally viewed unfavorably. How can one maximize the advantages of having a diverse plant community and still limit the adverse effects of competition? What monitoring techniques and instrumentation are best suited to studying root/shoot development of different species in a diverse plant community?

Methods of Studying Root Architecture and Development

Several different techniques are cited in the technical literature for studying plant root systems, for example, hydraulic excavation, sector excavations, borings, and trench profiling. Each of these methods has certain advantages and limitations. The suitability of these various methods for studying root architecture and development in soil bioengineering systems should be evaluated.

TOPIC 2: PREDICTIVE MODELS AND PERFORMANCE EVALUATION CRITERIA FOR BIOTECHNICAL STABILIZATION

Existing Problems or Deficiencies in Prediction and Evaluation

There are presently no clear, quantitative measurements of performance of biotechnical and soil bioengineering construction. Establishment of a vegetative cover (*slope greening*) is an important gauge of success but not a sufficient measure. A need exists to determine how the strength of both the living and inert components of the system vary with time and how they interact to affect the overall safety factor of the system over time. There are no generally agreed upon nor standard methods for conducting in situ tests on plant root systems, such as shear, pull out, tensile, and so on. Nor is sufficient information available about the change in properties with time and rooting environment.

Recommendations for Needed Research

(a) Monitor instrumented slopes and installations in representative environments and applications.

(b) Develop quantitative measurements of performance of soil bioengineering and biotechnical systems.

(c) Evaluate various in situ methods of strength testing of fibrous soil inclusions, including brushlayers, fascines, and so on, and develop standard methods for testing and interpretation.

(d) Apply analytical methods to evaluate various test methods (in situ, full scale tests, scaled models, etc.).

(e) Develop practical design methods from results in (a) through (d); make them available to practitioners.

TOPIC 3: COMPATIBILITY PROBLEMS BETWEEN STRUCTURAL/MECHANICAL AND VEGETATIVE COMPONENTS

Influence of Live and Inert Components on Each Other

To what extent will nonbiodegradable components or materials, such as synthetic nets and geogrids, adversely affect vegetation and vice versa? Plants can accommodate to the presence of synthetic materials but the extent to which this may weaken plant growth and vigor is unknown. Synthetic materials (nets and three-dimensional grids) can be manufactured that permit aperture enlargement with time to better accommodate stem and root growth without compromising the role and function of a geofabric. Some biotechnical systems employ metal components that may corrode and deteriorate over time. What effect will this metallic corrosion and/or deterioration have on vegetation?

Potential Areas of Conflict

What are the detrimental impacts, if any, of live plant materials that are introduced into inert structures such as gabions and crib walls? To what extent is the performance and function of engineered drains adversely affected by growth of vegetation? Drainage designs that employ geofabrics must be made with great care both to insure that plant roots will not interfere with drain function or efficiency, and conversely, so that drainage does not unduly deprive vegetation of water. How should appropriate levels of soil compaction be established to balance the need for improved engineering properties (e.g., strength) versus the needs for an enhanced plant growing environment?

Fate of Vegetation after Inert Structural Component Deteriorates

Will vegetation that has been established by mechanical means persist and flourish in harsh growing environments when the mechanical components deteriorate? This issue is particularly relevant to weathered rock slope cuts. Areas of specific inquiry include the following:

(a) What plant species can be established in very harsh environments, and what are the conditions under which they can be established?

(b) What criteria should be applied to harsh slopes to evaluate the appropriateness of using biotechnical systems and to select the best system?

(c) What are the impacts of snow and ice loads on these systems?

Transitions from Engineered Structures to Soil Bioengineering Systems

How can stable transitions be designed and constructed from soil bioengineering systems to structural retaining walls or hard armor revetments so that the interface between the two does not become a locus of weakness or other problems?

Surface Anchoring of Biotechnical Ground Covers

Rolled erosion control products are vulnerable to rilling and undermining in the absence of intimate contact with the ground surface. Additional research is needed to determine the best way of initially fastening or anchoring nets and blankets to the ground surface when used in conjunction with seedings and plantings.

Answers to many of these questions will emerge as more experience is gained with biotechnical and soil bioengineering stabilization. Some questions have already been addressed in this book; others still await answers. Users of this book may wish to take note of the questions and issues raised at the NSF Workshop and eventually contribute lessons learned from their own experience with biotechnical and soil bioengineering technology.

10.2 ANCHORED GEOTEXTILES AND GEONETS

10.2.1 Components and Principle of System

An anchored geonet is a good example of an emerging development in biotechnical stabilization. Unlike conventional rolled erosion control products and biotechnical ground covers, this technique provides a firm and intimate contact between the ground surface and the geonet, in addition to imparting a slight confining stress to the ground surface. Because of these attributes, anchored geonets are particularly well suited for the stabilization of sandy slopes and coastal landforms, which normally lack resistance to shear and tractive stresses near the surface.

An anchored geotextile or geonet consists basically of a fabric or net that is stretched over and pulled down tightly on the ground surface by means of anchors that are inserted through and fastened to the geotextile. Driving or pushing the anchors into the ground causes the geotextile to go into tension, and this in turn imparts a slight compressive force on the ground surface. The compressive force increases shear strength and resistance to sliding. This additional shear strength is particularly critical in the case of sandy soils, which have little resistance to sliding and particle detachment in the absence of any significant confining stress near the ground surface.

Anchored geotextile systems were first introduced by Koerner and Robbins (1986). A conceptualized version of an anchored geonet/geotextile system is shown in Figure 10-1. The main function of the geonet is to place the underlying soil into compression for reasons noted previously. The success of such a system depends upon a complex set of interactions, namely, the satisfactory transfer of load between the geotextile/geonet and the anchors, between the geotextile/geonet and soil, and between the anchor and soil. The nature and quantification of these interactions are still the subject of research, but enough is already known about the system to make rough performance estimates and preliminary recommendations with regard to installation.

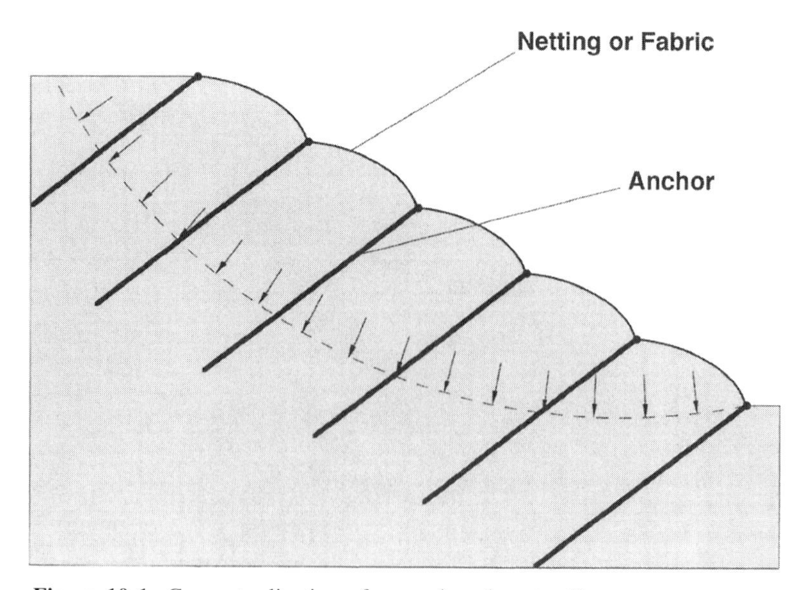

Figure 10-1. Conceptualization of an anchored geotextile or geonet system.

10.2.2 Load Transfer from Geotextile/Geonet to Soil

An incremental length of stretched fabric or netting on a soil surface is shown schematically in Figure 10-2. The normal stress between soil and fabric acting over the increment $\Delta\theta$ is given by

$$\sigma_n = \frac{(T_2 + T_1)}{2} \frac{\Delta\theta}{\Delta s} \tag{10-1a}$$

where Δs is the incremental curvilinear distance between points 1 and 2.

Since $\Delta\theta / \Delta s$ is the reciprocal of the radius of curvature (r), the normal stress at any point on the interface can be expressed simply as:

$$\sigma_n = \frac{T}{r} \tag{10-1b}$$

where T is the developed tensile load per unit length in the fabric over the increment of interest. Accordingly, the stresses transferred from an anchored geotextile to the soil at a given point on the interface are only a function of the tension in the fabric and its local curvature. This holds true for both plane (two-dimensional) curvature (from a line anchor) or axisymmetric (three-dimensional) curvature (from a geometric array of multiple anchor points).

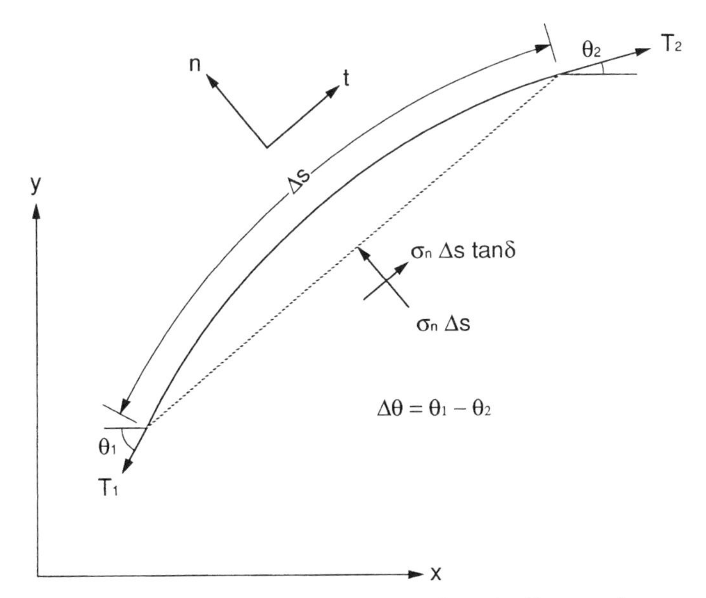

Figure 10-2. Load transfer between anchored geotextile and soil over an incremental fabric or netting length.

10.2.3 Contrast with Conventional Netting Installation

Instructions for the installation of erosion control netting and fabrics typically emphasize the importance of *not* stretching the fabric (Belton Industries, 1990). This warning is issued ostensibly to insure intimate contact with the ground surface and to avoid bridging over incipient rills and channels. The warning appears to be based on past experience with sites where rilling has been observed beneath tightly stretched netting.

This injunction not to stretch eliminates, however, the opportunity of exploiting the stabilizing mechanism from a geotextile or geonet that is tightly stretched and anchored securely to the ground. Furthermore, the problem of rilling and washout beneath a stretched fabric is largely a function of the type of netting and the manner in which it is anchored or fastened to the ground. Low tensile strength and modulus nettings should probably not be pulled down and stretched because they simply lack sufficient strength and stiffness to withstand the imposed tensile forces. According to Koerner (1990) geotextiles and geonets should be used that have a minimum tensile strength of 200 lb/inch (35 kN/m), that are relatively durable, and that have good creep resistance.

Some natural nettings (e.g., the higher density coir fabrics) have sufficient unit tensile strength and stiffness to impart a substantial confining stress on the ground surface as well as provide a water retention/mulching function that synthetic nets do not. The benefits of a surface confining stress will only be realized, however, provided the fabric is *properly stretched and anchored to the ground*. If properly anchored, the netting should improve not only the shallow mass stability of a sandy slope but also its resistance to surface erosion or particle detachment. The operative word here is "properly"—otherwise, the warnings about the adverse effects of stetching may be applicable.

Research by Hryciw (1991) and Vitton (1991) has shown that the effectiveness of an anchored geotextile is a function of the geometric arrangement, spacing, and inclination of the anchors in addition to the viscoelastic properties of the geotextile, soil-geotextile interface friction, and stetched shape of the geotextile. The stretched shape of the geotextile is an important factor that depends upon the pull generated by the anchors, the geometry of the anchor array, and the initial or prepared (groomed) shape of the ground surface.

Influence of Anchor Array or Pattern: As noted previously the amount of normal stress transferred to the ground surface by an anchored fabric (or geonet) depends largely on the amount of curvature developed in the fabric. Significant curvature tends to develop only in the immediate vicinity of the anchors. Accordingly, much of the soil between anchor points at shallow depths may be subject to very little confining stress increase. Furthermore, the tension in the fabric is also dissipated at increasing distances from the anchor by interface friction between the fabric and soil.

Different anchor point arrays are shown in Figure 10-3. A relatively wide erosion channel of width S_e may form beneath a stretched, but uncurved fabric

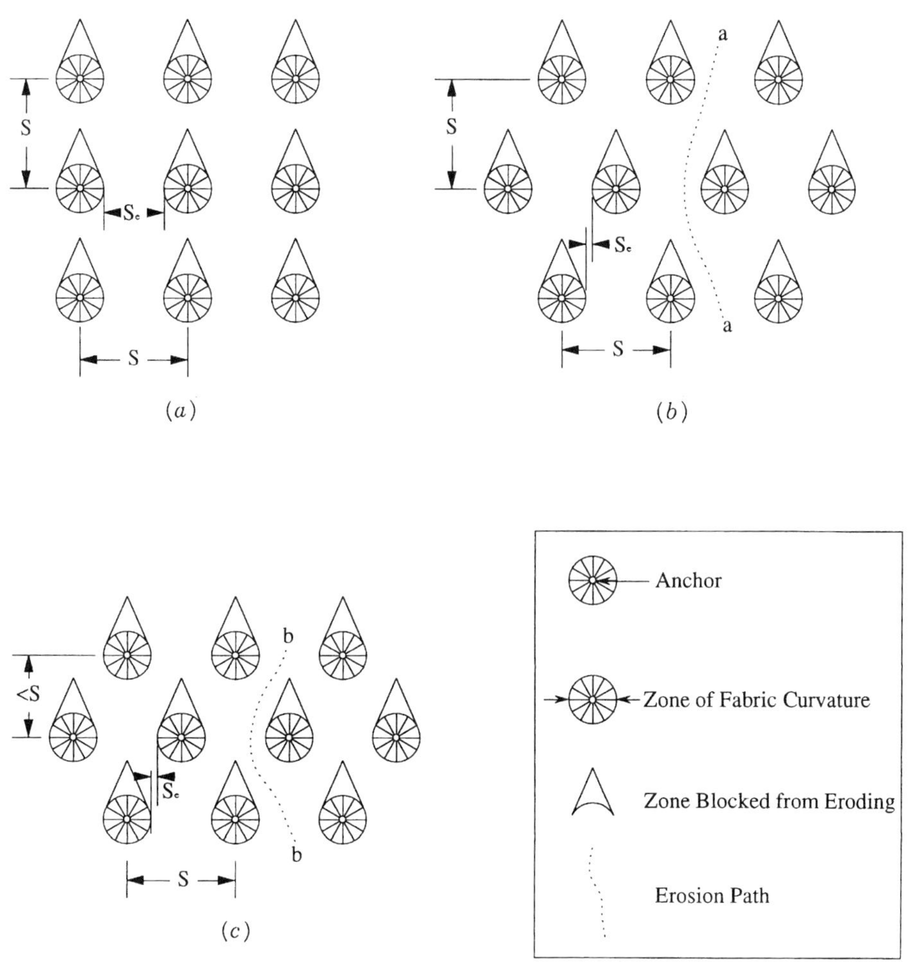

Figure 10-3. Anchor arrays for ground anchored netting systems. (*a*) Square anchor pattern. (*b*) Triangular anchor pattern. (*c*) Triangular anchor pattern with decreased row spacing. (After Hryciw, 1991.)

in a square pattern (Figure 10-3*a*). By offsetting alternate anchor rows (Figure 10-3*b*) into a triangular anchor pattern, the width of a direct downslope erosion channel can be reduced considerably. If in addition to offsetting alternate rows, the spacing between rows is reduced (Figure 10-3*c*), a potential meandering erosion channel must follow a more tortuous path, and therefore, erosion resistance will be increased still further.

Influence of Initial Ground Shape: The initial shape of the ground surface also affects the efficiency of stress transfer. Digging conical depressions in the ground in the case of a multiple point anchor array results in favorable curvature modification. Plowing the ground surface across a slope on contour, as shown in Figure 10-4, can enhance performance by favorable modification of curvature in the case of a line anchor system. This latter surface modification also improves erosion resistance by providing a "contour terrace" effect. Plowing furrows in the slope and mounding the soil on either (uphill and downhill) side enlarges the zone of curvature beneath the netting. Anchors in this case would be installed in the furrows, as shown schematically in Figure 10-4.

Influence of Anchor Inclination and Method of Emplacement: The inclination of the anchors and their method of emplacement also affect the stress transfer between the netting and the ground surface. A driven anchor system that relies on skin friction (e.g., ribbed metal rods) will respond differently than anchors that rely on end anchorage (e.g., duckbill anchors). Previous analytical studies (Hryciw, 1991) have also shown that a perpendicular direction to the slope is not the optimal anchor orientation. The optimal orientation depends upon the type of anchor, spacing, length, and other considerations.

10.2.4 Calculation of Required Anchor Loads

Homogeneous slopes of coarse-textured or cohesionless soils, such as dune sands, commonly fail by shallow surface sliding or sloughing. The thickness of the sliding mass in this case is generally small compared to the length of slope. Under these conditions an *infinite slope* analysis is appropriate. In such an analysis stability is governed by shear forces and resistance along the basal sliding surface and the boundary forces at the top and bottom ends of the slope are assumed to have negligible effects on the stability (see Chapter 2). The factor

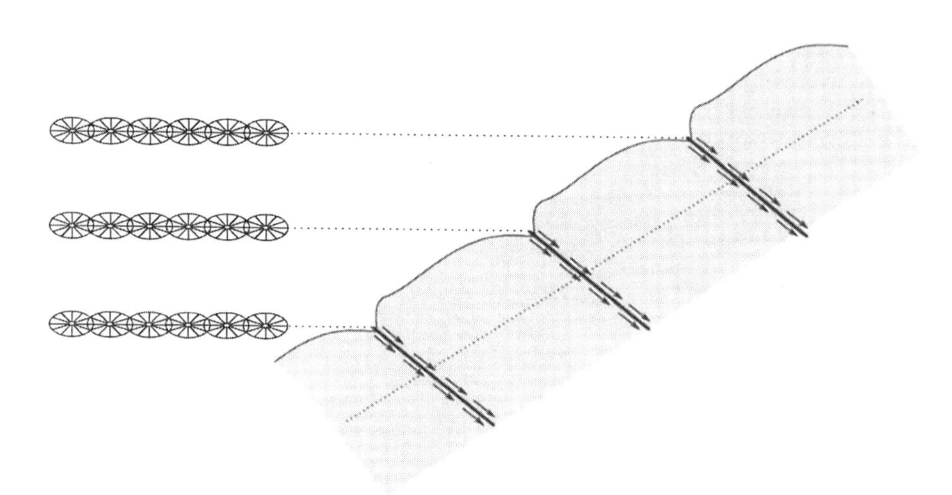

Figure 10-4. Preformed or furrowed surface for anchored geonet.

of safety against Coulomb sliding is derived from force-equilibrium equations for an element (or slice) of the infinite slope. Stresses on two vertical faces of an element in the finite slope are assumed to be equal, but opposite in direction, so that they cancel each other in the equilibrium equations. This analysis is identical for all elements along the slope length. In other words, all soil elements in the slope are subjected to identical forces due to gravity, seepage, and the forces exerted by an anchored geonet.

The presence of water can significantly affect the stability of sandy, cohesionless slopes. When water seeps parallel to the surface of such slopes, the stability drops by approximately one-half. The stability is decreased even further when seepage emerges from the slope (see Section 3.10). On the other hand, when water flows vertically downward, the stability is equivalent to that of a dry slope. In situations where water flows over the slope surface or emerges from the surface, surface erosion or piping can occur and either accompany or trigger shallow mass failure.

The stability of slopes against shallow sloughing and surficial erosion can be increased significantly by anchored geonets (Hryciw and Haji-Ahmad, 1992; Gray and Hryciw, 1993; Gray et al., 1996). A schematic diagram of an element in an infinite slope under gravity, anchored geonet, and seepage forces is shown in Figure 10-5. Since anchors are installed on contour, in preplowed furrows, fabric curvature develops only in the cross-sectional plane, as shown

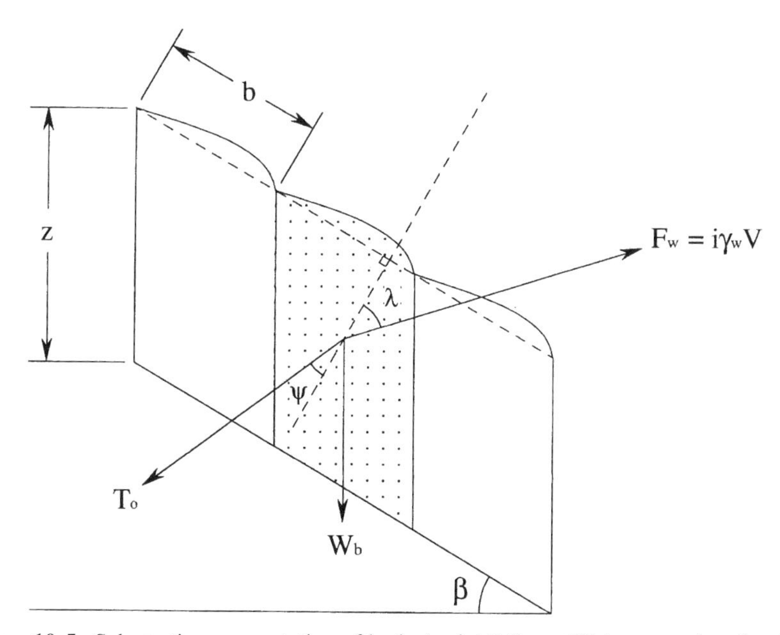

Figure 10-5. Schematic representation of body (weight) force (W_b), external anchor force (T_0), and seepage force (F_w) in an infinite slope with seepage acting in a variable direction ($0 < \lambda < 180 - \beta$).

in Figure 10-4. Accordingly, a two-dimensional analysis is appropriate for this problem.

The anchor force, T_0, in Figure 10-5 represents the resultant force exerted by the geosynthetic per slope area ($s \times 1$) where s is the spacing between rows of anchors. In general the direction of T_0 may be assumed to coincide with the anchor axis. The factor of safety against Coulomb sliding for this element can be written as:

$$FS = \frac{\cos\beta + \xi_b \cos\Psi - A_b \cos\lambda}{\sin\beta + \xi_b \sin\Psi - A_b \sin\lambda} \tan\phi \qquad (10\text{-}2)$$

where: $A_b = \gamma_w/\gamma_b$(the buoyant seepage force)
$i = \sin\beta/\sin\lambda$(the seepage gradient)
$\xi_b = T_0/zs\cos\beta\gamma_b$(dimensionless anchored geonet load ratio)
β = slope angle
ϕ = soil friction angle
Ψ = anchor orientation with respect to slope normal
λ = seepage direction with respect to slope normal
γ_b = buoyant unit weight of soil
γ_w = buoyant unit weight of water
s = spacing between anchor rows
T_0 = unit anchored geonet load (per "$s \times 1$" area of slope)
z = depth to failure surface (or thickness of sliding mass)

It is important to note that ξ_b, and therefore FS, decrease with depth, z. As z becomes large, ξ_b tends toward zero and FS in Equation 10-2 approaches the factor of safety of the slope prior to a geonet installation. Consequently, an anchored geonet provides its greatest benefit near the surface where sandy slopes are most unstable and vulnerable to erosion, while at the same time relying on a natural increase of strength with depth where confining stresses and shear strength are higher because of surcharge from the overlying weight of sand.

For efficiency in an anchored geonet installation, it is desirable to place the anchors at an orientation that maximizes the increase in stability under a given load. This optimum anchor orientation (Ψ_{op}) can be obtained by differentiating Equation 10-2 with respect to Ψ and setting the result equal to zero. The final expression for Ψ_{op} can be written in terms of the anchored geonet load ratio ξ_b as follows:

$$\xi_b = A_b \cos(\Psi_{op} - \lambda) - \cos(\Psi_{op} + \beta) \qquad (10\text{-}3)$$

If Equation 10-3 is substituted into Equation 10-2, the following simple relationship is obtained for the maximum factor of safety FS_{max} when the anchors are installed at their optimum orientation Ψ_{op}:

$$FS_{max} = \tan \Psi_{op} \tan \phi \tag{10-4}$$

Equation 10-4 can be used to compute the factor of safety for a given unit anchor load T_0, or alternatively, it can be used to calculate the unit anchor load required to achieve a specified factor of safety when the anchors are placed at their optimum orientation θ_{op}.

A schematic diagram and illustration of an experimental setup (Ghiassian et al., 1996) to evaluate the influence of anchor forces, spacing, and inclination on the stability of a saturated sand slope subjected to overtopping and seepage are shown in Figures 10-6 and 10-7, respectively.

10.2.5 Potential Applications of Anchored Geonets

Many coastal landforms (e.g., beaches, dunes, bluffs, levees) are composed of sandy material that is vulnerable to rainfall, wave, channel, and seepage erosion. The shear strength, or alternatively, the resistance to particle detachment and displacement, in coarse-textured, cohesionless soils depends upon the amount of confining or contact stress between particles. Near the surface, where the confining stresses are very low, the shear strength will likewise be very low. Accordingly, sandy slopes are very susceptible to surficial erosion and shallow mass wasting because in the absence of cohesion they offer little resistance to

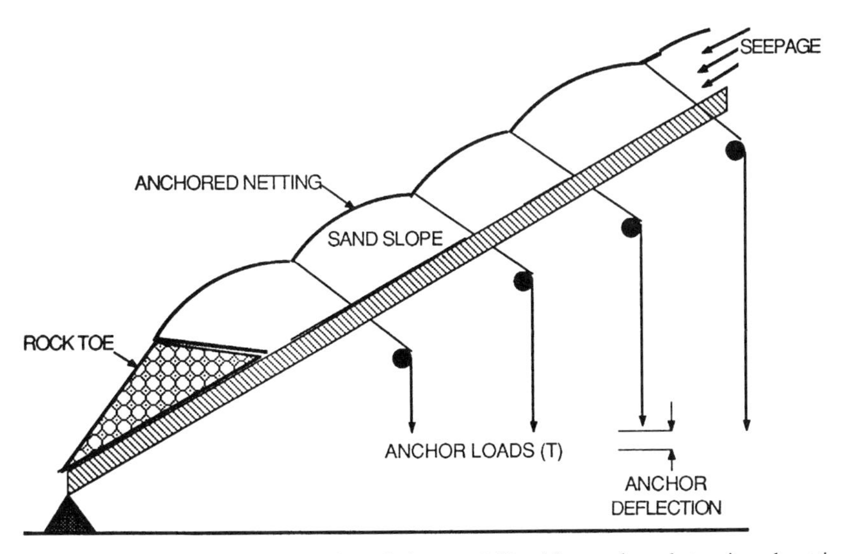

Figure 10-6. Schematic diagram of sand slope stabilized by anchored, tensioned netting. Anchor forces are developed by wires attached to rods that impart line loads to netting, which in turn causes tension in netting and compression in sand.

Figure 10-7. Photo of test facility for studying influence of anchored, tensioned netting on stability of sand slopes subjected to overtopping and seepage forces.

particle detachment and displacement under the influence of raindrop splash, overland flow, seepage, and wave action.

The barrier islands that lie offshore of much of the southeast coast of the United States are good examples of sandy, low-lying landforms that are very vulnerable to erosion. The islands are attacked periodically by high tides and waves during the winter and occasionally by hurricane driven storms. The main line of defense appears to be a wide beach and a vegetated dune field, as shown in Figures 10-8 and 10-9. Dune vegetation is pretty good at "building" dunes by trapping drifting, windblown sand. The vegetation is only minimally effective, on the other hand, at "protecting" the dunes and beach from direct wave action and seepage, as shown in Figures 10-10 and 10-11.

An anchored and tensioned fabric on the face of a sand dune in effect converts millions of loose, easily detached sand grains into a coherent, unitary mass that resist seepage and erosion forces. If geogrids or nets are used for this purpose, they have the additional advantage of permitting vegetation to be established in the interstices or openings of the net. Alternatively, a vegetative cover (or biocover) can be established on top of the anchored net. The vegetation traps drifting sand and develops a protective cover of sand and vegetation on top of the anchored netting system. The underlying anchored netting system provides long-term protection against wave attack and seepage erosion while the overlying biocover, composed of sand and dune grass, hides the anchored netting and shields it against UV radiation.

Figure 10-8. Coastal development buffered and protected by relatively wide beach and well-vegetated foredunes, Kiawah Island, South Carolina.

Figure 10-9. Close-up view of vegetated dune field. Specially adapted dune vegetation traps drifting sand and helps to maintain and build up the dunes.

Figure 10-10. Erosion of vegetated, coastal foredunes by waves and high tides, resulting in vertical scarp and damage to beach access structures.

Figure 10-11. Erosion of vegetated coastal dune by wave action. Note presence of salt-intolerant woody vegetation near erosion face or scarp, which is indicative of extent of erosion and loss of foredunes.

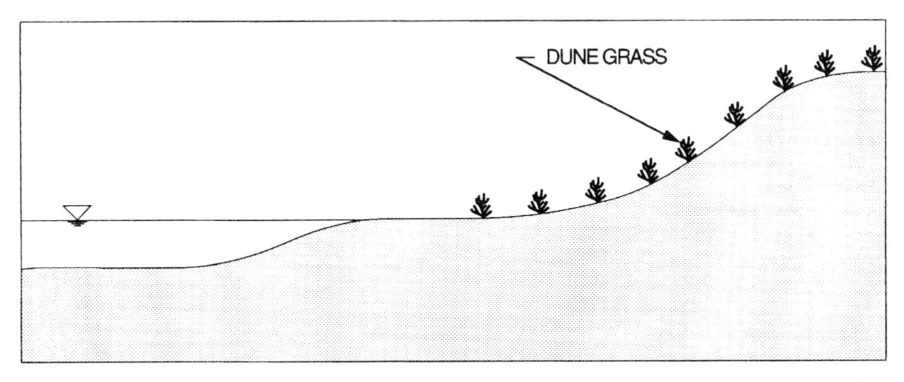

Figure 10-12. Sand dune protected by dune grass alone. Vegetation provides protection against wind erosion by trapping and restraining drifting sand. The vegetation helps to stabilize and build the dune over time.

This sand/vegetation veneer (or for that matter any unmodified, vegetated dune) is subject to erosion or washout during storms, but the core of the underlying portion of the dune that is protected by an anchored geonet would remain intact. This concept is illustrated schematically in Figures 10-12 through 10-15. The response of an unreinforced but vegetated dune is shown in Figures 10-12 and 10-13. Vegetation alone is unable to armor and protect a sand dune against seepage and wave forces. The response of the same dune with its core protected by an anchored geonet is shown in Figures 10-14 and 10-15. The sand/vegetation cover or veneer is subject to erosion or washout during storms, but the core of the dune remains intact. The sand/vegetation cover will reestablish naturally over time. Alternatively, the recovery process can be accelerated by importing and covering the protected part of the dune with a sand fill and

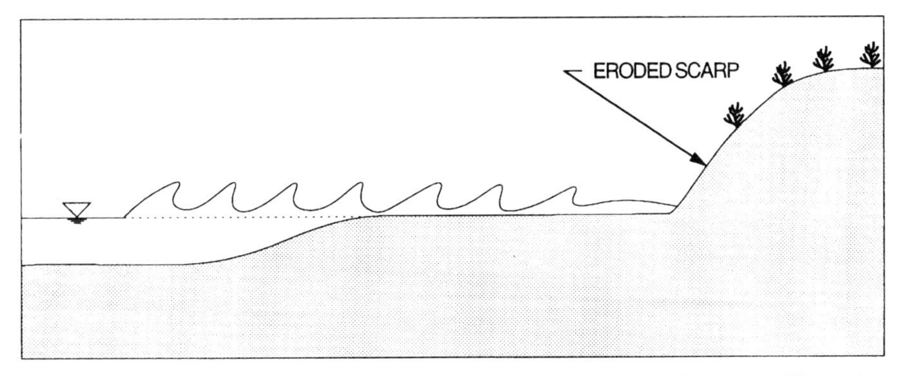

Figure 10-13. Erosion of coastal dune by wave action during a major storm. Vegetation alone is unable to armor and protect a sand dune against seepage and wave forces.

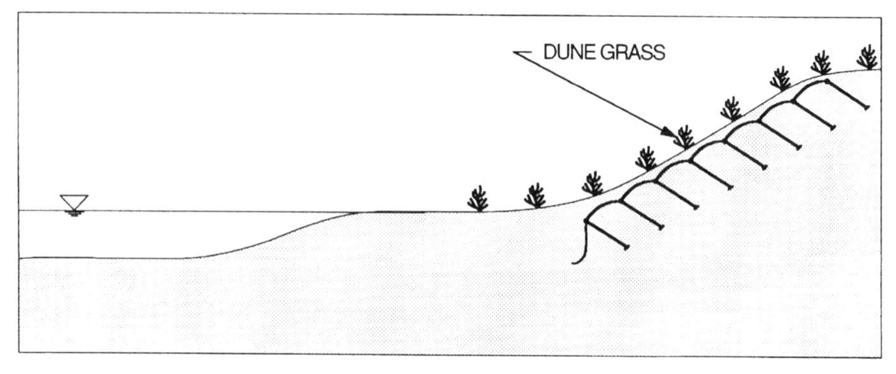

Figure 10-14. Sand dune protected by anchored geosynthetic and dune grass. Grass traps drifting sand and develops protective cover of sand and vegetation on top of the anchored netting system.

then planting it with dune grass and other suitable vegetation in the conventional manner.

An anchored geonet is a "soft armor" system that is visually and physically nonintrusive. This feature would be advantageous in coastal communities, which often have regulations preventing the use of structural armor systems, for example, rock riprap, in environmentally sensitive areas such as coastal dunes. The state of South Carolina, for example, passed a Beachfront Management Act that is designed to protect coastal areas and to regulate shoreline development. Significant features of the Beachfront Management Act include the establishment of setback distances from the shoreline and prohibition of "hard armor" erosion control structures, such as rock revetments and seawalls, in new coastal developments.

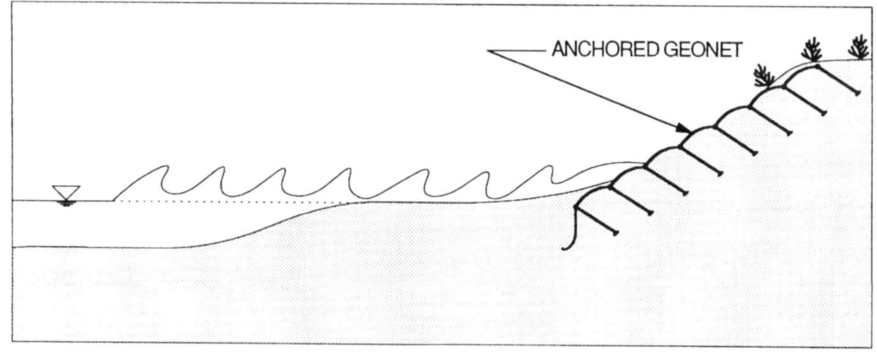

Figure 10-15. Sand dune protected and armored against storm waves by anchored geosynthetic system. Sand/vegetation cover is subject to erosion or washout during storm, but core of dune remains intact. Sand/vegetation cover will reestablish over time.

10.3 STABILIZATION WITH BIOLOGICAL ORGANISMS

Anyone walking on a sandy beach has no doubt noticed how firm the beach is underfoot where the sand is partially saturated. The shear strength of the sand has been increased by "apparent cohesion" or capillary stresses acting in pendular water rings surrounding the grain contacts. This is only one of several ways of developing apparent cohesion in an otherwise purely frictional or cohesionless sand. Another way of generating apparent cohesion in sand is via cementation with silica and carbonate precipitated at grain contacts from percolating groundwater or, alternatively, from the presence of fibrous inclusions or plant roots, as discussed in Chapter 3. Still another way is from microbiological binding from the presence of fungal and bacterial colonies in a sand. Bacteria produce extracellular polymeric material, which tends to bind sand particles together in a "biological glue," whereas the fungal colonies produce microscopic, hair-like hyphae, which at small scale have an effect similar to plant roots (Meadows et al., 1994).

A little bit of cohesion or apparent cohesion makes a big difference to the stability of sandy soils and slopes. This can be demonstrated in several ways. The effect of small amounts of apparent cohesion in preventing shallow surface sloughing in sand slopes was illustrated by means of slope stability analyses in Chapters 2 and 3. Another way of demonstrating the significance and importance of cohesion to stability is to calculate the amount of cohesion required to maintain a vertical, self-supporting slope face or cut. The height of a vertical, unsupported cut is given by the following equation (Huang, 1983):

$$H_{\text{crit}} = (Kc_{\text{avail}}/\gamma)\tan(45 + \phi/2) \qquad \text{(10-5a)}$$

or

$$c_{\text{req'd}} = [H\gamma/K\tan(45 + \phi/2)] \qquad \text{(10-5b)}$$

where:
H_{crit} = the critical or maximum allowable height of cut
K = a constant
γ = soil density
$c_{\text{req'd}}$ = the required cohesion
c_{avail} = the available cohesion
ϕ = the angle of internal friction

The constant K varies from 2 to 4 depending upon whether an upper or lower bound solution is invoked. Observations of stable, vertical cliffs or bluffs in loessial soils, which have a slight cohesion as a result of light carbonate cementation at their grain contacts, can be used to determine an appropriate value for the constant K. Vertical banks or bluffs in loessial soils in western Iowa have been observed (Lambe and Whitman, 1969) to have an average height H_{av} of

14.6 feet, c_{av} = 187 psf, γ_{av} = 76 pcf, and ϕ_{av} = 25 degrees. Substitution of these values into Equation 10-5 yileds a value for the constant K of 3.8. This suggests that an upper bound value for the constant (K = 4) is appropriate as a first approximation. Accordingly, we may can write an expression for the minimum required cohesion as follows:

$$c_{req'd} = [H\gamma/4 \tan(45 + \phi/2)] \qquad (10\text{-}6)$$

This equation can be used to calculate the required cohesion for any vertical cut or slope. For purposes of illustration assume a 6-foot high, vertical, unsupported cut in a sandy soil with a density of 100 pcf and friction angle of 30 degrees. Substituting these values into Equation 10-6 yields a value of $c_{req'd}$ = 1.8 psi (12.1 kPa), which is well within the range of cohesion that can be supplied by plant roots (see Table 3-5 in Chapter 3). It is also within range of the apparent cohesion or microbiological binding that can be supplied by bacterial and fungal colonies growing in sand.

The potential use of microbiological organisms for stabilizing sediments and slopes has recently been studied by several investigators (Muir Wood et al., 1995; Meadows et al., 1994). Samples of quartz sand were placed in containers, submerged in a synthetic sea water, and inoculated with either the bacterium *Pseudomonas atlantica* or the fungus *Penicillium chrysogenum*. All inoculated specimens were left to incubate for three weeks before the angles of avalanche failure and angles of repose were determined. These angles were measured by slowly tipping the containers, as shown schematically in Figure 10-16, and noting the angle at which the surface of the sand first started to slide (angle of avalanche) and the equilibrium angle of the sand surface after failure (angle of repose).

The cohesive binding produced by the bacterial and fungal colonies in the sand were estimated from these measurements. An order-of-magnitude value

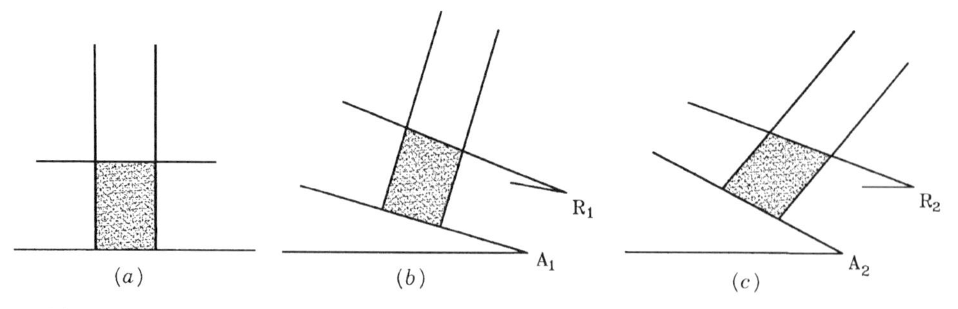

Figure 10-16. Stability measurements in sand specimens with bacterial or fungal colonies. (*a*) Initial orientation. (*b*) First angle of avalanche (A_1) and repose (R_1). (*c*) Second angle of avalanche (A_2) and repose (R_2). (From Muir Wood et al., 1995. Used with permission of the Institution of Civil Engineers.)

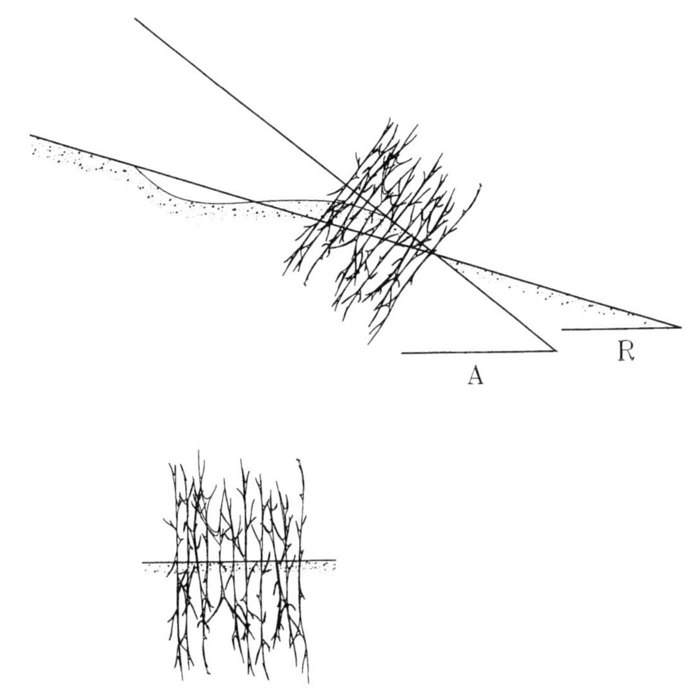

Figure 10-17. Schematic illustration of fungal colony producing microscopic, hair-like hyphae that extend into sloping, sediment surface and provide reinforcing effect (angles of avalanche, *A*, and response, *R*). (From Muir Wood et al., 1995. Used with permission of the Institution of Civil Engineers.)

of c = 0.7 psi (5 kPa) was deduced for the bacteria. The extracellular polymeric material produced by the bacteria acted like a "biological glue," effectively supplying cohesion near the surface, which increased the stability of the sandy sediment against shallow sloughing or sliding as it was tilted. Similar results were obtained with the fungal hyphae. In this case the action of the microscopic, hair-like hyphae growing on the surface of the sandy sediment (shown schematically in Figure 10-17) was similar to the reinforcement effect of plant roots, but on a much smaller scale.

10.4 CONCLUDING REMARKS

Biotechnical and soil bioengineering methods greatly expand our range of options for stabilizing and protecting slopes against erosion and shallow mass movement. More to the point they allow us to carry out this mission in a cost-effective, visually attractive, and environmentally sensitive manner. Questions still remain with regard to quantification and performance evaluation;

but answers to these questions will emerge as more experience is gained with biotechnical and soil bioengineering stabilization. Guidelines are provided in this book for the installation of a variety of soil bioengineering and biotechnical stabilization methods that have met the test of time and that have ultimately resulted in stable slopes with a diverse habitat and natural appearance.

Biotechnical stabilization is also a dynamic and evolving field as attested to by a brief review of recent developments in this concluding chapter. The use of biological organisms for biotechnical stabilization purposes, namely fungal and bacterial colonies, appears to be a promising new avenue of approach. As Muir Wood et al. (1995) have presciently noted, "There is continuous spectrum of biological influences on slope stability extending from microorganisms at one end to the effect of plant roots at the other." It remains for us to explore and capitalize on these influences in a rational and productive manner.

10.5 REFERENCES CITED

Barker, D. H. (1994). The way ahead: Continuing and future developments in vegetative slope engineering or ecoengineering. *Proceedings*, International Conference on Vegetation and Slopes, Institution of Civil Engineers, University Museum, Oxford, England, September 29–30.

Belton Industries, Inc. (1990). Anti-wash Geojute, installation instructions.

Ghiassian, H., R. D. Hryciw, and D. H. Gray (1995). Laboratory testing apparatus for slopes stabilized by anchored geosynthetics. *Geotechnical Testing Journal*, GTJODJ 19(1): 65–73.

Gray, D. H., and R. D. Hryciw (1993). Biotechnical stabilization of sandy coastal landforms. *Proceedings*, International Rip Rap Workshop, Fort Collins, CO, July 12–16, Vol. 1, pp. 341–356.

Gray, D. H., and A. T. Leiser (1982). *Biotechnical Slope Protection and Erosion Control*. New York: Van Nostrand Reinhold.

Gray, D. H., R. D. Hryciw, and H. Ghiassian (1966). Protection of coastal sand dunes with anchored geonets. *Proceedings*, Conference on Beach Preservation Technology, St. Petersburg, FL, Jan 24–26.

Hryciw, R. D. (1991). Load transfer mechanisms in anchored geosynthetic systems. University of Michigan Research Report to the Air Force Office of Scientific Research, Grant No. 88-0166, 59 pp.

Hryciw, R. D., and K. Haji-Ahmad (1992). Slope stabilization by anchored geosynthetic systems: Anchorage optimization. *Proceedings*, ASCE Specialty Conference on Slopes and Embankments, Berkeley, CA, pp. 1464– 1480.

Huang, R. (1983). *Stability of Earth Slopes*. New York: Van Nostrand Reinhold.

Koerner, R. M. (1990). *Designing with Geosynthetics*. Englewood Cliffs, NJ: Prentice-Hall.

Koerner, R. M., and J. C. Robbins (1986). In-situ stabilization of slopes using nailed geosynthetics. *Proceedings*, Third International Conference Geosynthetics, Vienna, Austria, pp. 395–400.

Lambe, T. W., and R. V. Whitman (1969). *Soil Mechanics*. New York: Wiley.

Meadows, A., P. S. Meadows, D. Muir Wood, and J. H. M. Murray (1994). Microbiological effects on slope stability: An experimental analysis. *Sedimentology* **41** (to appear).

Muir Wood, D., A. Meadows, and J. H. M. Murray (1995). Effect of fungal and bacterial colonies on slope stability: In *Vegetation and Slopes*, edited by D. H. Barker, Institution of Civil Engineers, *Proceedings*, International Conference held at the University Museum, Oxford, England, September 29–30, pp. 46–51.

National Science Foundation (1991). *Proceedings*, Workshop on Biotechnical Stabilization, edited by D. H. Gray, The University of Michigan, Ann Arbor, MI, August.

Vitton, S. J. (1991). An experimental investigation into the load transfer mechanism in anchored geosynthetic systems. Ph.D. diss., University of Michigan.

APPENDIX 1
Soil Bioengineering Plant Species

This appendix contains a list of plants suitable for use with soil bioengineering or biotechnical stabilization. The salient characteristics of each plant (availability, habitat value, size/form, root type, root propagation from cuttings) are noted.

Name	Location	Availability	Habitat Value	Size/Form	Root Type	Rooting Ability from Cuttings
Acer negundo Box elder	N, NE	Common	Excellent	Small tree	Moderately deep spreading	Poor to fair
Alnus rubra Red alder	NW	Very common	Excellent	Large tree	Shallow spreading	Poor
Baccharis glutinosa Water wally	W	Common	Very good	Medium shrub	Fibrous	Good
Baccharis halimifolia Eastern baccharis	S, SE	Common	Poor	Small to medium shrub	Fibrous	Fair to good
Baccharis pilularis Coyotebrush	W	Very common	Good	Medium shrub	Fibrous	Good
Baccharis viminea Mule fat	W	Very common	Very good	Medium shrub	Fibrous	Good
Betula papyrifera Paper birch	N, E, W	Common	Good	Tree	Fibrous shallow	Poor
Betula pumila Low birch	N.E. W	Common	Very good	Medium shrub	Fibrous	Poor
Cornus amomum Silky dogwood	N, SE	Very common	Very good	Small shrub	Shallow fibrous	Very good
Cornus racemosa Gray dogwood	NE	Common	Very good	Medium to small shrub	Shallow	Good
Cornus rugosa Roundleaf dogwood	NE	Common	Very good	Medium to small shrub	Shallow fibrous	Fair to good
Cornus sericea ssp. stolonifera Red osier dogwood	N, NE, NW	Very common	Very good	Medium to small shrub	Shallow	Very good
Crataegus Sp. Hawthorn	SE	Uncommon	Good	Small dense tree	Tap root	Fair to good
Elaeagnus commutata Silverberry	N. central	Very common	Poor	Medium shrub	Shallow	Fair to good

Name	Location	Availability	Habitat Value	Size/Form	Root Type	Rooting Ability from Cuttings
Ligustrum sinense Chinese privet	S, SE	Common	Fair to good	Small to medium shrub	Shallow fibrous	Good
Lonicera involucrata Black twinberry	E	Common	Poor to fair	Small shrub	Shallow	Good
Physocarpus capitatus Pacific ninebark	NW, W	Common	Fair	Small	Fibrous	Good
Physocarpus opulifolius Common ninebark	NE	Common	Good	Medium to high shrub	Shallow lateral	Fair to good
Populus angustifolia Arrowleaf cottonwood	W	Common	Good	Tree	Shallow	Very good
Populus balsamifera ssp. trichocarpa Black cottonwood	NW	Common	Good	Tree	Shallow fibrous	Very good
Populus deltoides Eastern cottonwood	MW, E	Very common	Good	Large tree	Shallow	Very good
Populus fremontii Freemont cottonwood	SW	Very common	Good	Tree	Shallow fibrous	Very good
Populus tremuloides Quaking aspen	NW	Very common	Good	Large tree	Shallow	Fair
Robinia pseudoacacia Black locust	NE	Common	Very poor	Tree	Shallow	Good
Rubus allegheniensis Allegheny blackberry	NE	Very common	Very good	Small shrub	Fibrous	Good
Rubus spectabilis Salmonberry	SW, NW	Very common	Good	Small shrub	Fibrous	Fair to good
Rubus strigosus Red raspberry	N, NE, W	Very common	Very good	Small shrub	Fibrous	Good
Salix exigua Coyote willow	NW	Fairly common	Good	Medium shrub	Shallow suckering	Good
ssp. interior Sandbar willow	N, SE	Common	Good	Large shrub	Shallow to deep	Fair to good
Salix amygdaloides Peachleaf willow	N, S	Common	Good	Very large shrub	Shallow to deep	Very good
Salix bonplandiana Pussy willow	W, MW	Very common	Good	Medium shrub	Fibrous	Very good
Salix eriocephala ssp. ligulifolia Erect willow	NW	Common	Good	Large shrub	Fibrous	Very good
Salix gooddingii Goodding willow	SW	Very common	Good	Large shrub to small tree	Shallow to deep	Excellent
Salix hookeriana Hooker willow	NW	Common	Good	Large	Fibrous dense	Very good
Salix humilis Prairie willow	N, NE	Very common	Good	Medium shrub	Fibrous	Good
Salix lasiolepis Arroya willow	W	Common	Good	Medium shrub	Fibrous	Very good

Name	Location	Availability	Habitat Value	Size/Form	Root Type	Rooting Ability from Cuttings
Salix lemmonii Lemmon willow	W	Common	Good	Medium shrub	Fibrous	Very good
Salix lucida Shining willow	N, NE	Very common	Good	Medium to large shrub	Fibrous	Very good
ssp. lasiandra Pacific willow	NW	Very common	Good	Large shrub to small tree	Fibrous	Very good
Salix lutea Yellow willow	W	Very common	Good	Medium to large shrub	Fibrous	Very good
Salix nigra Black willow	N. SE	Very common	Good	Large shrub to small tree	Shallow to deep	Excellent
Salix purpurea Streamco	N, S, E, W	Very common	Good	Medium shrub	Shallow	Very good
Salix scouleriana Scoulers willow	NE	Very common	Good	Large shrub to small tree	Shallow	Very good
Salix sitchensis Sitka willow	NW	Common	Good	Very large shrub		Very good
Salix X cotteti Bankers willow	N, S, E, W	Uncommon	Good	Small shrub	Shallow	Very good
Salix discolor Red willow	N, NE	Very common	Good	Large shrub	Shallow	Very good
Sambucus cerulea Blueberry elderberry	W	Common	Very good	Medium shrub	Fibrous	Poor
Sambucus canadensis American elderberry	NE, SE	Very common	Very good	Medium shrub	Fibrous	Good
Sambucus racemosa Red elderbery	NW	Common	Good	Medium shrub		Good
ssp. pubens Scarlet elder	NE	Common	Very good	Medium shrub	Deep laterals	Fair to good
Spiraea alba Meadowsweet spirea	N, E	Common	Good	Small dense tree	Dense shallow lateral	Fair to good
Spiraea douglasii Douglas spirea	NW	Common	Fair	Dense shrub	Fibrous suckering	Good
Spiraea tomentosa Hardhack spirea	NE	Common	Good	Small shrub	Dense shallow	Fair
Symphoricarpos albus Snowberry	N, NW, E	Common	Good	Small shrub	Shallow fibrous	Good
Viburnum alnifolium Hubbiebush viburnum	NE	Fairly common	Good	Large shrub	Shallow fibrous	Good
Viburnum dentatum Arrowwood viburnum	E	Common	Good	Medium shrub	Shallow fibrous	Good
Viburnum lentago Nannyberry viburnum	S, SE	Fairly Common	Good	Large shrub	Shallow	Fair-good

APPENDIX 2
Plant Tolerance

This appendix contains a list of plants suitable for use with soil bioengineering or biotechnical stabilization. The tolerance to various adverse site conditions (salt, drought, flooding, and deposition) is noted for each plant.

Name	Location	Availability	Tolerance to Deposition[a]	Tolerance to Flooding[b]	Tolerance to Drought[c]	Salt Tolerance[d]
Acer negundo Box elder	N, NE	Common	High	High	High	Medium
Alnus rubra Red alder	NW	Very common	High	Medium	Low	Low
Baccharis glutinosa Water wally	W	Common	Medium	High	Medium	Low
Baccharis halimifolia Eastern baccharis	S, SE	Common	Medium	High	Medium	Medium
Baccharis piluaris Coyotebrush	W	Very common	Medium	Medium	High	Medium
Baccharis viminea Mule fat	W	Very common	High	HIgh	High	Medium
Betula Papyrifera Paper birch	N, E, W	Common	Medium	Medium	Medium	Medium
Betula pumila Low birch	N, E, W	Common	Low			Low
Cornus amomum Silky dogwood	N, SE	Very common	Low	Medium	Medium	Low
Cornus racemosa Gray dogwood	NE	Common	Medium	Medium	High	Low
Cornus rugosa Roundleaf dogwood	NE	Common				
Cornus sericea ssp. stolonifera Red osier dogwood	N, NE, NW	Very common	Low	High	Medium	Low
Crataegus Sp. Hawthorn	SE	Uncommon	Medium	Low	High	Low
Elaeagnus commutata Silverberry	N, central	Very common	High	Low	High	Medium
Ligustrum sinense Chinese privet	S. SE	Common	High	Medium	Medium	Low

Name	Location	Availability	Tolerance to Deposition[a]	Tolerance to Flooding[b]	Tolerance to Drought[c]	Salt Tolerance[d]
Lonicera involucrata Black twinberry	E	Common	Medium	Medium	Low	Low
Physocarpus capitatus Pacific ninebark	NW, W	Common	Low	Medium	Low	Low
Physocarpus opulifolius Common ninebark	NE	Common	Low	Medium	Medium	Medium
Populus angustifolia Arrowleaf cottonwood	W	Common	Medium	Medium	High	Medium
Populus balsamifera ssp. trichocarpa Black cottonwood	NW	Common	Medium	Medium	Medium	Medium
Populus deltoides Eastern cottonwood	MW, E	Very common	Medium	High	Medium	Low
Populus fremontii Freemont cottonwood	SW	Very common	Medium	Medium	Medium	Medium
Populus tremuloides Quaking aspen	NW	Very common	Medium	Low	Medium	Medium
Robinia pseudoacacia Black locust	NE	Common	Medium	Low	High	High
Rubus allegheniensis Allegheny blackberry	NE	Very common	Medium	Medium	Medium	Low
Rubus spectabilis Salmonberry	SW, NW	Very common	Medium	Medium	Medium	Low
Rubus strigosus Red raspberry	N, NE, W	Very common	Medium	Low	Medium	Low
Salix exigua Coyote willow	NW	Fairly common	High	High	Medium	Low
ssp. interior Sandbar willow	N, SE	Common	High	High	Low	High
Salix amygdaloides Peachleaf willow	N.S.	Common	High	High	Low	High
Salix bonplandiana Pussy willow	W, MW	Very common	Medium	Medium	Low	
Salix eriocephala ssp. ligulifolia Erect willow	Common	HIgh	High	Medium	Low	
Salix gooddingii Goodding willow	SW	Very common	High	Medium	Medium	Low
Salix hookeriana Hooker willow	NW	Common	High	High	Low	Medium
Salix humilis Prairie willow	N, Ne	Very common	Medium	Medium	High	Low
Salix lasiolepis Arroya willow	W	Common	High	High	Medium	Low
Spiraea alba Meadowsweet spirea	N, E	Common	Low	Medium	Medium	
Spiraea douglasii Douglas spirea	NW	Common	Medium	Medium	Medium	Low
Spiraea tomentosa Hardback spirea	NE	Common	Medium	Medium	Medium	Medium

Name	Location	Availability	Tolerance to Deposition[a]	Tolerance to Flooding[b]	Tolerance to Drought[c]	Salt Tolerance[d]
Symphoricarpos albus Snowberry	N, NW, E	Common	Low	Low	High	High
Viburnum dentatum Arrowwood viburnum	E	Common	Medium	Medium	Medium	Low
Viburnum lentago Nannyberry viburnum		Fairly common	Medium	Low	Medium	Low

[a]Tolerance to deposition: Regrowth from shallow coverage by stream deposits or soil slips, high, medium, or low.
[b]Tolerance to flooding: High (severely damaged after 10 to 30 days of flooding), medium (severely damaged after 6 to 10 days of flooding), or low (severely damaged after 1 to 5 days of flooding).
[c]Tolerance to drought: Resistance to drought (relative to native vegetation on similar sites) is high, medium, or low.
[d]Salt tolerance: Tolerance (relative to salt-tolerant native vegetation on similar sites) is high, medium, or low.

APPENDIX III
Nomenclature and Symbols

ENGLISH NOTATION

A = cross-sectional area (also annual computed soil loss from USLE)
A_s = cross-sectional area of fibers
A_r = cross-sectional area of roots
A_1 = cross-sectional area of fill above phreatic surface
A_2 = saturated cross-sectional area of fill below phreatic surface
B = width of retaining wall
C = vegetation management or cover factor
C = average or weighted vegetation cover factor
D = root diameter
E_R = tensile modulus of root fiber
FS = factor of safety
H = height or vertical thickness of slope or wall (also design wave height)
H_w = height of piezometric surface above sliding surface
K = soil erodibility factor
K_A = coefficient of active earth pressure
K_D = stability or roughness coefficient
L = slope length factor (also length of root fiber)
L_E = effective or "grip" length of imbedded reinforcement
M_O = overturning moment
M_R = resisting moment
N = normal bearing force at base of retaining wall
P = erosion control (or surface condition) factor
P_A = active earth force per unit length of wall
Q = flow discharge
R = hydraulic radius
S = steepness factor (also slope of water surface or channel gradient)
T = tensile load per unit width of fabric
T_S = mobilized tensile stress in root at incipient slippage
T_r = root tensile strength
T_i = tensile strength or roots in size class i
T_{ro} = initial root tensile strength

T_{rt} = root tensile strength after elapsed time t
T_{allow} = allowable unit tensile load in geogrid reinforcement
V = mean flow velocity
V_p = permissible flow velocity
W = weight of wall (also weight of individual armor stones in a revetment)
a = empirical constant
b = empirical constant
a_i = mean cross-sectional area of roots in size class i
c = soil cohesion or cohesion intercept
c_R = pseudo root cohesion
c_u = undrained cohesion (or shear strength)
c' = effective cohesion
c'_r = residual cohesion
d = required vertical spacing between reinforcements (also depth of flow)
e = eccentricity of load
f = coefficient of friction between soil and wood
i = seepage gradient
$\boldsymbol{n}$ = Manning's roughness coefficient
n, m = empirical constants
n_i = number of roots in size class i
m = empirical constant
q = discharge per unit with of channel
q_o = uniform vertical surcharge stress
q_{ULT} = ultimate bearing capacity of soil
q_{AVE} = average bearing stress exerted by wall
s = shear strength
Δs = shear strength increase (also incremental curvilinear length)
s_R = specific weight of armor stone in a revetment
t = elapsed time
t_R = mobilized tensile stress of root fibers per unit area of soil
x = shear displacement
x = distance of normal force reaction from toe of wall
$x_{1,2,3}$ = moment arm distances
z = shear zone thickness (also depth below ground surface)

GREEK LETTERS

α = wall batter angle (with respect to horizontal direction)
β = slope angle of natural ground or backfill
ϕ = angle of internal friction
ϕ_b = angle of internal friction of base soil (beneath wall)
ϕ_W = angle of wall friction

ϕ' = effective angle of internal friction
ϕ'_r = residual angle of internal friction
γ = moist or total unit weight of soil
γ_F = unit weight of root fibers
γ_R = unit weight of rock
γ_w = unit weight (density) of water
γ_{BUOY} = buoyant density of soil
γ_{SATD} = saturated density of soil
λ = seepage direction with respect to slope normal
μ = interface friction coefficient between soil and reinforcement
ρ_R = biomass of roots per unit volume of soil
σ = normal stress
τ = average shear or tractive stress
τ_b = bond or interface friction stress between root and soil
τ_p = permissible shear or tractive stress
τ_d = maximum shear or tractive stress
θ = angle of shear distortion (also the seepage direction or angle with respect to horizontal direction)
Ψ = anchor orientation with respect to slope normal

INDEX